Electron Spin Resonance
Volume 13B

A Specialist Periodical Report

Electron Spin Resonance
Volume 13B

A Review of Recent Literature to Mid-1992

Senior Reporter
M. C. R. Symons, FRS, *Department of Chemistry, University of Leicester*

Reporters
J. M. Baker, *University of Oxford*
A. Bencini, *Università degli Studi di Firenze, Italy*
G. R. Hanson, *University of Queensland, St. Lucia, Queensland, Australia*
P. H. Rieger, *Brown University, Providence, Rhode Island, USA*
C. Zanchini, *Università degli Studi di Firenze, Italy*

ISBN 0-85186-911-4

Copyright © 1993
The Royal Society of Chemistry

All Rights Reserved
No part of this book may be reproduced or transmitted in any form or by any means—graphic, electronic, including photocopying, recording, taping or information storage and retrieval systems—without written permission from the Royal Society of Chemistry

Published by The Royal Society of Chemistry,
Thomas Graham House, The Science Park, Cambridge CB4 4WF

Preface

As before, this volume follows on from the previous 'B' volumes, and is devoted to purely inorganic and bio-inorganic esr (epr) studies. The 'A' volumes cover organic aspects, both chemical and biological, and the contents of Volume 13A can be found at the end of the 'Contents' section of this volume.

The esr *versus* epr controversy remains, and some have tried to solve the difficulty with yet other forms of nomenclature. Probably both will continue, and when clarification is essential both will be used, as above, or *vice versa*. Probably the best compromise would be EMR to form closer links with NMR. I always thought that, for me, the 'Chemical Society' would always remain the 'Chem. Soc.' - but it has now become the 'R.S.C.'. Times change!

This volume contains standard, on-going reviews on transition metal ions by Bencini and Zanchini and on metalloproteins by Hanson (together with my own review on inorganic radicals). I wish to express my deep appreciation to these authors, who have, for many years, produced invaluable reviews that form the core of these B volumes.

Then there are outstanding reviews on the lanthanides, by Baker, and on the art of computer simulations of spectra, by Rieger - both leading authorities in these fields. I am sure that we are all most grateful for the production of these interesting and very timely reviews.

Finally, a general 'thank you' to all those who have supported me so strongly and expertly in the compilation of these volumes over many years. I have greatly enjoyed working with you, and have learned - or tried to learn - a great deal from you. This is my last volume, so I wish my successor the same help, advice and pleasure that I have had.

I also thank Alan Cubitt and the Royal Society of Chemistry for putting up with me and for giving me so much encouragement and support.

Martyn C.R. Symons

Contents

CHAPTER 1 Transition Metal Ions   1
By Alessandro Bencini and Claudia Zanchini

1 Introduction   1

2 General   2
   Theory   4
   Analysis of the Spectra and Computing   6
   Phase Transitions   10
   Jahn-Teller Effect   13
   Paramagnetic Ligands   14
   Mixed Valence Systems   18
      d_2^2-d^3 Systems   18
      d^3-d^4 Systems   18
      d_2^4-d_c^5 Systems   19
      d^5-d^6 Systems   19
   Binuclear and Oligonuclear Complexes   21
   Extended Systems   26
      d_5^9 Configuration   27
      d^5 Configuration   28
   Other Systems   29
   Semiconductors   29
   Superconductors   30

3 $S_i = 1/2$ 33
   d_3^1 Configuration   33
   d_5^3 Configuration   39
   d_7^5 Configuration   40
   d_9^7 Configuration   43
   d^9 Configuration   46
   d^{10} Configuration   53

4 $S_i = 1$ 53
   d_4^2 Configuration   53
   d_8^4 Configuration   54
   d^8 Configuration   54

5 $S_i = 3/2$ 54
   d_7^3 Configuration   54
   d^7 Configuration   56

6 $S_i = 2$ 56
   d^6 Configuration   56

7 $S_i = 5/2$ 57
   d^5 Configuration   57

References   61

CHAPTER 2 Metalloproteins 86
By G.R. Hanson

1 Introduction 86

2 Copper Proteins 86
   2.1   Type I Copper Proteins 86
   2.2   Type II Copper Proteins 87
   2.3   Multicentred Copper Proteins 89

3 Iron Proteins 91
   3.1   Non-Heme Iron Proteins 91
   3.2   Heme Iron Proteins 96

4 Iron Sulfur Proteins 101
   4.1   Simple Iron Sulfur Proteins 101
   4.2   Complex Iron Sulfur Proteins 103

5 Nickel Containing Enzymes 105

6 Molybdenum and Tungsten Enzymes 107
   6.1   Mononuclear Oxomolybdenum Enzymes 107
   6.2   Molybdenum Iron Enzymes 108
   6.3   Tungsten Enzymes 109

7 Vanadium Proteins 110

8 Manganese Proteins 111

9 Chromium Substituted Proteins and Toxicity 111

10 Paramagnetic Metal Substituted Proteins 112

11 Mitochondrial Enzymes 113
   11.1   Complex I - NADH Ubiquinone Oxidoreductase 113
   11.2   Complex II - Succinate Ubiquinone Oxidoreductase 114
   11.3   Complex III - Ubiquinol Cytochrome c Oxidoreductase 114
   11.4   Complex IV - Cytochrome c Oxidase 116

12 Photosynthetic Proteins 117
   12.1   Photosystem II Reaction Centre 117
   12.2   Oxygen Evolving Complex 119
   12.3   The Cytochrome b_6-f Complex 121
   12.4   Photosystem I Reaction Centre 121

References 122

CHAPTER 3 EPR and ENDOR in the Lanthanides 131
By J.M. Baker

1 Introduction 131

2 Properties of the Lanthanide Ions 131

3 The Crystal Field 134
   3.1   Orbit-Lattice Interaction 136

4	Electron Paramagnetic Resonance	138
4.1	The Spin Hamiltonian	139
4.2	The Spin-Lattice Hamiltonian and the Effect of External Stress	141
4.3	Line Width	143
4.4	Hyperfine Structure	144
4.5	Transferred (Super-)hyperfine Structure	146
4.6	Non-Kramers' Ions	148
5	Electron Nuclear Double Resonance	151
5.1	Self ENDOR	152
5.2	Ligand ENDOR	152
5.3	ENDOR of Distant Nuclei	154
6	Materials Investigated	154
6.1	Measurements on Lanthanide Compounds	156
6.2	Insulating Hosts Which are not Lanthanide Salts	160
6.3	Structural Phase Transitions	162
6.4	Sites of Low Symmetry	164
6.4.1	Crystals of Low Symmetry	165
6.4.2	Glasses	165
7	Interactions between Paramagnetic Ions	167
8	Conclusion	172
	References	173

CHAPTER 4 Simulation and Analysis of ESR Powder Patterns 178
By Philip H. Rieger

1	Introduction	178
2	The Anatomy of a Powder Pattern	178
2.1	'Extra' Features in Powder Spectra	179
2.2	Non-coincident Principal Axes	181
2.3	Powder Pattern Component Line Shapes and Linewidths	183
3	Computer Simulation of Powder Patterns	186
3.1	Analytical Solutions	186
3.2	Numerical Methods	187
3.3	The Taylor Method	188
3.4	Limitations and Improvements	190
3.4.1	Number of Orientations	190
3.4.2	Variable Transition Probabilities	191
3.4.3	Variable Component Widths	191
4	Analysis of Powder Spectra	192
4.1	Comparison of Experimental and Simulated Spectra	193
4.2	Fitting of Spectral Features	194
	References	198

CHAPTER 5 Inorganic and Organometallic Radicals   200
By Martyn C.R. Symons

1 Introduction   200
   1.1   Books and Reviews   200
   1.2   Techniques   201

2 Trapped and Solvated Electrons   202
   2.1   Introduction   202
   2.2   Electrons in Fluids   203
   2.3   Electrons in Solids   203

3 Atoms and Monatomic Ions   204
   3.1   Introduction   204
   3.2   Hydrogen Atoms   205
   3.3   Copper, Silver, and Gold   206
   3.4   Aluminium, Gallium, and Thallium   209
   3.5   Lead and Bismuth   210
   3.6   Oxygen and Sulphur Centres   210

4 Diatomic and Related Radical Systems   211
   4.1   Introduction   211
   4.2   Hydrides   211
   4.3 σ-A-B Radicals   211
   4.4 π-Radicals   212
   4.5   Diatomics Containing One Transition Metal   215
   4.6   Diatomics Containing Two Equal Transition Metals   215
   4.7   Diatomics with Different Transition-metal Atoms   216

5 Triatomic AB_2 Radicals   216
   5.1   Introduction   216
   5.2   Hydrogen and Alkyl Derivatives (AH_2, $AH(R)$, AR_2)   216
   5.3   HAB Radicals   217
   5.4   Nine Electron Radicals   218
   5.5   11, 13, and 15 Electron Radicals   219
   5.6   17 Electron Radicals   219
   5.7   19 Electron Radicals   220
   5.8   Transition Metal Centres   220

6 Tetra-atomic AB_3 Radicals   220
   6.1   Introduction   220
   6.2   Carbon Centred AB_3 Radicals   220
   6.3   Nitrogen Centred AB_3 Radicals   222
   6.4   Sulphur and Selenium AB_3 Radicals   222
   6.5   Phosphorus Centred AB_3 Radicals   224

7 Penta-atomic AB_4 Radicals   225
   7.1   Introduction   225
   7.2   Electron Loss AB_4 Centres   225
   7.3   Electron Gain Centres   226
   7.4   Transition Metal AB_4 Centres   227

8	Other Inorganic Radicals	228
	8.1 Introduction	228
	8.2 The IO_5^- Centre	228
	8.3 σ^* Radical Dimers	228
	8.4 Other σ^* Radicals	229
	8.5 Carbon and Carbon-Sulphur Centres	230
	8.6 Nitrogen and Sulphur Centres	231
	8.7 Metal Clusters	231
9	Paramagnetic Centres in Inorganic Materials	233
	9.1 Introduction	233
	9.2 Carbon	233
	9.3 Crystalline Silicon	233
	9.4 Amorphous and Hydrogenated Silicon	234
	9.5 Silica Centres	235
	9.6 III-V Compounds	235
	9.7 II-VI Materials	236
	9.8 I-VIII Materials: Silver Halides	236
10	Spin Trapping of Inorganic Radicals	236
	10.1 Introduction	236
	10.2 Trapping Inorganic Radicals in Chemical Systems	237
	10.3 Biological Uses of Spin Traps	237
11	Transition Metal 'Radical' Centres	238
	11.1 Introduction	238
	11.2 Some Chromium Complexes	238
	11.3 Some Mo, W, and Re Complexes	238
	References	239

Author Index   244

Summarised Contents of Volume 13A

CHAPTER 1 Organic Radicals in Solution

By B.J. Tabner

1) Introduction; 2) Carbon-centred Radicals;
3) Nitrogen-centred Radicals; 4) Oxygen-centred
Radicals; 5) Nitroxyl Radicals; 6) Radical
Cations; 7) Radical Anions; 8) CIDEP; References

CHAPTER 2 Organic Radicals in Solid Matrices

By C.J. Rhodes

1) Introduction; 2) Radical Cations; 3) Radical
Anions and Dissociative Electron Capture;
4) Radicals Relevant to Biological Systems;
5) Peroxyl Radicals; 6) Carbon-centred Radicals;
7) Silicon-centred Radicals; 8) Nitrogen-centred
Radicals; 9) Phosphorus-containing Radicals;
10) Radicals in Heterogeneous Systems; References

CHAPTER 3 Organic Radicals in Solids

By D.M. Close

1) Introduction; 2) Recent Developments;
3) Spectroscopic Aspects; 4) Molecules of
Biological Interest; References

CHAPTER 4 Fluorescence Detected Magnetic Resonance

By D.W. Werst and A.D. Trifunac

1) Introduction; 2) FDMR Detection of
Radical Ions; 3) Structure and Dynamics of
Radical Cations; 4) Condensed-phase Ion Chemistry
- An Overview; 5) Summary; References

CHAPTER 5 Theoretical and Physical Aspects of ESR

By A. Hudson

1) Introduction; 2) Time Domain and Related
Methods; 3) Numerical Methods and Spectral
Analysis; 4) Applications of Quantum Chemistry;
5) Condensed Phases and Molecular Dynamics;
6) CIDEP and Related Phenomena; 7) Miscellany;
References

CHAPTER 6 Applications of ESR in Polymer Chemistry
By D.J.T. Hill, J.H. O'Donnell, and P.J. Pomery

1) Introduction; 2) Polymer Degradation;
3) Polymerization; 4) Polymer Structure;
References

CHAPTER 7 Industrial Applications of ESR Spectroscopy
By S.A. Fairhurst

1) Introduction; 2) Films and Coatings;
3) Fossil Fuels; References

AUTHOR INDEX

1
Transition Metal Ions

BY ALESSANDRO BENCINI AND CLAUDIA ZANCHINI

1 Introduction

E.s.r. spectroscopy has been applied in a number of research fields mainly as a diagnostic tool for determining the spin and/or the oxidation state of the transition metal ion. A large number of papers therefore appeared, dealing with straightforward applications of e.s.r., while in other cases e.s.r. was applied to systems which were found to be too complicated to permit a complete spectroscopic characterization. An example of these latter systems are the transition-ion containing clusters found in living organisms (usually formed with iron and manganese) and their synthetic analogues, whose synthesis and crystallographic characterization have received much attention in the last few years. In order to rationalize this harvest of data we follow (with minor modifications) the classification scheme adopted in issue 12B.[1] In part 1 we will report review articles of general interest or books; in part 2 attention will be given to "areas of interest" like studies on phase transitions, mixed valence systems, superconductors, *etc.*; in part 3 and subsequent sections the systematic of e.s.r. is presented classifying the systems according to the spin of the transition metal ion. Only articles written in English will be generally reviewed and we will try to put emphasis on papers reporting on rather uncommon spin systems or papers in which a characterization as detailed as possible has been performed, especially when data coming from more than one spectroscopic and physico-chemical technique were interpreted together. In order to avoid as much as possible the overlap among different areas of interest we have separated complexes in different sections. Oligonuclear complexes containing mixed valences (generally delocalized) have been grouped in the Mixed Valence Systems section, while those exhibiting long range exchange interactions have been reported in the Extended Systems section. Magnetic exchange interactions will be represented with the spin hamiltonian H = J_{ij} S_i • S_j. A negative value of J_{ij} indicates a ferromagnetic interaction. Readers chiefly interested in the nature of the transition ion or in high nuclearity systems must therefore check more than one section. .

The book from J.R. Pilbrow entitled *Transition Ion Electron Paramagnetic Resonance* gives an up to date account of the applications of the e.s.r. spectroscopy to transition metal ions. The book covers the fundamental aspects of e.s.r. of transition metal ions and also presents the principles of both CW and pulsed techniques and zero-field e.s.r. [2] One chapter is also devoted to the principles and techniques of simulation of e.s.r. spectra.

Several review articles dealing with various applications of e.s.r. spectroscopy and related techniques appeared. Particularly interesting is the article by Trautwein *et al.* in which Mössbauer, e.s.r., and magnetic susceptibility studies on iron-containing proteins and their synthetic analogues are reviewed.[3] Applications of e.s.r. to study the structure of impurities in insulators[4] and in antiferromagnetic solids[5] and to investigate the surface structure of catalysts.[6] An extensive review article outlines the nature of the information which can be obtained from the subset of spin-spin interactions that yield additional lines in the e.s.r. spectra.[7] Ferromagnetic e.s.r. is considered in ref. 8 and in ref. 9 NMR and e.s.r. in two-dimensional magnets has been presented. Optically detected magnetic resonance has been discussed in ref. 10 and applied to clarify the structure of intrinsic and impurity centres in ionic solids.[11]

The principles and applications of pulse e.s.r.[12] and pulse electron-nuclear multiple resonance have been reviewed.[13,14] A recently developed technique based on pulsed e.s.r. methods for dynamic polarisation of nuclear spins in semiconductors is reviewed.[15]

Modern techniques in electron paramagnetic resonance, such as two-dimensional and Fourier-transform e.s.r., far infrared e.s.r, imaging, spin-echo, and high field e.s.r. are presented in three review articles.[16-18]

2 General

A number of new e.s.r. spectrometers have been described and the apparatuses are under patent licence. Some instruments, however, are rather peculiar and will be mentioned here. A CW e.s.r spectrometer working at the magnetic earth-field with a frequency of 1.845 MHz with optimized sensitivity is described.[19] Only very narrow lines can of course be detected (maximum sweep 125 µT). A modular low frequency spectrometer with sensitivity of 8×10^{21} spins T^{-1} has been constructed using commercially available components.[20] Coupling spins to other degrees of freedom in order to obtain dynamical polarization and/or greater detection sensitivity constitutes rather new approaches to magnetic resonance. Several of these techniques have been described.[21] A reduction of the overall receiver noise figure of 24.6 dB has been obtained for a Varian E-110 spectrometer working at Q-band frequency by the use of a low-noise Gunn diode oscillator.[22] Simple modification of a Varian E-line series of spectrometers is suggested in order to perform fast direct detection of the e.s.r. signal.[23] An e.s.r. signal from a sample of irradiated quartz which can be easily saturated was used to monitor the performance of an automatic frequency control (AFC) circuit, measuring the different saturation behaviour of the absorption and dispersion signals.[24] Modulation of the magnetic field to detect the e.s.r. signal is still under development,[25,26] while electron paramagnetic rotary resonance is being studied as an alternative method.[27]

A software package containing routines for data acquisition, processing and graphic representation of e.s.r. (and other) spectra has been described.[28] The data acquisition system allows one to perform kinetic studies as well and includes spectra accumulation.

Pulse e.s.r. technique is receiving much attention and we mention in particular the following advances from Arthur Schweiger and co-workers: suppression of artefacts and undesired echoes by selection of coherence-transfer pathways;[29] detection of small hyperfine and quadrupole interactions;[30] generation of electron-spin-echo-envelope modulations (ESEEM) in paramagnetic systems that do not contain nonsecular hyperfine interactions;[31] resolution of an e.s.r. spectrum in a second dimension.[32]

Electron-spin transient nutation spectroscopy has been studied by Astashkin and Schweiger in order to find a way of simplifying complicated solid state e.s.r. spectra. The technique is based on the physical fact that the nutation frequency of a spin system precessing around a magnetic field is given by the product of the transition matrix element and the strength of the magnetic field causing the nutation. In solid state spectra these matrix elements may vary from line to line and the nutation frequency is therefore different from line to line. In a transient nutation experiment the microwave magnetic field is suddenly switched on at t=0 and the nutation intensity is measured at any fixed time as a function of the static magnetic field flux density. Examples of applications to the resolution of single crystal spectra of copper(II) and manganese(II) samples have been furnished.[33]

E.s.r. imaging using the microwave scanning technique has been applied by Furusawa and Ikeyato to detect the distribution of nitrogen and nickel impurities in diamond crystals,[34] and the same authors also describe a method of producing high-quality linear field gradients,[35] which constitutes the more common method of e.s.r. imaging. High fields (5 T) have been used in imaging with spatial resolution up to 1 µm in one dimension.[36] The perspectives in 3D e.s.r. imaging using CW methods and electron spin-echo (ESE) detection have been analysed.[132]

Photo-e.s.r. has been applied to characterize deep levels in cadmium telluride.[37] Laser-flash excitation following by ESE-e.s.r. has been applied to characterize excited states in chromate(VI) doped dipotassium sulphate.[38]

Zhong and Pilbrow have given a simple interpretation of the saturation-transfer e.s.r. spectroscopy based on the effects of spectral diffusion induced by molecular motion on adiabatic rapid-passage responses.[39]

Very low temperatures, as low as 0.4 K, have been reached with a X-band e.s.r. spectrometer in which the sample enclosed in a small-helix resonator inside a ^{3}He cryostat is reached by the microwave radiation through a semirigid coaxial cable and a small loop. Detection of a minimum of 1.5×10^5 spins T^{-1} was reported.[40]

Quantitative detection of the number of spin and preparation of the standards for intensity measurements have been described.[41,42]

Theory. Numerical calculations of the zero-field splitting (ZFS) parameters for high spin d^5 ions in cubic and tetragonal symmetries have been performed by Wan-Lu and Rudowicz. The calculations of the second (b_2^0) and fourth degree (b_4^0, b_4^4) terms of the zero-field splitting spin hamiltonian, written using Stevens operators,[43] have been performed in the crystal field framework including spin-orbit coupling as a perturbation on the crystal field and electrostatic interactions. All the 252 states of the d^5 configuration have been included in the calculations and the perturbation procedure was extended to fifth order. The parameter b_2^0 was found to be mainly determined by second and third order perturbation terms, i.e. interactions among the ^6S ground state and the excited quartet states, while the fourth degree ZFS parameters were affected by both quartet and doublet levels. Relationships between the zero-field splitting parameters and the crystal field, electrostatic, and spin-orbit coupling parameters have been examined. Agreement with the experimental data is obtained for manganese(II) and iron(III) in fluoroperovskites.[44] The energies of the electronic transitions and the zero-field splitting of high spin iron(III) (S = 5/2) in cubic fields, which is parameterized by $a = b_4^0 = 2b_4^4/5$, have been computed including all the 252 states of the d^5 configuration and e.s.r. data for iron(III) in yttrium gallium garnet have been reproduced.[45] A paper dealing with the calculation of the e.s.r spectra of high spin iron(II) (d^6, S = 2) in C_{2v} symmetry based on the crystal field theory appeared. The formalism developed seems to consider only $^5A_1$ as the ground state and presents the calculation of Zeeman splitting between adjacent states. The computed zero-field splitting, $D = b_2^0 = 8.8035$ cm^{-1} proves to be axial, but the actual value seems to depend on the nature of the states chosen for the calculation.[46] Electronic transitions and e.s.r. spectrum of nickel(II) (d^8, S=1) in lithium niobate(V) have been calculated using the crystal field formalism assuming a C_{3v} symmetry.[47]

$X\alpha$-SW molecular orbital calculations of the isotropic and dipolar contribution to the hyperfine tensor have been performed for copper(II) (d^9, S = 1/2) in a $[CuCl_4(NH_3)_2]^{2-}$ centre to clarify the observed dependence of the superhyperfine splitting on the metal-ligand distances.[48] Calculations of the spin hamiltonian parameters on iridium(II) (d^7, S = 1/2),[49] copper(II) complexes,[50] metal phthalocyanines,[51] and metal-nitroxide[52] complexes using various MO methods have been reported.

G. Shen, C. Xu, and G. Bai[53] analyzed the e.s.r. spectrum of the binuclear complex $[Ph_4AsCuCl_3]_2$ since, they stated, "few theoretical analyses to interpret the results" (for binuclear copper(II) systems) "have been made". Apparently, however, the authors are not aware of a bulk of scientific literature: useful references can be 54 and

55 . Their theoretical treatment is based on a rather peculiar hamiltonian which uses a crystal field hamiltonian which includes the Dirac-Van Vleck-Heisenberg isotropic exchange spin-hamiltonian and the conclusions showed a relationship between the zero-field-splitting of the triplet state and the isotropic exchange coupling constant, contrary to well established literature results.[54,55]

Analysis of the Spectra and Computing. - A fundamental aspect in the calculation of e.s.r. spectra is the correct formulation of the lineshape function. In particular, a few questions have arisen in recent years about the equivalence or non-equivalence of the field and frequency modulation schemes and about the inclusion of the $1/g$ factor in the simulation of S = 1/2 spin systems. Zhong and Pilbrow[56] worked out an expression of the e.s.r. absorption in single crystals and polycrystalline powders, which we briefly reported here. The basic equation, for single crystal absorption, is

$$S(B,\omega,\Omega) = C\,\eta Q_0 \frac{2N\hbar\omega}{ZkT} \sum_{\alpha,\beta} |\langle\alpha|\mu_x|\beta\rangle|^2 F_\omega[\omega - \omega_0(B,\Omega),\sigma_\omega] \quad (1)$$

where as usual η is the sample filling factor, Q_0 is the unloaded quality factor of the sample resonant cavity (or loop-gap resonator), C is an instrumental constant, α and β are the states (eigenvalues of the spin hamiltonian) between which the transition, caused by the interaction of the microwave magnetic field with the spin system (μ_x), occurs. F_ω is the lineshape function explicitly expressed as a function of the flux density of the static magnetic field, B, of the microwave frequency, ω, and of the phenomenological linewidth, σ, in frequency units. Z is the partition function of the Boltzmann distribution. The fourth term in (1) represent, in fact, the difference in thermal population between the two levels in the high temperature approximation, $h\nu \ll kT$. The resonance centre of the absorption, ω_0, is defined by

$$\omega_0(B,\Omega) = \frac{E_\beta - E_\alpha}{\hbar} \quad (2)$$

$\Omega = \Omega(\theta,\varphi)$ being the angular variable. Equation (1) explicitly includes a Boltzmann factor as a frequency dependent quantity. It is a common practice, when dealing with field sweep spectra, to work with a lineshape function expressed using field variables, F_B, instead of F_ω. A simple relationship connects F_ω and F_B:

1: Transition Metal Ions

$$\frac{\partial \omega_0}{\partial B} F_\omega(\omega-\omega_0,\sigma_\omega) = F_B(B-B_0,\sigma_B); \quad \sigma_\omega = \frac{\partial \omega_0}{\partial B}\sigma_B \qquad (3)$$

Equation (3) is rigorously valid for a Dirac-δ distribution of states and its use in e.s.r. was proposed by Aasa and Vanngard[57] and accepted for symmetric lineshapes with finite widths. In general the lineshape function F has the following property:

$$\frac{\partial F_\omega}{\partial \omega} = -\frac{\partial F_\omega}{\partial \omega_0} \qquad (4)$$

This allows one to write down the connection between the frequency and field modulations as

$$\frac{\partial F}{\partial B} = -\frac{\partial \omega_0}{\partial B}\frac{\partial F}{\partial \omega} \qquad (5)$$

Neglecting the field dependence of the transition probability and of the linewidth, σ_ω, in (1) a simple form of the e.s.r. spectrum is obtained:

$$\frac{\partial S(B,\omega,\Omega)}{\partial B} \cong \sum_{\alpha,\beta} G \frac{\partial F_\omega}{\partial B} \qquad (6)$$

where G represent all the terms in equation (1) other than F_ω. Using (5) and (4) one gets

$$\frac{\partial S(B,\omega,\Omega)}{\partial B} \cong \sum_{\alpha,\beta} G \frac{\partial F_\omega}{\partial \omega_0}\frac{\partial \omega_0}{\partial B} \qquad (7)$$

or, using (3), the more familiar expression

$$\frac{\partial S(B,\omega,\Omega)}{\partial B} \cong \sum_{\alpha,\beta} G \frac{\partial F_B}{\partial B}\frac{1}{\frac{\partial \omega_0}{\partial B}} \qquad (8)$$

as already found by van Veen.[58] For spin systems in which the resonance condition can be expressed as $\hbar\omega = g\mu_B B$, equation (8) takes the simpler form

$$\frac{\partial S(B,\omega,\Omega)}{\partial B} \cong \sum_{\alpha,\beta} G \frac{\partial F_\omega}{\partial \omega_0}\frac{\hbar}{g\beta} \qquad (9)$$

The expression for the e.s.r. signal of polycrystalline powders can be easily found by integrating (7) or (8) over the angular variables Ω, or equivalently

$$\frac{\partial I(B,\omega)}{\partial B} = -\int \sum_{\alpha,\beta} G \frac{\partial F_\omega}{\partial \omega}\frac{\partial \omega_0}{\partial B}\partial\Omega \qquad (10)$$

A general treatment of the lineshape functions to be used in e.s.r. spectral simulations and the effect of the asymmetry of the lineshape functions can be also found in ref. 2.

The computer simulation of powder spectra is important in a large number of situations in which crystalline solids cannot be obtained, and this is one reason why a part of the e.s.r. literature is still dealing with developing new algorithms for an efficient simulation of polycrystalline powder spectra. It must be stressed, however, that not all the computer codes agree on the exact form of the resonance absorption function and expressions different from (10) or not including the $\partial \omega_0 / \partial B$ part are still used. Differences can thus be found in the literature on the actual simulations. The actual form of the lineshape function used, in general, does not severely affect resonances which are symmetric around the transition frequency, for which equation (3) applies, but its effects become observable when asymmetric resonances are considered (g strain and A strain effects). A list of computer programs developed to analyze and fit e.s.r. spectra can be found in Appendix T of ref. 2. Gribnau, van Tits and Reijerse reported a procedure to speed up the calculation of e.s.r. spectra in their program MAGRES.[59] The program is based on matrix diagonalization and can in principle handle any spin system; the suggested procedure consists of accumulating a large number of single crystal spectra, but only 2-10% of these spectra are computed by exact diagonalization. All the other points necessary for a meaningful spatial integration have been computed via a fitting procedure. The authors claim that the computer time is reduced by a factor of 10 without apparent deterioration of the quality of the spectra.[60] Simultaneous calculation of e.s.r. and ENDOR powder spectra has been obtained by Kreiter and Hüttermann with the program package MSPEN/MSGRA. The program is based on direct diagonalization of the spin hamiltonian.[61] A FORTRAN computer program, FILMCAL, has been announced to be developed to simulate e.s.r. spectra of transition metal ions in thin layer of phthalocyanine films[62].

Computer simulations of e.s.r. spectra in amorphous thiomolybdates(V) (S=1/2) taking into account distribution of g values arising from random local structural fluctuations[63] and of polycrystalline powder spectra of high-spin iron(III) (S=5/2) in hexacoordinated chromophores based on full matrix diagonalization[64] have been reported.

A fitting procedure for extracting the spin hamiltonian parameters from powder spectra of copper(II) complexes has been reported.[65]

Hyde et al. described an algorithm, called *pseudomodulation*, to transform a digitized spectrum and to filter the noise. The methodology simulates the effect of an additional field modulation on an experimental spectrum. The authors suggest applying their method to simulated spectra prior to fitting a model to data.[66]

The lineshape function to be used in simulation of the e.s.r. spectra can be neither pure Gaussian, G(B), nor pure Lorentzian, L(B). In many cases a convolution of these two functions is used

$$I(B) = \int_{-\infty}^{\infty} G(t)L(B-t)dt \qquad (11)$$

called a Voigt profile and representing a Lorentzian spectral line broadened by Gaussian effects. Since the integral in (11) cannot be analytically solved a number of investigations dealt, in the past, with its numerical calculation. A Mexican group suggested a method to deconvolute the Voigt profile[67] and its improvements.[68] Applications to enhance the resolution of the e.s.r. spectrum of molybdenum-phthalocyanine in solution of fuming sulphuric acid have been reported.[69]

An analytical method has been applied to calculating powder patterns which reduces the patterns themselves to a closed form through an integral over one variable. This method was previously applied to the calculation of powder spectra of ^6S ions in cubic crystal fields and to axially symmetric S=1/2 spectra[70,71] and it is now applied to the calculation of the powder pattern of the central $\Delta M_s = \pm 1/2$ line of paramagnetic ions in sites of orthorhombic symmetry.[72]

Theoretical treatments of the e.s.r. lineshapes in randomly oriented, anisotropic systems undergoing internal motion,[73] for spins performing a random walk in 1-, 2-, and 3-dimensional lattices,[74] and in the regime of fast modulation of the zero-field splitting[75] have been presented. The effect of isotropic Heisenberg exchange interaction on the e.s.r. lineshape and intensity for S=1/2 spin systems under saturation have been also examined.[76] Spin-lattice relaxation of nickel(II) in potassium trifluoromagnesite and potassium trifluorozincate crystals were calculated form e.s.r. and optical absorption spectra using a parametric relationship between the spin-lattice coupling and the stress dependence of the zero-field splitting.[77]

Measuring the spin hamiltonian parameters from single crystal spectra requires a fitting procedure to the experimental transition fields and a number of them have been reported in the past. The fitting procedure is often based on least-squares minimization, either linear or non-linear. A distinctive feature of the various procedures is the method adopted to compute the transition fields, either perturbation or diagonalization of the spin hamiltonian matrix. A non-linear least squares fitting procedure for extracting the spin hamiltonian parameters from single crystal spectra is presented by Hrabanski and Lech. The procedure is based on the minimization of the chi-squared function

$$\chi^2 = \sum_{k=1}^{n} \frac{\left[B_k^0 - B_k\right]^2}{\sigma_k^2} \qquad (12)$$

where B_k^0 and B_k are the experimental and computed transition fields and σ^2_k are effective weight factors for the kth line. The transition fields are computed through a third-order perturbation procedure. No procedure to estimate the standard deviations on the parameters and of their correlation is given. The method was applied to fit the room temperature X-band e.s.r. spectra for manganese(II) (S=5/2) doped single crystals of [Mg(H$_2$O)$_6$][SnCl$_6$] on a microcomputer.[78]

The spin hamiltonian to be used in the interpretation of S = 5/2 spin systems originating from ions having $^6A_1$ ground states, like high-spin iron(III) and manganese(II) in octahedral environment, sometimes includes fourth order terms. An analysis of the effects of these terms on the e.s.r. spectra of glasses containing impurities of manganese(II) has been performed; in particular the condition for observing $g_{eff} \approx 4.3$ and $g_{eff} \approx 3.3$ resonances in the presence of these terms have been assessed.[79]

Phase Transitions. Transition metal ions have been used for a long time to monitor structural changes in solids, induced by temperature or pressure, by relating the variation of the spin hamiltonian parameters to the geometrical deformation of the environment of the ions. The symmetry of the e.s.r. spectra also yielded information on the geometry of the lattice. The applications of e.s.r. to study structural phase transitions have been reviewed by Müller and Fayet.[80]

Particular interest has been devoted in the last two years to transitions to a ferroelectric or ferroelastic phase and/or to incommensurate phases. In the ferroelectric phase the domains in the crystal have non-zero total dipole moment oriented co-operatively along a specific direction; in ferroelastic phases the solid is unaffected by the electric fields, but changes under the effects of applied stresses. Condensation of phonons whose wavelength is incommensurable with the structure of the crystal at a given temperature results in a transition to an incommensurable phase, in which the lattice translational symmetry is lost and all the molecules become magnetically inequivalent. This latter effect can be revealed by e.s.r. since the spectrum results in a continuous distribution of resonance lines. In this sense the e.s.r. spectrum of an incommensurate crystal resembles that of a powder. Splitting of the resonance lines and extra lines can also result from the incommensurability of the phase. A clear account of the e.s.r. spectra in incommensurate phases has been presented by Zwanenburg, Michiels and de Boer who investigated the phase transitions in copper(II) doped K_2SeO_4.[81] In Figure 1 the temperature variation of one e.s.r. line of the spectrum is shown.

The transition to the incommensurate phase occurs at T_I = 130 K but it is still observable at 144 K depending on the way of varying the temperature. Below 135 K hyperfine structure due to interactions with protons of lattice water molecules is observed. The splitting of the resonances line was found to obey the following law:

$$B = B_0 + B_1 \cos\phi(x) + \frac{1}{2}B_2 \cos^2 \phi(x) \qquad (13)$$

where B_0 is the resonant field in the high temperature phase, $\phi(x)$ is the phase of the modulation inducing the transition. B_1 and B_2 are adjustable parameters with the following temperature dependence

$$B_1 = [(T_I - T)/T_I]^\beta \, ; \, B_2 = [(T_I - T)/T_I]^{2\beta} \qquad (14)$$

with β = 0.51 determined from B_1.

The phase transitions in single crystals of chromium(III) doped potassium scandium molybdate, $K(Sc,Cr)(MoO_4)_2$, have been studied following the temperature dependence of the resonance line of the transition $|1/2\rangle \leftrightarrow |-1/2\rangle$ of chromium(III) impurities. The crystals undergo two second-order ferroelastic phase transitions from the room temperature trigonal phase to a monoclinic below 307 K and to a triclinic one below 288 K. The crystallographic symmetry lowers from $P\bar{3}m1$ to $C2/m$ to $P\bar{1}$.[82] Below 240 K the appearance of additional splitting of the e.s.r. line was ascribed to an incommensurate phase transition.[83-85] Incommensurate phases were observed also for chromium(III) doped $RbIn(MoO_4)_2$,[86] for manganese(II) doped Rb_2ZnCl_4,[87] and for manganese(II) doped $[Mg(H_2O)_6][SiF_6]$.[88,89]

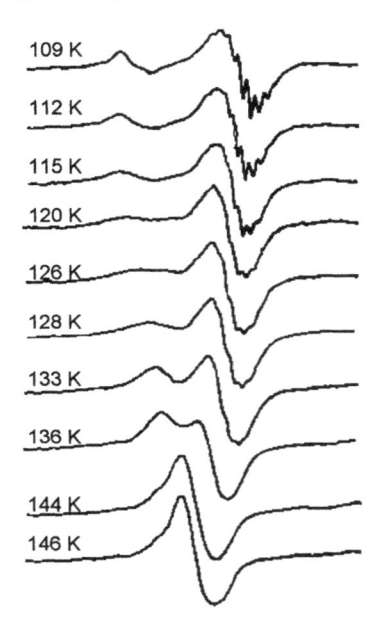

Figure 1. The temperature dependence of one line of the e.s.r. spectrum of K_2SeO_4 in the incommensurate phase. The extra features due to the incommensurable modulation are visible in the spectra up to ≈ 142 K.[81]

$RbCdF_3$ undergoes a transition from a cubic to a tetragonal phase at T_c ≈124 K. The nature of the transition has been monitored by measuring the e.s.r. spectra of $Rb(Cd,Ni)F_3$ single crystals. The spectra in the cubic phase are isotropic with g = 2.34(1); at 77 K the spectra have been interpreted using an axial S = 1 spin hamiltonian with $g_{\parallel}$ = 2.321(5), $g_{\perp}$= 2.295(5), and D = 0.408(1) cm^{-1}. The zero field splitting parameter, D, below T_c has a temperature dependence of the type

$D = D_0|T-T_0|^{2\beta}$ \hfill (15)

where D_0 = 0.0584 cm^{-1}, T_0 = 124 K, and β = 0.25.[90]

Transition-metal halide complexes of formula [NR$_4$]$_2$[MX$_4$] (M = Zn, Co, Mn; X = Cl, Br; R = Me, Et) undergo a number of phase transitions as the temperature is lowered from room to liquid nitrogen: paraelectric ↔ incommensurate ↔ commensurate ↔ ferroelectric (or ferroelastic) ↔ paraelectric. The phase transitions in [NEt$_4$]$_2$[MnBr$_4$], [NMe$_4$]$_2$[(Zn,Mn)X$_4$], [NMe$_4$]$_2$[(Co,Mn)Br$_4$], [NEt$_4$]$_2$[(Zn,Mn)X$_4$], [NEt$_4$]$_2$[(Co,Mn)Br$_4$] have been investigated through e.s.r. spectroscopy and the phase transition was found to depend on the nature of both the metal halides (MX$_4$) and the organic groups.[91]

Phase transitions have been studied also in copper(II) and vanadyl(II) doped (NH$_4$)$_3$H(SeO$_4$)$_2$,[92] (Fe,Mn)SiF$_6$·6H$_2$O,[93] (Fe,Cu)SiF$_6$·6H$_2$O,[94] Cd$_2$Nb$_2$O$_7$,[95] manganese(II) doped NH$_4$HC$_2$O$_4$·0.5H$_2$O,[96] iron(III) doped BaTiO$_3$[97] and SrTiO$_3$,[98] Fe$_4$V$_2$Mo$_3$O$_{20}$,[99] CsNiBr$_3$,[100] RbNiBr$_3$,[100] CsMnI$_3$,[100] Tb$_x$Y$_{3-x}$Fe$_5$O$_{12}$ (x<0.5).[101]

Jahn-Teller Effect. Bersuker et al. have presented a theoretical description of the coupling of the electronic E ground state to an e vibrational mode in a trigonal two-centre system and derived the form of the potential-energy surface.[102,103] The model applies to a couple of metal impurities in the geometrical arrangement shown in Figure 2.

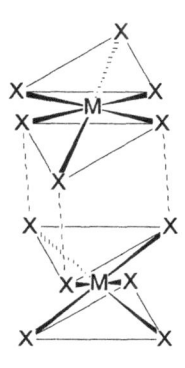

Figure 2. The two-centre system M$_2$X$_{12}$ (M = Cu, Ag, Ni, ...; X = F, Cl, O, ...) with common trigonal symmetry axis.[102]

In this model system no common atoms in the nearest neighbourhood of the metal ions have been considered and only the linear Jahn-Teller effect of the $E \otimes e$ type has been included. With these approximations, the lowest sheet of the adiabatic potential-energy surface possesses a one-dimensional trough in the space of local e vibrations and gives an in-phase reorientation of the local nuclear distortions at the Jahn-Teller centres. As a consequence of the strong vibronic coupling, the ground state of the system is non-degenerate and is well separated by the upper sheets by a large energy gap. The effect of the electron-electron interaction between the paramagnetic centres in the form of the Heisenberg-Dirac-Van Vleck spin hamiltonian was also included as a smaller effect and the computed vibronic-energy spectrum was used to compute the effective magnetic moment, μ_{eff}, and g values. The most striking effects have been computed on μ_{eff} which showed a non-monotonic temperature dependence.

Copper(II) ion is still widely used as a Jahn-Teller system. In octahedral complexes the Jahn-Teller interaction $E_g \otimes e_g$ is operative. This interaction has been studied in copper(II) complexes with terpyridine (terpy) and related ligands. In Cu(terpy)$_2$(PF$_6$)$_2$ the room temperature e.s.r. spectrum is axial with $g_{/\!/}$ = 2.023 and $g_{\perp}$ = 2.191 and becomes orthorhombic below 77 K. At 4.2 K the g values are g_x = 2.26, g_y = 2.129, g_z = 2.020. The high temperature structure is characterized by an averaged compressed octahedral structure, and the low temperature structure contains elongated octahedra in an antiferrodistortive order.[104] Static-dynamic Jahn-Teller effects have been observed by Misra et al. for copper(II) impurities in potassium oxalate monohydrate,[105] and in MgNa$_2$(SO$_4$)$_2$·4H$_2$O and CoNa$_2$(SO$_4$)$_2$·4H$_2$O.[106]

Other systems in which Jahn-Teller interactions have been found to be operative concern chromium(III) in gallium arsenide,[107] vanadium(IV) in strontium titanate,[108] silver(II) in sodium fluoride,[109] RBa$_2$Cu$_3$O$_{7-x}$,[110] platinum complexes of defect in silicon.[111,112]

Paramagnetic Ligands. A number of papers dealt with complexes of transition metal ions with catecholate-semiquinone-quinone systems. The redox relationship between the three forms is shown in Figure 3. Generally a rather strong chemical bond is formed between the ligand and the metal giving rise to rather strong magnetic coupling when both the ligand and the metal possess unpaired electrons. A problem which always occurs in these systems is the correct assignment of the formal oxidation state of the metal ion. It is important to mention that the bonding interaction with the metal involves the frontier orbital of the organic moiety and therefore affects its electrochemical behaviour. A number of applications of e.s.r. spectroscopy, which have been done in a rather routine fashion to see whether the unpaired electron is metal or ligand centred, will not be explicitly mentioned in the following.

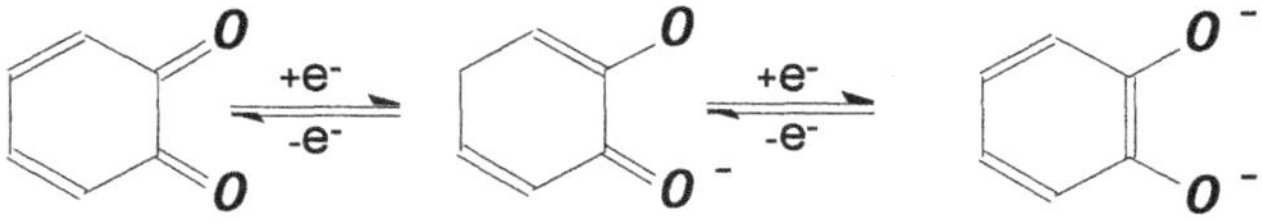

Figure 3. The electrochemical relationship between a quinone (diamagnetic), semiquinonate (paramagnetic, S = 1/2), and catecholate (diamagnetic).

Antiferromagnetic coupling between nickel(II) and semiquinonate ion has been observed in [Ni(Me$_3$[12]N$_3$)(DTBSQ)]ClO$_4$ (Me$_3$[12]N$_3$ = 2,4,4-trimethyl-1,5,9-triazacyclododec-1-ene; DTBSQ = 3,5-di-*ter*-butyl-*o*-semiquinonate) for which a ground S = 1/2 state, arising from the magnetic coupling between S = 1 of nickel(II) and S = 1/2 of the semiquinonate, has been reported. The X-band e.s.r. polycrystalline powder e.s.r. spectrum showed an isotropic signal at g = 2.26, practically independent on the temperature. This g value was interpreted assuming a strong isotropic exchange coupling between the magnetic electrons using the relationship

$$g = \frac{4g_{Ni} - g_{SQ}}{3} \qquad (16)$$

with g_{SQ} = 2.00 and g_{Ni} = 2.2. In the analogous copper(II) complex the magnetic coupling was found to be ferromagnetic leading to the stabilization of a triplet state.[113] High field (hv = 8.17 cm^{-1}) e.s.r. polycrystalline powder spectra at 4.2 K have been interpreted with an S = 1 spin hamiltonian with g_x = 2.01, g_y = 2.19, g_z = 2.01, D = 1.09 cm^{-1}, and E/D = -0.240. A molecular model of the exchange mechanism between the metal and the ligand is also discussed.[114] Antiferromagnetic coupling has been observed also in tetraoxolene complexes with iron(III),[115] chromium(III)[115] and nickel(III)[116] ions. The iron and chromium derivatives have formulas of the type [M$_2$(CTH)$_2$(DHBQ)$_2$]Y$_2$ (M = Cr, Fe; CTH = *dl*-5,7,7,12,14,14-hexamethyl-1,4,8,11-tetraazacyclotetradecane; H$_2$DHBQ = 2,5-dihydroxy-1,4-benzoquinone; Y = ClO$_4$, PF$_6$)

and the ground state was found to correspond to S = 5/2 and S = 9/2 for chromium(III) and iron(III) derivatives respectively.

Antiferromagnetic exchange interactions between semiquinonates molecules which are adjacent to each other in the solid state in [(tetramethylenediamine) (9,10-phenantrenequinone) (9,10-phenantrenediolato) copper(II)] gave rise to the e.s.r. spectrum of the copper(II) ion in a square pyramidal environment with $g_{/\!/}$ = 2.225 and $g_\perp$ = 2.07. The parallel direction was found to coincide with the perpendicular to the equatorial plane.[117]

Other selected systems in which the formation of the paramagnetic species was followed with e.s.r. spectroscopy can be found in references 118-129.

Apart from quinones also porphyrins and porphyrin-like systems have been commonly used as ligands towards transition metal ions. Some of the most relevant complexes, especially those having ligand centred or unusual oxidation state, are reported in this section, while other examples can be found in the section devoted to the appropriate metal. Iron(III) and iron(II) form of a porphyrinato-acetato complex, [Fe$^{III/II}$(Ac)(TPpivP)]$^{0,-}$ (TPpivP = pivalamide-picket-fence porphyrin), have been studied through Mössbauer and e.s.r. spectroscopies. Both the iron(II) and the iron(III) complexes are high-spin with S = 2 and S = 5/2 ground spin states respectively. The zero field splitting in the ferrous complex was found to be highly orthorhombic with D = -0.9 cm^{-1}, while was found to be nearly axial with D = 7.5 cm^{-1} in the ferric one.[130] Cobalt(II) tetraphenylporphyrins, bearing nitrogen containing substituents on one of the phenyl rings, have been found to form binuclear species in solution and to bind dioxygen to cobalt(II).[131] E.s.r. and ENDOR spectra of cobalt(III) and magnesium(II) octaethylporphyrins have been analyzed in term of a predominantly $^2A_{1u}$ ground state[132] while axial ligand field interaction causes an increasing mixing of $^2A_{2u}$.[133] E.s.r. and magnetically perturbed Mössbauer spectroscopy have been applied to characterize oxoferryl, OFe(IV), porphyrin cation radicals, in some complexes, in frozen solutions.[134,135] The magnetic interaction between the iron(IV) ion and the ligand radical have been taken into consideration and hyperfine and zero-field splitting parameters have been reported.[134] The formation of a dimeric ferryl porphyrin species has been also reported.[136]

The nitronyl nitroxide radical ligand NITpPy = 2-(4-pyridyl)-4,4,5,5-tetramethylimidazoline-1-oxyl 3-oxide), shown in the scheme below, binds three distinct copper(II) ions in the complex [Cu(hfac)$_2$]$_3$(NITpPy)$_2$ (hfac = hexafluoroacetylacetonato).

The copper(II) ions bound to the NO groups are ferromagnetically coupled (J ≈ − 80 cm^{-1}) and are antiferromagnetically interacting with the one bound to N (J' ≈15 cm^{-1}). The overall behaviour is that of a one-dimensional antiferromagnet. The room temperature single crystal e.s.r. spectra have been interpreted using a S = 1/2 spin hamiltonian with g_1 = 2.16, g_2 = 2.08, g_3 = 2.06 and are almost temperature independent. A (3cos^{2}θ-1)^n dependence of the peak-to-peak linewidth with the orientation of the magnetic field has been observed.[137] A rather uncommon S = 4 spin state has been found to arise from the magnetic exchange interaction in [Mn(F$_3$bzac)$_2$(NITMe)]$_2$ (F$_3$bzacH = benzoyltrifluoroacetone; NITMe = 2,4,4,5,5-pentamethyl-Δ^2-imidazolyl-1-oxyl-3-oxide).[138]

A number of transition metal complexes have been used as counterions in solids containing TCNQ$^-$ radical anions (TCNQ = tetracyanoquinodimethanide) yielding both conducting[139,140] and insulating[141,142] solids. In most cases the e.s.r. signals are attributable to the organic radicals which stack one over the other and eccitonic behaviour could also be observed. The single crystal e.s.r. spectra of Cu(phen)$_3$(TCNQ)$_2$ (phen = 1,10-phenanthroline) have been interpreted using an S = 1/2 spin hamiltonian and have been found to be temperature dependent with g values g_x = 2.037(3), g_y = 2.052(3), g_z = 2.104(3) and g_x = 2.052(3), g_y = 2.057(3), g_z = 2.180(3) at room temperature and 4.2 K respectively. This temperature dependence has been attributed to interactions of the copper(II) ion unpaired electron with the mobile triplet excitons of the organic radicals.[142] A series of iron, ruthenium and rhodium halogen complex salts with TTF = tetrathiafulvalene has been also investigated[143]. A triplet excitonic spectrum has been observed in tris(N,N,N',N'-tetramethyl-p-phenylendiamine) bis(4,5-dimercapto-1,3-dithiole-2-thionato)nickelate which was attributed to the existence of N,N,N',N'-tetramethyl-p-phenylendiamine dimers. No evidence of interaction with the metal complex has been reported.[144]

Mixed Valence Systems. E.s.r. spectroscopy can give invaluable information on the localization of the valences on the metal centres, especially when the measurement of the hyperfine splitting is possible. E.s.r. spectroscopy has been used to characterize [LMo^{V}OCl(μ-O)Mo^{VI}O$_2$L] (L = hydrotris(3,5-dimethyl-1-pyrazolyl)borate), a d^1-d^0 system, in which the unpaired electron was found to be localized on the d^1 centre,[145] while a complete delocalization of the single unpaired electron at 4 K was claimed to occur in [Cu$_2$LH](ClO$_4$)$_4$ (L = N(CH$_2$CH$_2$N:CHCH:NCH$_2$CH$_2$)$_3$N), a d^9-d^{10} system.[146] The observation of 15 hyperfine lines in the 120 K frozen solution spectrum of [Cu^{II}Co^{III}(dpb)$_4$]$^+$ and [Cu^{II}Co^{III}(dpb)$_4$]$^-$ (dpb = N,N'-diphenylbenzamidinate ion) suggested that the unpaired electron is delocalized over the two cobalt centres. Similar results were obtained for the analogous rhodium complexes.[147] Trinuclear systems d^1-d^1-d^1 (V(IV)-V(IV)-V(IV))[148] and d^2-d^1-d^2 (Mo(IV)-Mo(V)-Mo(IV))[149] have been also investigated.

d^2-d^3 **systems.** Single crystal e.s.r. spectra of the mixed valence (PPh$_4$)$_2$[Cl$_3$W^{III}(μ-Cl)-(μ-SePh)$_2$W^{IV}Cl$_3$]·2CH$_2$Cl$_2$ dimer doped into the diamagnetic W(IV)-W(IV) analogue have been interpreted using a S = 1/2 spin hamiltonian with largely anisotropic g values: g_1 = 2.317, g_2 = 1.901, g_3 = 1.777. Hyperfine interactions with [183]W and [35,37]Cl have been also observed in some orientations.[150]

d^3-d^4 **systems.** Several manganese(IV)-manganese(III) systems containing a Mn$_2$O$_2$ moiety have been synthesized and their electronic structure has been investigated with magnetic susceptibility measurements and e.s.r. spectroscopy.[151-153] In any case an antiferromagnetic interaction was found to yield a doublet ground state well isolated from the excited quartet, sextet, and octet spin states which are at energies 3J/2, 8J/2, and 15J/2 respectively. The coupling constants, J, range form 230 to 440 cm^{-1}. The e.s.r. spectra are characterized by a 16 line hyperfine pattern with $A \approx 80 \times 10^{-4}$ cm^{-1} and g = 2.00. Manganese(IV)-manganese(III) couple is present also in the hexanuclear complex [Mn$_6$(3,2,3-tet)$_4$O$_6$(OAc)$_3$]I$_5$.[154] Other investigated systems are reported in references 155-157.

***d⁴-d⁵* systems**. A [Mn$_2$O$_2$] moiety containing a manganese(III)-manganese(II) couple has been found in the complex (HNEt$_3$)$_2$[Mn(biphen)$_2$(biphenH)] (biphenH$_2$ = 2,2'-biphenol). The coordinations around the manganese(II) and manganese(III) ions are distorted square pyramidal and octahedral respectively. A weak ferromagnetic coupling, J = -1.72 cm^{-1}, was found to be operative to yield a S = 9/2 ground state while generally a weak antiferromagnetic coupling was found in other similar systems to yield a S = 1/2 ground state. Magnetic data have been interpreted with a spin hamiltonian including axial zero field splitting of the single ion centres with g = 1.83, D(Mn^{II}) = 1.0 cm^{-1}, D(Mn^{III}) = -1.6 cm^{-1}. The X-band polycrystalline powder e.s.r. spectrum at 8 K showed broad resonances in the region $g \approx 2$ and $g \approx 5$, and a weak feature at g $\approx$ 12 appeared in frozen solution spectra. The transition fields as a function of the axial zero field splitting parameter have been computed.[158] These plots can be useful for the interpretation of the spectra and are reported in Figure 4.

A mixed valence couple manganese(III)-manganese(II) has been also found in the heptanuclear system [Mn$_7$(trien)$_2$(dien)$_2$O$_4$(OAc)$_8$](PF$_6$)$_4$·2H$_2$O yielding an e.s.r. spectrum with a signal centred at $g \approx 2$ and another one at $g \approx 3.9$.[159] These transitions are similar to those already observed in antiferromagnetically coupled systems and could arise from transition within the S = 1/2 ground state and within the S = 3/2 next excited state.

***d⁵-d⁶* systems.** Considerable attention is always given to tetranuclear complexes with the cubane structures. A telluride containing the core [Fe$_4$Te$_4$]$^+$ has been synthesized and characterized through magnetic susceptibility, e.s.r., and 1H-NMR. This complex formally contains three high-spin iron(II) and one high-spin iron(III) centres. A ground state S = 3/2 has been suggested.[160] A S = 3/2 ground state has been identified for the mixed-metal cubane core [NiFe$_3$Se$_4$]+, whose 3 K e.s.r. spectrum show resonances at

g_1 = 4.6, g_2 = 3.3 and g_3 = 1.95[161]. A S = 1/2 spin ground state has been found in the cluster core [Fe$_3$CoS$_4$]$^{2+}$.[162] An asymmetric Fe$_4$S$_5$ core has been found in [Fe$_4$(µ$_3$-S$_2$)(µ$_3$-S)$_3$(MeCp)]$^+$ to yield a S =1/2 e.s.r. spectrum at 4.2 K with g_1 = 2.128, g_2 = 2.008, g_3 = 1.981.[163]

A hexanuclear cluster [Fe$_6$S$_5$(µ-SPh)(n-Bu$_3$P)$_4$(SPh)$_2$] formally containing five iron(II) and one iron(III) centres gave e.s.r. signals at 77 K assigned to a S = 1/2 spin state with $g_{//}$ = 2.063, $g_\perp$ = 2.023.[164] No magnetic results are presented to complement this observation.

E.s.r. spectroscopy was further applied to characterize mixed valence systems in references 165,166.

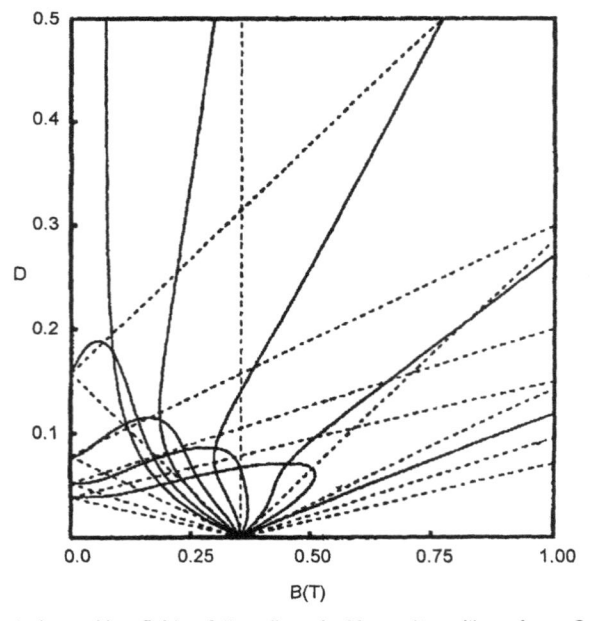

Figure 4. Computed transition fields of the allowed ΔM = ± 1 transitions for a S = 9/2 spin system as a function of the axial ZFS parameter D with hν = 0.314 cm^{-1}. Solid and dashed lines refer to perpendicular and parallel orientations of the static magnetic field respectively.[158]

Binuclear and Oligonuclear Complexes. - The compounds are grouped according to the number of interacting transition metal ions as binuclear (two metals), tri- and tetranuclear, and higher nuclearity complexes.

E.s.r. spectroscopy has been widely applied in the last ten years to characterize both the spin of the ground state and the nuclearity of oligonuclear transition metal ion complexes. Binuclear systems are, of course, the simplest systems to be studied and magneto-structural correlations have been suggested in a number of cases, particularly for copper(II) binuclear complexes.[54,55] Recent applications of e.s.r. moved from simple copper(II)-copper(II) dimers to complexes containing a larger number of electrons like iron(III)-iron(III) dimers and/or to systems with nuclearity higher than two. On increasing the number of magnetically interacting electrons, the complexity of the e.s.r. spectra increases and their analysis becomes therefore much more complicated. The applications of the e.s.r., as a consequence, result for the moment limited to attempts of detecting the exact nature of the spin state, of confirming the analysis of magnetic and/or Mössbauer data, or of determining the number of the magnetically interacting nuclei, when structural data are not available. More detailed spectroscopic and theoretical interpretations have been often postponed. In the following we will try to evidence the most relevant characteristics of the observed spectra and the most straightforward applications of e.s.r. are only mentioned without further comments.

Binuclear complexes. The largest number of binuclear complexes studied by e.s.r. spectroscopy are copper(II) compounds, in which the isotropic magnetic interaction generates S = 1 and S = 0 spin states. Since in most cases the zero field splitting D of the triplet state is smaller than hv, the triplet state can be easily recognized by e.s.r.. Single crystal room temperature and variable temperature polycrystalline e.s.r. spectra have been reported for ferromagnetically coupled Cu - Cu pairs in the [Zn(dnpdtc)$_2$]$_2$ (dnpdtc = *N,N*-di-*n*-propyldithiocarbamato) complex. The measured zero-field splitting increases on lowering the temperature, and this behaviour has been attributed to a thermal contraction of the host lattice.[167]

Selected complexes for which polycrystalline powder spectra have been measured are: [Cu(tn)Cu(pba)]·4H$_2$O, [Cu(bipy)Cu(pba)]·H$_2$O and

[Cu(dien)Cu(pba)]·H$_2$O[168] (pba = 1,3-propanediylbis(oxamato), tn = 1,3-diaminopropane, bipy = 2,2'-bipyridine, dien = diethylenetriamine), [(mpym)$_2$ (H$_2$O)$_2$(NO$_3$)$_2$ Cu$_2$(C$_2$O$_4$)]·2H$_2$O (mpym = 4-methoxy-2-(5-methoxy-3-methyl-1H-pyrazol-1-yl)-6-methylpyrimidine),[169] [Cu(pAb)(phen)(H$_2$O)]$_2$(NO$_3$)$_2$ ·2pAbH·2H$_2$O and [Cu(DPP)(4,7-phen)(H$_2$O)]$_2$(NO$_3$)$_2$ · 2H$_2$O (pAbH = p-aminobenzoic acid; DPPH = 3,4-dimethoxyhydrocinnamic acid; phen = 1,10-phenanthroline; 4,7-phen = 4,7-dimethyl-1,10-phenanthroline),[170] [(TmenCu)$_2$Tae(ClO$_4$)$_2$·H$_2$O] and [(TmenCu)$_2$Tae· H$_2$O](BPh$_4$)$_2$ (Tmen = N,N,N',N'-tetramethylethylenediamine, Tae = tetraacetylethanato(2–)),[171] [Cu$_2$Cl$_3$(C$_7$H$_6$N$_2$)$_5$]Cl·4H$_2$O,[172] [Cu$_2$(bipy)$_2$(OH)$_2$(CF$_3$SO$_3$)$_2$][173] (bipy = 2,2'-bipyridine) and CuX$_2$(H$_2$O)$_2$ and CuX$_2$(py) (X = 2-fluoro-6-chlorobenzoate).[174]

Solution spectra have been used to characterize dicopper(II) complexes formed with the 3,12-bis(carboxymethyl)-6,9-dioxa-3,12-diazatetradecanedioic acid,[175] the binucleating ligand 2,6-bis[(bis ((1-methylimidazol-2-yl)methyl)amino)methyl]-4-methylphenol,[176] the tetradentate ligand 1,2,4,5-tetrakis(4,5-dihydroimidazol-2-yl)benzene, (BTIM), [Cu$_2$(BTIM)$_2$Cl$_2$]Cl$_2$·7H$_2$O and [Cu$_2$(BTIM)$_2$(N$_3$)$_2$](N$_3$)$_2$·5H$_2$O,[177] and the N-substituted salicylideneamines (X-sal-R) (X = H, 5-Cl, 5-Br, R = N-alkyl).[178]

E.s.r. spectra of binuclear copper(II)[179,180] and nickel(III)[181] complexes formed by spiro macrobicycles ligands with aza and thia-aza donor set have been also reported.

The μ-oxo dimeric species [O=Mo^V(TPP)]$_2$O (TPP = tetraphenylporphyrin), obtained by spontaneous decomposition of the photoproduct of [O=Mo^{IV}(TPP)], has shown a triplet state e.s.r. spectrum with D = 3.35·10^{-2} cm^{-1} and |E| = 7.6·10^{-4} cm^{-1}, clearly distinguishable from that of the precursor.[182]

The bimetallic chains complexes MM'(EDTA)·6H$_2$O, [MM'], built up with alternated hydrated, M, and chelated, M', metallic sites, with M = Co and M' = Co, Cu have a high degree of alternation in the coupling constants, and behave as quasi-isolated dimers. The [CoCo] complex is e.s.r. silent, while the single crystal spectra of [CoCu] have be interpreted with a triplet spin hamiltonian, the triplet state arising from the coupling between the quasi-isotropic doublet of copper(II) and the anisotropic lowest Kramers doublet of the cobalt(II) ion.[183] The hyperfine structure of ^{59}Co in cobalt(II) doped CsMgCl$_3$ single crystals has been used to evidence the formation of binuclear moieties and to characterize them.[184]

X-band frozen solution spectra have been reported for a series of paramagnetic [Rh$_2$]$^{3+}$ complexes[185] and for a [Rh$_2$]$^{5+}$ species.[186] In complexes of the type [Ru$_2$Cl$_5$(PR$_3$)$_4$] (with R = Bu, Me or R$_3$ = Me$_2$Ph), formally containing a [Ru$_2$]$^{5+}$ species, both symmetrical and asymmetrical coordination around the two tri-µ-chloro bridged metal ions has been observed. The delocalization of the unpaired electron in the two kinds of compound has been discussed on the basis of structural, e.s.r. and electrochemical data.[187] The frozen solution e.s.r. spectra of the asymmetrical tri-µ-chloro bridged [Ru$_2$Cl$_6$(PEt$_3$)$_3$][187] and bi-µ-chloro bridged [Ru$_2$Cl$_6$(PR$_3$)$_4$][188] (R = Et, Bu) diruthenium(III) complexes have been interpreted as due to two non-equivalent low-spin d^5 ions.[188] E.s.r. spectra of trinuclear tri-µ-chloro bridged rhodium compounds have been also reported.[189,190]

Magnetic interactions between ions having S > 1/2 often generate a number of states having different spin multiplicity, which can be thermally populated even at very low temperature. This is a difficulty which adds to the intriguing problem of the interpretation of the e.s.r. spectra of spin systems with S > 1, both integer and non integer, and their interpretation can be a matter of controversy[191,192]. In [Ni$_2$(BPMP)(O$_2$CR)$_2$]$^{2+}$ (HBPMP = 2,6-bis[[bis(2-pyridylmethyl)-amino]-methyl]-4-methylphenol; R = Me, Et) a S = 3/2 ground state well isolated from other ones, with $g_{//}$ = 2 and $g_\perp$ = 5.7, was found. It could originate from ferromagnetic coupling between high-spin nickel(II) (S = 1) and low-spin nickel(III) (S = 1/2).[193] In [(LH$_2$)$_2$Mn$_2$(µ-OH)]$^+$ (LH$_2$ = 4,7-bis(2-hydroxybenzyl)-1-oxa-4,7-diazacyclononane) the manganese(II) ions are antiferromagnetically coupled and the e.s.r. spectrum showed a complicated pattern with signals belonging to different spin multiplets.[194]

The complexes [Fe$_2$(BPMP)(OPr)$_2$](BPh$_4$)·0.8CH$_2$Cl$_2$ and [Fe$_2$(BPMP)(OBz)$_2$](BPh$_4$)·0.8CH$_2$Cl$_2$ (BPMP = 2,6-bis[[bis(2-pyridylmethyl)-amino]-methyl]-4-methyl-phenolato, OPr = µ-O,O'-propionato, OBz = µ-O,O'-benzoato), where a [Fe$_2$O(carboxylate)$_2$] moiety is present, contain iron(II)-iron(II) couples which have been characterized through Mössbauer, NMR and e.s.r. spectroscopies. The iron(II) centres are high-spin (S = 2). The e.s.r. spectra, recorded in frozen solution at 3K, showed only low field resonances at g = 16, taken as indicative of a ground state with integer spin.[195] A similar spectrum has been observed also in the asymmetric, triply carboxylate bridged diiron(II) compound [Fe$_2$(BIPhMe)$_2$(O$_2$CH)$_4$] (BIPhMe = 2,2'-bis(1-methyl-imidazolyl)-phenyl-methoxy-methane). In this complex magnetic susceptibility data indicated the absence of a significant magnetic exchange coupling between the iron(II) ions.[196] These last examples show that e.s.r. spectra which look similar can arise from rather different spin states. A signal at $g_{eff} \approx 16$ was also observed in the fully reduced form of methane monooxygenase hydroxylase[197] and this observation originated efforts in the synthesis of possible synthetic models of the enzyme.

Antiferromagnetic coupling between iron(III) (S = 5/2) and nickel(II) (S = 1) in [Fe^{III}Ni^{II}(BPMP)(OPr)$_2$](BPh$_4$)$_2$ yields a S = 3/2 ground state whose e.s.r. spectra have been interpreted with D = 0.7 cm^{-1} and E/D = 0.32.[198]

The g = 12 e.s.r. signal observed for the iron(III)-copper(II) monobridged couple in [Fe^{III}Cu^{II}(BPMP)Cl)$_2$](BPh$_4$)$_2$ has been attributed to the integer spin of the ground state of the couple. Magnetic measurements are not available.[199]

The complex [Mn^{III}Mn^{IV}(L)$_2$(μ-O)$_2$](ClO$_4$)$_3$ (L = tris(2-methylpyridyl)amine, and 1,4,8,11-tetraazacyclotetradecane) contains an antiferromagnetically coupled valence trapped manganese(III)-manganese(IV) dimer with S = 1/2 ground state for which ENDOR spectra have been reported.[200]

Tri- and Tetra-nuclear complexes - A number of papers have been dealing with the synthesis and characterization of polynuclear complexes containing manganese ions in order to find structural models of naturally occurring polymanganese moieties and mimic the fundamental processes of photoactivation in photosystem II of green plants.[200-203] A strong antiferromagnetic coupling was found to be operative between the metal ions in the trinuclear complex [(bpy)$_4$Mn^{IV}$_3$(μ-O)$_4$(H$_2$O)$_2$]$^{4+}$(J_{12} = 182 cm^{-1} and J_{13} = J_{23} = 98 cm^{-1}) and results in a S = 1/2 ground state. The frozen solution e.s.r. spectrum has been interpreted using g = 1.965, A_3 = 66 · 10^{-4} cm^{-1} and A_{12} = 61 · 10^{-4} cm^{-1} (65.3 G). A comparison has also been made with previously reported assignments.[201] The photolysis process of the tetranuclear [Mn$_2$^{II}Mn$_2$^{III}(μ-O)$_2$(O$_2$CCPh$_2$)$_6$(OEt$_2$)$_2$] complex, both in absence and presence of strongly donating ligands, has been followed through solution e.s.r. spectra.[202] For the distorted cubane mixed valence manganese complexes, [Mn^{IV}Mn$_3$^{III}(μ-

$O)_3(\mu\text{-Cl})(O_2CR)_3Cl_{6-x}(L)_x] \cdot nCH_3CN^{203}$ (R = Me, Et; L = imidazole, pyridine; x = 1, 3; n = 1.5, 2.5) the complicated polycrystalline and frozen solution e.s.r. spectra have been suggested to be consistent with the ground S = 9/2 state evidenced by magnetic susceptibility measurement, with D ≈0.3 cm⁻¹.

Low temperature single crystal e.s.r. spectra of the $\{Cu_3[C_2S_2(NCH_2CH_2CH_2SCH_2CH_2OH)_2]_2\}X_2$ (X = ClO_4^-, NO_3^-) complexes have been measured. The trinuclear core is formed by a central CuS_4 and two terminal $CuS_2N_2O_2$ chromophores. Intermolecular exchange interactions between trinuclear units resulted in exchange narrowed spectra.[204] This situation has been often met in the past for trinuclear copper(II) complexes and has always prevented any detailed analysis of the e.s.r. spectra.

Tetrametallic clusters M_4O_{16} (M = Cu and Co), encapsulated in heteropolyoxotungstate complexes derived from Keggins and Dawson-Wells structures, have been studied. Antiferromagnetic interactions lead to a triplet ground state in the copper clusters, evidenced through e.s.r. polycrystalline powder spectra.[205]

E.s.r. spectra have been reported for tribromo[4-((ethylthio)methyl)-5-methylimidazole]copper(II)copper(I)[206] and for the tetranuclear $(\mu\text{-}Y)(DN)_4Cu_2^{II}Cu_2^{I}X_4$ (DN = N,N-diethylnicotinamide; Y = 3,4,5,6-tetrachlorocatecholate; X = Cl, Br, I) complexes containing mixed valence copper(II)-copper(I) couples.[207]

E.s.r. spectra of heteropolymetallic systems like the heterotetranuclear complexes $[\{Cu(oxpn)\}_3M]^{2+}$ (M = cadmium(II), nickel(II) and manganese(II))[208] and trinuclear Cu-M^{II}(dmg)-Cu, (dmg = dimethylglyoximato, M = Ni, Pd) with 1,4,7-trimethyl-1,4,7-triazacyclononane or N,N,N',N'-tetramethylethylenediamine as capping ligand have been reported.[209]

E.s.r. active silver clusters in zeolites rho have been generated by reduction of silver exchanged samples with molecular hydrogen. The stable Ag_4^{n+} (n = 1 or 3) species show a five-lines multiplet centred at g = 1.973 with an isotropic hyperfine constant A = 140 G indicative of the interaction of a single unpaired electron with four equivalent nuclei with I = 1/2. A comparison with the results obtained for other silver exchanged zeolites have been reported.[210]

Higher nuclearity complexes - A S = 1/2 ground state has been assigned to the hexamolybdenum complexes [(Mo$_6$X$_7$Y)X'$_6$]$^{2-}$ (X = X' = Cl, Br; Y = S, Se) whose electronic structure has been discussed on the basis of frozen solution e.s.r. data.[211] The electronic structure of the hexanuclear octahedral cluster [Co$_6$(μ_3-S)$_8$(PEt$_3$)$_6$]BPh$_4$ has been studied through e.s.r, ^{1}H-NMR spectroscopies and magnetic susceptibility measurement. The complex formally contains five diamagnetic cobalt(III) centres and one cobalt(II) ion (S = 1/2). The single crystal e.s.r. spectra have been found to be extremely temperature dependent, and at 4.2 K have been interpreted with g_1 = 1.95(1), g_2 = 2.04(1), g_3 = 2.35(1).[212] A S = 10 ground state has been evidenced in [Mn$_{12}$O$_{12}$(CH$_3$COO)$_{16}$(H$_2$O)$_4$]·2CH$_3$COOH·4H$_2$O. No e.s.r. spectra have been detected at X- and Q-band frequency, while polycrystalline powder spectra at 4.2 K have been obtained with a high frequency spectrometer operating with a far infrared laser from 17.5 cm^{-1} to 8.2 cm^{-1}. The spectra were interpreted with a D value of ≈ 0.3 cm^{-1}.[213]

Extended Systems. In the last few years the interest of researchers for linear chains of non-integer spins increased, due to the fact that, in these systems, a non-magnetic ground state separated in zero field from magnetic excited states is anticipated[214,215]. The energy separation between the diamagnetic ground state and the next excited state is often called the "Haldane gap". One of the most thoroughly investigated systems is the Heisenberg linear chain antiferromagnet [Ni(en)$_2$NO$_2$]ClO$_4$ (en = ethylenediamine) whose magnetic susceptibility data have been reasonably well understood within the antiferromagnetic S = 1 linear chain model including a zero-field splitting |D| ≈ 8.5 cm^{-1}.[216] From high field magnetization studies a Haldane gap of ≈ 10 cm^{-1} was estimated.[217] This complex gave an e.s.r. signal in the 50 GHz region under the pulsed magnetic field which is temperature dependent. In the temperature range 300-35 K, a broad magnetic resonance with g ≈ 2 and line shape intermediate between Gaussian and Lorentzian has been observed and assigned as an exchange-mixed paramagnetic line where the fine structure components are modulated by the intrachain exchange interactions. Below 20 K a new resonance appeared whose intensity decreases with temperature. This resonance has been assigned to a locked

1: Transition Metal Ions   27

cluster of two spins, with resultant spin S = 1, immersed in a chain, where the resonance occurs between the sub levels of the bound state. The angular dependence of the resonance field was fitted using a rhombic S = 1 spin hamiltonian with D = 7.5 cm^{-1} and E = 0.7 cm^{-1}.[218] Other studies on this and related systems have been reported by Date and Kindo.[219] Upon doping [Ni(en)$_2$NO$_2$]ClO$_4$ with a small amount of copper(II) a new signal whose intensity increases with temperature below 10 K was observed at X band frequency.[220-222] The observed signal was attributed to a magnetically interacting copper(II)-nickel(II) couple. The intensity of the signal, I, has been reproduced with the equation (17)

$$I = I_0\{\exp(E_G/kT)-1\}\{1-\exp(-\delta/kT)\} \quad (17)$$

where E_G is the Haldane gap and the third term comes from the Boltzmann distribution among the energy levels giving rise to the e.s.r signal. A reasonable fit with experiment was obtained with E_G = 9.0 cm^{-1} and $\delta \approx 0.35$ cm^{-1}. E.s.r. experiments on samples doped with diamagnetic impurities showed spectra corresponding to the fractional spin S = 1/2 states at the "open" ends of the nickel(II) chains adjacent to the non magnetic substitutions.[223] CsNiF$_3$ is another system under current investigation using e.s.r. spectroscopy.[224]

d^9 configuration. Bivalent copper. Calvo et al. continued a systematic study on exchange coupled systems formed by copper(II) and several amino acids, which often showed low-dimensional (mono- and bi-dimensional) features. They report X-ray structures, magnetic susceptibility measurements, and single crystal e.s.r. results. In the case of exchange coupling between magnetically inequivalent copper(II) ions, A and B, linked by carboxylate bridges the angular dependence of the peak-to-peak linewidth ΔB_{pp} of the exchange-collapsed e.s.r. line was found to follow equation (18)

$$\Delta B_{pp}(\theta,\phi) = a_1 \sin^2\theta \cos^2\phi + a_2 \sin^2\theta \sin^2\phi + a_3 \cos^2\theta + a_4 \sin 2\theta \cos\phi + \\ + a_5[g_A(\theta,\phi) - g_B(\theta,\phi)]^2 + a_6 \cos^4\theta \quad (18)$$

where θ and φ are the angular co-ordinates of the static magnetic field, the coefficients a_i with $i \leq 5$ represent contributions from magnetic dipolar, hyperfine interactions and anisotropic and antisymmetric exchange terms, and a_6 is related to the intercopper isotropic exchange interaction, J, by

$$a_6 = \frac{v^2 h^2 \sqrt{2\pi/3}}{8|J|\mu_B g^3} \qquad (19)$$

In (19) v is the Larmor frequency and g is the averaged gyromagnetic factor.[225,226] The angular variation of the g^2 values has been found to be influenced also by fourth-order contributions.[227] The exchange coupling between inequivalent magnetic sites in the one-dimensional solid [(2-aminopyridinium)CuCl$_4$] was estimated from the angular dependence of the linewidth of the e.s.r. signal.[228] Also in (2,2'-bipyridine-3,3'-dicarboxylic acid)dichlorocopper(II) monohydrate intramolecular exchange interactions were found to have sizeable effect on the e.s.r. spectra.[229]

Effect of low dimensionality on the e.s.r. spectra of some copper(II) aminoacid complexes can be found in references 230-234.

The single crystal e.s.r. spectra of [(4-picolinium)Cu$_3$Cl$_8$] and [(1,1,4-trimethylpiperazinium)Cu$_3$Cl$_8$], a one-dimensional system formed by a linear stacking of Cu$_3$Cl$_8$ moieties, have been reported.[235] The temperature dependence of the e.s.r. linewidth in one-dimensional systems formed by copper(II) trimers, like [(4-picolinium)Cu$_3$Cl$_8$], was computed to be approximately proportional to $\sqrt{T}$ when the intratrimer isotropic exchange is the dominant contribution.[236] In KCuF$_3$, a one-dimensional Heisenberg antiferromagnet, the Dzyaloshinski-Moriya antisymmetric exchange caused a temperature dependence of the resonance field observed in the direction perpendicular to the chain axis.[237]

Other investigated systems are [Cu(mpym)(H$_2$O)(C$_4$O$_4$)]·2H$_2$O (mpym = 4-methoxy-2-(5-methoxy-3-methyl-1H-pyrazol-1-yl)-6-methylpyrimidine; C$_4$O$_4{}^{2-}$ = squarato),[238] [BEDT-TTF]$_8$[CuBr$_4$]Br$_3$ (BEDT-TTF = bis(ethylenedithio) tetrathiafulvalene),[239] Cu(1-phenylpyrazole)$_2$Cl$_2$[240] and [Co(en)$_3$]$_2$[Cu$_2$Cl$_8$]Cl$_2$·2H$_2$O.[241] In these last two compounds the effect of pressure on the exchange coupling constants was studied.

d^5 **configuration.** *Bivalent manganese.* Only a few polynuclear manganese(II) containing systems have been studied in the last two years. In the most studied system (Me$_4$N)MnCl$_3$ (TMMC) the chain of manganese ions behaves as an ideal S = 1/2 spin chain, while in the chain complex [Mn$_2$(EDTA)·9H$_2$O] the angular dependence of the linewidth of the e.s.r. signal showed observable deviations from $(3\cos^2\theta - 1)^n$ (n = 4/3, 2) which have been attributed to the zero-field splitting of the manganese(II) ion arising from the low symmetry of the co-ordination site.[242,243]

Other investigated systems are $KNi_xMn_{1-x}F_3$,[244] $Rb_2Mn_xCr_{1-x}Cl_4$,[245] $[MnL_2L'(H_2O)_2]$ (L = imidazole; H_2L' = croconic acid),[246] and manganese(II) doped $CoCl_2 \cdot 2H_2O$.[247]

Other systems. A number of investigations on the nature of the magnetic ordering and on the effect of doping extended systems with paramagnetic ions have been performed. Among these we cite $Mg_{1-x}Fe_xCl_2$,[248] $FeTiO_3$,[249] two dimensional iron(II) polymers,[250] $[M(chxn)_2Br]Br_2$ (M = Pd, Pt, Ni; chxn = 1R,2R-cyclohexanediamine),[251] β-$LiVOPO_4$,[252] graphite intercalated $CrCl_3$,[253,254] two-dimensional $CrCl_3$,[255] $Cr_2V_{4-x}Mo_xO_{13+x/2}$,[256] $[Mo_2(O_2CCF_3)_4(9,10\text{-anthraquinone})]$,[257] layered metal phosphates intercalated Mo_2^{5+}.[258]

Single crystal e.s.r. spectra of the ferrimagnetic alternating spin chain compounds $[Cu(hfac)_2NITiPr]$ (hfac = hexafluoroacetylacetonate; NITiPr = 2-isopropyl-4,4,5,5-tetramethyl-4,5-dihydro-1H-imidazolyl-1-oxyl-3-oxide)[259] and $[Mn(hfac)_2NITPhOMe]$ (NITPhOMe = 2-(4-methoxyphenyl)-4,4,5,5-tetramethylimidazoline-1-oxy-3-oxide)[260] have been reported. In the last compound a *g* shift is observed parallel and perpendicular to the chain axis as a function of temperature. This shift is one of the characteristics of ferro-, antiferro-, and one-dimensional ferrimagnetic systems. Feeble magnetic interactions not observable with magnetic susceptibility measurements have been detected using e.s.r. spectroscopy in the chain compound $[MnCu(Br_4obbz)(H_2O)_3]\cdot 5H_2O$ (Br_4obbz = oxamido-N,N'-bis(3,5-dibromobenzoato)).[261]

Following the temperature dependence of the e.s.r. signal a neutral soliton state has been found to occur in $[Pt(en)_2][Pt(en)_2Cl_2](ClO_4)_4$ in which dimeric states of platinum(II) ions optically excited are mobile.[262]

Semiconductors. The characterization of the current carriers in semiconducting solids obtained from molecular systems is surely the most common application of the e.s.r. spectroscopy to semiconductors. Ligand centred oxidation has been found to occur in [Au(dmit)$_2$]$^{n-}$ and [Au(dsis)$_2$]$^{n-}$ (H$_2$dmit = 4,5-dimercapto-1,3-dithiole-2-thione; H$_2$dsis = 4,5,-di(hydroseleno)-1,3-diselenole-2-selone),[263] and in [Cu$_4$(dmit)$_3$]$^{n-}$ and [Cu$_4$(dsis)$_3$]$^{n-}$ which form semiconducting solids.[264] Ni(dmit)$_2$ forms a semiconducting solid with the organic π donor tetraphenyldithiapyranylidene. The e.s.r. spectrum can be interpreted with a S = 1/2 spin hamiltonian with g_x = 2.003, g_y = 2.035, and g_z = 2.046. Since both the donor and the acceptor have one single unpaired electron, this indicates that some exchange coupling is operative.[265] Exchange interactions between itinerant electrons and localized spins caused a linear relationship between ΔB_{pp} and temperature in (perylene)$_2$[M(S$_2$C$_2$(CN)$_2$)$_2$] (M = Pd, Pt)[266] and in copper-phthalocyaninato systems.[267] Further studies of the exchange interactions between conduction electrons and electronic spins localized on transition metal complexes are reported in references 268, 269.

The electrical conduction in the mixed metal complex K[Pt$_{1-x}$Au$_x$(mnt)$_2$]·2H$_2$O (mnt = maleonitriledithiolene) increases of almost two order of magnitude when x = 0.04 with respect to x = 0 and a decrease in g anisotropy is observed.[270] The effect of the composition on the e.s.r. linewidth has been studied in the semimagnetic semiconductor Pb$_{1-x-y}$Sn$_y$Mn$_x$Te,[271] in cobalt copper chromium sulphide selenide,[272] and in ruthenium sulphide selenide.[273] Narrowing of the e.s.r. line in TiBe$_2$ in the temperature range 1.55-1.8 K has been ascribed to exchange interactions arising from a spontaneous weak thermal ferromagnetism.[274]

Transition metal ions are often added as impurities to semiconductors and used as probes to investigate the nature of defects and to determine the position of the conduction band with respect to the energy levels of the transition ion.[275-280] The limits of the application of e.s.r. spectroscopy (and other spectroscopic techniques) as compared to positron annihilation in studying defects in semiconductors have been reviewed.[281]

Superconductors. A large number of papers have been dealing with the characterization of the high-T$_c$ superconductors obtained from copper oxides and extensive use of e.s.r. and, in general, of microwave absorption has been made. Two review articles dealing with the application of e.s.r. and non-resonant microwave reflection or absorption appeared.[282,283] The most recent one,[283] which is specifically dealing with microwave absorption, gives 277 references. The ceramic cuprate superconductors have been classified as type II superconductors although the combined effect of thermal fluctuations and impurities leads to phase diagrams and transport properties in magnetic fields that are radically different from those of low-T_c type II superconductors.[284] As already done in the previous issue[1] we will mention in this section only some of the most relevant applications of e.s.r. to the characterization of superconductors following the areas of application we have outlined there.

It is well established that the transition to the superconducting state in the ceramic cuprate superconductors is accompanied by an intense field dependent absorption of microwave energy at low magnetic field. The appearance of this signal is always accompanied by severe changes of the resonance frequency and of the Q-factor of the e.s.r. cavity. Magnetically modulated microwave absorption (MAMMA), which observes only magnetic field dependent changes, has been applied to characterize superconducting transitions using a conventional e.s.r. spectrometer. The possible inconveniences arising from microwave responses due to Josephson junctions, flux-trapping, etc..., at grain boundaries have also been discussed.[285] The absorption of microwave radiation due to the Josephson effect (JMA), which arise between the grain boundaries in the ceramic superconductors, can be also measured with a standard e.s.r. apparatus and application to the determination of the critical temperature have been reported.[286,287] Magnetically modulated microwave reflection (MMR) can be measured using a modified e.s.r. spectrometer and has been used to measure the critical temperature. The MMR signal is in fact proportional to changes in the Q-factor of the cavity.[288]

E.s.r. has been used to measure the angular variation of the upper critical field, H_{c2}, in single crystals of $YBa_2Cu_3O_7$ by sweeping the temperature across T_c at a constant and very low magnetic field.[289]

Together with the critical temperature and the values of the upper and lower critical fields, type II semiconductors are characterized by the knowledge of the magnetic flux distribution in the mixed state. The broadening of the signal of radicals (diphenylpicrylhydrazil,[290-295] bisphthalocyaninelutetium[296]) absorbed on the surface of the material has been related to the inhomogeneities in the magnetic flux through the sample, allowing the determination of the London penetration depth, λ, in $YBa_2Cu_3O_7$ and related systems with an estimated accuracy $\approx$ 4% or better.[297] In $YBa_2Cu_3O_7$ these measurements ruled out the Abrikosov vortex lattice as the main source of the broadening, which is therefore attributed to the grain structure of the material.[298] Kevan et al. showed that some complications in the measurement of the penetration depth can arise from the fact that the broadening depends also on the parallel or perpendicular alignment of the paramagnetic probe in the magnetic field.[299] The application of conduction e.s.r. (CESR) to the characterization of superconductors has been considered and a comparison between conventional and high T_c superconductors has been presented.[300]

E.s.r. spectroscopy has been used to investigate the nature of the magnetic interactions in insulators whose structure is related to those of the superconducting solids, such as CuO,[301-303] Bi_2CuO_4,[304] $La_4Cu_2Cu_2O_{10}$,[305] $YBa_2Cu_3O_{7-\delta}$,[306] Eu_2CuO_4,[307] La_2CuO_4,[308,309] $Bi_2Sr_2Ca_{n-1}Cu_nO_{2n+4}$ (n = 1,2,3).[310] The e.s.r. spectrum of Y_2BaCuO_5 has been studied[311,312] and this compound has been assigned as a common impurity in YBaCuO superconductors.[312]

The strong microwave absorption at low field (LFMA) observed in the ceramic cuprate superconductors is still being extensively investigated in order to clarify its origin[313-316] and the possibility of using this absorption to characterize the superconductors.[317-322] Low field absorptions have been measured also in a number of rare earth cuprates of the type R_2CuO_4[323] and their similarities and differences with those observed in superconducting oxides have been discussed.[324] Selected studies on LFMA concern the effect of the microwave power,[325,326] hysteresis in applied magnetic fields,[327,328] temperature and magnetic field dependence,[329,330] magnetic field modulation,[331] anisotropy.[332] On cooling ceramic cuprates superconductors in zero magnetic field and using appropriate values of the temperature and of the scan rate, a large number of equally spaced lines can be seen in the e.s.r. spectrum. These lines are attributed to tunnelling of the current through the grains which exhibit flux quantization, so that flux quantized oscillations arise.[333,334] Time dependent effects related to trapped magnetic flux in LaSrCuO systems have been studied.[335]

At temperatures higher than T_c resonances can be often observed at $g \approx 2.0$. A large number of papers report on these signals and sometimes different mechanisms are invoked to explain their origin. Selected references report e.s.r. in BiPbSrCaCuO,[336-340] MBa$_2$Cu$_3$O$_{7-\delta}$.[341-348]

Other studies report on the e.s.r. spectra in YBa$_2$Cu$_3$O$_7$ and Bi$_2$Sr$_2$CaCu$_2$O$_8$,[349,350] in YBa$_2$Cu$_{3-x}$Sn$_x$O$_{7-\delta}$,[351] RBaCuO (R = Y, Eu, Gd),[352] MnVMoO systems.[353] Several systems have been doped with either diamagnetic or paramagnetic transition metal ions and the effect of doping on the superconducting properties and on the e.s.r. spectra have been studied. Selected references are 354-361.

3 S = 1/2

d^1 **configuration.** - *Tervalent titanium and zirconium.* The electron paramagnetic resonance has been used to analyze the environment and the reactivity of titanium (III) centres in solids having a relevant catalytic and technological interest such as silica substrates,[362] ZSM-5 zeolites[363] and Ti(OR)$_4$-AlR'$_3$ (R = Bu, Et and R' = Et, Me) systems.[364,365] Single crystal studies have been performed on hydrothermally grown potassium titanyl phosphate[366], KTiOPO$_4$, and on the [TiO$_4$/Na]$_A^0$ centre of X-irradiated α-quartz.[367] E.s.r. and ENDOR spectroscopies have been employed to characterize titanium(III) sites in titanium(III) doped LaMgAl$_{11}$O$_{19}$.[368,369] The most populated sites have been identified in the ENDOR experiments and a defect model has been proposed for the charge compensation of the Mg(II) ions substituting Al(III) in sites adjacent to Ti(III).[368] The e.s.r. spectra of the ilmenite/corundum type solid solution series Mg$_{1-x}$Ti$^{IV}_{1-x}$Ti$^{III}_{2x}$O$_3$ and of the spinel-type compounds Mg(Mg$_{1-x}$Ti$^{IV}_{1-x}$Ti$^{III}_{2x}$)O$_4$ have been compared in order to analyze the influence of the structure and of the composition on the local environment of the Ti(III) ions.[370]

In the X-ray irradiated yttria-stabilized zirconia at least three distinct types of paramagnetic defects have been identified using e.s.r.. For one of them, exhibiting axial geometry, the authors proposed a model of intrinsic defect consisting of an electron trapped in a 4d orbital of a zirconium(III) ion in a strongly distorted crystal field generated by seven or six oxygen atoms in presence of one or two vacancies, respectively. Crystal field calculations have been performed to interpret the e.s.r. results.[371]

Quadrivalent vanadium and niobium. Due to the high sensitivity of its e.s.r. spectra to the local environment, vanadyl ion has been extensively used as probe to study defect properties and structural changes of a number of host lattices such as single crystals of triammonium citrate[372] and potassium-zinc sulphate,[373] and alkali cadmium borosulphate glasses.[374] Single crystal studies have been performed on VO^{2+} doped in cesium oxalate monohydrate (COMH),[375] where the vanadyl ion enters in the lattice interstitially, with an overall C_{4v} symmetry of oxygen atoms,[375] on vanadyl doped in zinc maleate tetrahydrate,[376] and on a series of $VOCl_2L_2$ complexes (L is a N,N'-alkyl substituted urea) doped into the corresponding trichloro indium(III) compounds,[377] for which angular overlap calculation have been also performed. E.s.r. studies have been performed on a series of vanadyl doped Tutton salts $M(NH_4)_2(SO_4)_2 \cdot 6H_2O$, with M = Zn^{2+} (ZASH),[378] Cd^{2+} (CASH)[379] and Mg^{2+} (MASH),[379] Fe^{2+} (FASH)[380] and Co^{2+} (COASH).[381] When the vanadyl ion substitutes for a metallic ion in the Tutton salts lattices, it expels one of the water molecules of the co-ordination environment of the host ion, forming a species adequately described by a $[VO(H_2O)_5]^{2+}$ model complex. The V=O bond can orient itself along one of the three possible directions of the tetragonally distorted octahedral symmetry, yielding three physically equivalent but magnetically non-equivalent sites, occupied with different population. For all the Tutton salts a superhyperfine splitting has been observed and analyzed at different temperatures for the two more populated sites (for ZASH the third site is studied[375]). E.s.r. single crystal data on VO^{2+}- doped $Mg(ND_4)_2(SO_4)_2 \cdot 6D_2O$ have confirmed unequivocally that the superhyperfine splitting is due to the four protons of the two nearest-neighbour water molecules.[382] The densities of the unpaired electron of VO^{2+} at the ligand protons have been also estimated.[378,379] For the paramagnetic FASH (M = Fe) and COASH (M = Co) host the VO^{2+} - Fe^{2+} exchange interaction has been estimated on the basis of the observed low temperature *g*-shift and the spin lattice relaxation times of Fe^{2+} and Co^{2+} ions have been derived from the analysis of the vanadyl e.s.r. linewidths.[380,381]

The e.s.r. spectrum of vanadyl ethylenediamine tetraacetate complex (VO-EDTA) dissolved in polyvinyl alcohol (PVA) films has been investigated as a function of the film stretching.[383] Low frequency L-band (1.2 GHz) aqueous solutions e.s.r. spectra have been studied for vanadyl and manganese(II) sulphate and the frequency dependence of the hyperfine coupling and the line-broadening have been analyzed.[384] E.s.r. spectra of oxovanadium(IV)-D-aspartic acid and oxovanadium(IV)-D-aspartic acid α-benzylester complexes have been used to clarify the binding mode of the ligands.[385]

Vanadium(IV) has been used as a probe in systems of catalytic interest like ZSM5/silicate zeolite,[386] highly dispersed V_2O_5/SiO_2 mixed gels catalysts,[387] and in water adsorbed onto silica gels.[388] Low-temperature oxygen chemisorption has been applied to characterize a series of V_2O_5/ZrO_2 catalysts.[389] The V_2O_5/TiO_2 system has attracted the interest of a number of researchers: the dependence of the surface structure, reactivity and amount of paramagnetic vanadium(IV) species on the crystallographic nature and on the conditions of the heat treatment has been investigated,[390,391] and e.s.r. and high resolution electron microscopy have been reported for anatase, rutile and titania(B) supported vanadia samples treated with aqueous ammonia.[392] The e.s.r. analysis of *in situ* reactions has been revealed useful in clarifying the mechanism of the redox intercalation reactions between alkylammonium iodides and the layered host α-$VOPO_2 \cdot 2H_2O$.[393]

The field of metallomesogens have recently attracted great interest due to the possibilities of generating liquid crystalline materials exhibiting unusual structures and properties.[394] E.s.r. spectra of vanadium(IV) are sensitive to the variation in the co-ordination environment of the metal ions or to the local magnetic interactions, and, therefore, can give information on the phase transitions and on the orientational order in the smectic and nematic phases. Oxovanadium(IV) probes dissolved in the analogue bis[4-((4-alkoxybenzoyl)oxy)-N-*n*-propylsalicylaldiminato]nickel(II) nematic host have been investigated and the central chelate cores of these molecules have been found to be strongly oriented.[395] A detailed e.s.r. study has been performed on the two mesogenic complexes bis[N-pentyl-4-(4-decyloxybenzoyloxy)-salicylaldiminato]oxovanadium(IV) and bis[N-(4-pentyl-oxyphenyl)-4-(4-decyloxybenzoyloxy)-salicyl-aldiminato]oxovanadium(IV), for which nematic phases result to be stable between 110-131 °C and 225-266 °C, respectively. The evolution of the spectra during the thermal treatment, made both in the presence and in the absence of external magnetic field, have been discussed in terms of phase transitions and of molecular orientations induced by the magnetic field.[396]

The synthesis and spectroscopic characterization have been reported for 5-(4-N-hexadecylpyridiniumyl)-10,15,20-triphenylporphyrin bromide and its vanadyl, copper(II) and nickel(II) metalated derivatives, in order to examine in detail the monomeric complexes in both protic and aprotic non-aqueous media and the conditions in which micelles are formed in mixed solvents.[397] Similar studies have been performed on the spirocyclic imidophosphinato complex [V(OPPh$_2$NPPh$_2$O)$_2$O], whose crystal structure has been also solved.[398]

Paramagnetic impurities have been seen to influence significantly the scintillator performance of zinc tungstate single crystals and several ions with high positive charge might occupy tungsten(VI) sites. Niobium enters the host lattice mainly as diamagnetic niobium(V), but four different niobium(IV) centres have been observed by e.s.r. after thermal reduction or electron irradiation. A charge-compensating anion defect model has been proposed to explain the observed behaviour.[399]

Quinquevalent Chromium and Molybdenum. E.s.r. spectroscopy has been commonly used to study chromium(V) species: paramagnetic centres have been thermally generated in chromate(VI) doped (NH$_4$)$_3$H(SeO$_4$)$_2$ single crystals;[400] the redox reactivity of nitridochromium(V) porphyrins was found consistent with porphyrin centred processes;[401] the nature of the surface species formed before and after adsorption of oxygen and nitric oxide has been studied in CrO$_x$/ZrO$_2$ catalyst,[402,403] and the electron density and the formal oxidation state in chromium(V)-oxo complexes of macrocyclic tetraamido-N ligands have been determined.[404] The chemical behaviour of the complexes bis(2-ethyl-2-hydroxybutanoato(2-))-oxochromate(V), [Cr(ehba)$_2$O]$^-$,[405] bis(1,2-ethanediolato(2-))-oxo-chromate(V), [Cr(ed)$_2$O]$^-$,[406] and

Na[trans-Cr(ehba)$_2$O]·H$_2$O[407] has been widely investigated in solutions using u.v.-visible, e.s.r. and NMR spectroscopies and electrochemistry. Reactive chromium(V) species, generated by redox processes within the cells, have been suggested to contribute to the genotoxicity of chromium(VI), the oxidation state (V) being stabilized by the formation of complexes with ligands such as thiols[408], diols[406] and α-hydroxyacids.[405] Formaldehyde, as reducing agent in water, was found to favour the formation of chromium(V) complexes by reaction of potassium chromate with adenosine and other purine nucleosides.[409]

Molybdenum is know to have an important role in biological electron-transfer processes and great interest has been devoted to the analysis of the electronic structure and electrochemical behaviour of naturally occurring enzymes and of synthetic model complexes having sulphur atoms in the co-ordination sphere of the metal ion. Molybdenum(V) compounds [Mo(S$_2$CNR$_2$)$_n$(acda)$_{4-n}$]Br (acda = 2-aminocyclopent-1-ene-1-carbodithiolate, R = Et, Pr and n = 2,3) have been obtained by the corresponding diamagnetic eight co-ordinated mixed ligand molybdenum(IV) complexes by one electron oxidation;[410] a tetragonal pyramidal geometry has been suggested for the MoOS$_4$ chromophore in (NEt$_4$)[MoOL$_2$] (H$_2$L = α,2-toluenedithiol);[411] one electron reduction of the imido [Mo(NC$_6$H$_4$S)Cl(S$_2$CN(C$_2$H$_5$)$_2$)$_2$], having a MoS$_5$NCl chromophore, yields to the e.s.r.-active molybdenum(V) analogue,[412] and a mononuclear molybdenum(V) species (Bu$_4$N)[MoO(O$_2$CC(S)Ph$_2$)$_2$], for which structural data indicate a O(S$_2$O$_2$) donor set, has been used as model for the active centre of sulphite oxidase or nitrate reductase.[413]

The nature of the molybdenum(V) centres of milk xanthine oxidase has been investigated by A.G. Wedd and co-workers using multifrequency electron spin resonance. ^{95}Mo hyperfine tensor has been measured for the enzyme and for the synthetic species [MoOXL]$^-$ and [MoO(XH)L] (X = O and S; LH$_2$ = N,N'-dimethyl-N,N'-bis(2-mercaptophenyl)-1,2-diaminoethane). ^{33}S superhyperfine tensors have been derived for the thio- and mercapto ligands and the centres [MoO(SH)], [MoO(SH)(OH)], [MoO(OH)] and [MoOS] were found to be responsible for the observed e.s.r. signals. A mechanism for the minimal catalytic cycle at the molybdenum site of xanthine oxidase has been also suggested.[414]

The spectroscopic and electrochemical properties of the mononuclear oxomolybdenum(V) centre of the molybdenum cofactor of the oxo-type molyibdoenzymes have been studied in two series of model complexes of the type LMoO[E(CHR)$_n$E'] (for n = 2 - 4: E = E' = O, S and R = H ; for n = 2: E = O, E' = S, R = H, Me) and LMoOY(XR) (L = hydrotris(3,5-dimethyl-1-pyrazolyl)borate; for Y = OR and X = O: R = Me, Et and ^{n}Pr; for Y = Cl and X = O: R = Me, Et and ^{n}Pr; for Y = Cl and X = S: R = Et , ^{n}Pr and ^{i}Pr).[415,416]

The reaction of the cyclometallaphosphazene [Mo(NPPh$_2$NPPh$_2$N)Cl$_3$] with ^{t}BuOH in thf yields the spirocyclic molybdenum(V) compound [Mo(OPPh$_2$NPPh$_2$O)$_2$OCl], whose e.s.r. spectra do not show ligand hyperfine splitting, and are in agreement with a 4d^1(d$_{xy}$) configuration with nearly axial **g** and **A** tensors.[417] In the mononuclear oxomolybdenum(V) complexes [MoO(R$_2$C$_{22}$H$_{20}$N$_4$)][C$_3$(CN)$_5$] (R = H, Me, Cl; R$_2$C$_{22}$H$_{22}$N$_4$ is the substituted dibenzotetraaza[14]annulene) the spin hamiltonian parameters indicated a strong delocalization of the unpaired electron onto the ligands.[418] Amorphous molybdenum sulphides and in particular the co-ordination environment of the paramagnetic molybdenum(V) centres have been also investigated.[419,420]

A variety of molybdenum(V) and molybdenum(III) paramagnetic species have been recognized in the e.s.r spectra of MoO$_3$ catalysts obtained by thermal decomposition *in vacuo* of hexakisammonium heptamolybdate.[421] The dominant component belongs to an extremely reactive, strongly unsaturated tetraco-ordinate molybdenum(V) surface species with $g_{//}$ = 1.770 and $g_\perp$ = 1.920. The co-ordination sphere of the paramagnetic centres[422] and their reactions with carbon monoxide and methanol[423] in silica supported molybdenum(V) have been investigated using conventional[423] and microscopic imaging[422] e.s.r. spectroscopies.

Sexivalent Technetium and Rhenium. In the last years the interest in the co-ordination chemistry of the first man-made element, Tc, has increased due to the wide use of the isotope ^{99}Tc, I = 9/2, in diagnostic nuclear medicine, and, in particular, the analysis of nitridotechnetium complexes has attracted the attention of a number of researchers.[424-426] E.s.r. spectroscopy has been successfully used for the detection of the paramagnetic species obtained by the oxidation of (Ph$_4$As)$_2$[Tc^{V}N(dto)$_2$] (dto = dithiooxalato) by Cl$_2$ and Br$_2$, [Tc^{VI}NCl$_4$]$^-$, [Tc^{VI}NBr$_4$]$^-$ and other Br/dto mixed ligand complexes,[424,427] by the oxidation of [Tc^{V}NCl$_2$(EPh$_3$)$_2$] (E = P, As) by SOCl$_2$ via the intermediate [Tc^{VI}NCl$_3$(EPh$_3$)$_2$],[428] and by the reactions of (Bu$_4$N)TcNX$_4$ (X = Cl, Br) and (Bu$_4$N)TcOCl$_4$ with sodium azide in acetone solutions, [TcNX$_n$(N$_3$)$_{4-n}$]$^-$ (X = Cl, Br; n = 0-4).[429,426,425] Single crystal e.s.r., ENDOR and ESEEM studies have been performed on (Ph$_4$As)[TcNCl$_4$], diluted in the isoelectronic diamagnetic oxotechnetate(V) (Ph$_4$As)[TcOCl$_4$], whose structure has also been reported.[430] Superhyperfine splittings due to ^{35,37}Cl and ^{15}N (in the enriched complex) have been observed at 4.2 K and 2.1 K and analyzed through ENDOR and ESEEM spectroscopies. The isotropic ^{14,15}N hyperfine coupling was found to be mainly determined by spin polarization effects.[431] Single crystal e.s.r. and ^{15}N powder ENDOR studies have been reported for (Bu$_4$N)[TcNBr$_4$], doped in the isostructural oxotechnetate(V) (Bu$_4$N)[TcOBr$_4$]. The fraction of the unpaired electron delocalized on the ligand orbitals is larger than in the corresponding chloro complex.[432] The application of e.s.r. spectroscopy to the analytical chemistry of technetium complexes in both II and VI oxidation states has been reviewed by Abram and Kirmse.[433]

The *trans*-dioxo complex *trans*-[ReO$_2$(dmap)$_4$](PF$_6$)$_2$ (dmap = 4-(dimethylamino)-pyridine) has been isolated and its X-ray structure determined. The 50% DMSO solution e.s.r. spectra showed large hyperfine and quadrupole coupling constants.[434]

d^3 **Configuration.** - *Quinquivalent Ruthenium and Osmium.* The syntheses of hypervalent ruthenium(V) and osmium(V) oxo-complexes with α-hydroxy carboxylate and α-amino carboxylate ligands have been reported. The *d^3* ions are low spin in a trigonal bipyramidal environment. E.s.r. has been used to characterize the isolated ruthenium(V) species (g_x = 2.208, g_y = 2.062, g_z = 1.902) and to follow their reactions with organic substrates. A comparison with the analogous oxochromium(V) complexes has been made.[435]

d^5 **Configuration.** - *Univalent chromium and Tervalent Iron, Ruthenium and Osmium*. Many efforts have been directed to the synthesis of low-spin chromium(I) complexes, but the e.s.r. analysis has always remained at the stage of the characterization of the spin state. Cyanonitrosyl complexes have been actively studied by Maurya and Mishra,[436] an e.s.r. study of the photo substitution of the CN⁻ ligand by H_2O in $[Cr(CN)_5NO]^{3-}$ and $[Cr(CN)_2(H_2O)_3NO]$ has been reported,[437] and the solution e.s.r. spectra of the oxidation products of *mer*-$[Cr(CO)_3(\eta^1$-ape$)(\eta^2$-ape$)]$ and *fac*-$[Cr(CO)_3(\eta^1$-dae$)(\eta^2$-dae$)]$ (ape = $Ph_2PCH_2CH_2AsPh_2$; dae = $Ph_2AsCH_2CH_2AsPh_2$) gave evidence of phosphorus and arsenic hyperfine splitting.[438]

In order to obtain a better understanding of the electronic structure of iron(III)-cyanide complexes, information about the interaction of the unpaired electron with both ^{13}C and ^{14}N for all six CN ligands have been obtained through the single crystal e.s.r.,[439,440] ENDOR and ESEEM spectra[440] of $Fe(CN)_6^{3-}$ doped in the $NaCl^{439}$ and $KCl^{439,440}$ lattices. The 11 K e.s.r. spectrum has revealed the presence of a large number of centres, due to different configurations of the charge compensating cation vacancies. Among them, the two well differentiated I_a and I_b centres have been further characterized with the aid of ENDOR and ESEEM spectroscopies.

The complex $[Fe(diammac)](ClO_4)_3$, where diammac is the sexidentate polyamine macrocycle 6,13-dimethyl-1,4,8,11-tetraaza-cyclo-tetradecane-6,13-diamine, has provided an ideal system to investigate the relationship between the **g** tensor and the metal ligand bonding parameters of low-spin iron(III) compounds, due to the strong and purely σ - bonding character of the ligand.[441] Single crystal spectra at 130 K have been interpreted with g_1 = 1.52, g_2 = 2.58 and g_3 = 2.80, which have been satisfactorily reproduced within the angular overlap model of the ligand field. The *g* anisotropy was related to angular distortion in the FeN_6 chromophore.[442]

The scientific community is devoting considerable efforts in determining the factors affecting the physical, chemical and magnetic properties of iron proteins, in both low and high spin states, and in gaining a better understanding of how these properties may relate to each other. E.s.r., electronic and Mössbauer spectra have been used to characterize a wide number of low-spin iron(III) complexes. Among them we cite the strongly oxidizing iron(III) complexes of the ligand bis(difluoro-

(dimethylglyoximato)borate, (dmgBF$_2$), (Et$_4$N)[Fe(dmgBF$_2$)X$_2$], (X = Cl, Br) which appeared to provide low-spin ferric species useful as bioinorganic models for bleomycin and for non-heme iron proteins;[443] the bis(nitro)[444] [Fe(NO$_2$)$_2$(TpivPP)]$^-$, the nitropyridine[445] [Fe(NO$_2$)(py)(TpivPP)] and the nitro-imidazole[445] [Fe(NO$_2$)(Him)(TpivPP)] adducts of the picket-fence meso-α,α,α,α-tetrakis-(o-pivalamido-phenyl)-porphyrin, (H$_2$TpivPP); the ammonia-co-ordinated tetraphenylporphyrin (TPP) complex Fe(NH$_3$)$_2$(TPP)SO$_3$CF$_3$;[446] and the iron(III) complexes of etiporphyrin I and of electron deficient porphyrins with one, two or four highly electron-withdrawing CF$_3$ groups at pyrrole β positions of the porphyrins.[447] K. Tajima added a new chapter to the analysis of the reaction leading to peroxide adducts of porphyrin-iron complexes, studying the formation of Fe(OEP)(OH)(OOH) (OEP = octaethylporphyrinato) by reduction of Fe(OEP)O$_2$ with ascorbic acid sodium salt.[448]

The actual spin state of iron(III) is well known to depend on the nature of the co-ordinated ligands. For example the axial ligand in iron(III) porphyrinates can cause the spin state to change from pure low spin (S = 1/2), with strongly and moderately basic pyridines, to intermediate spin states (S = 3/2 ↔ S = 5/2), with lower basicity pyridines, or to thermal spin equilibrium complexes. Magnetic susceptibility, Mössbauer and e.s.r. spectroscopies are the most widely used techniques to characterize these complexes. The [Fe(OEP)(3-CNPy)$_2$]ClO$_4$ complex, whose e.s.r. spectrum at 77 K has $g_{\parallel}$ = 1.97 and $g_{\perp}$ = 4.28, has been assigned as an intermediate S = 3/2 spin complex and the actual spin state has been related to the axial ligand orientation with respect to the equatorial ligand.[449] Low spin iron(III) with g_x = 1.630, g_y = 2.275 and g_z = 2.818 at 77 K has been found in complexes [Fe(TMP)(4-NMe$_2$Py)$_2$]ClO$_4$, [Fe(OEP)(4-NMe$_2$Py)$_2$]ClO$_4$ and [Fe(TMP)(1-MeIm)$_2$]ClO$_4$ (TMP = meso-tetramesitylporphyrinato; OEP = octaethylporphyrinato 4NMe$_2$Py = 4-methylaminopyridine; 1-MeIm = 1-methylimidazole).[450] A more complicated spectrum has been observed in the bis(1-vinylimidazole) adduct of the sterically hindered porphyrin complex (meso-tetrakis-(2,6-dichloro-phenyl)porphyrinato)iron(III), [Fe(T2,6Cl$_2$PP)(1-VinIm)$_2$]ClO$_4$, for which the presence of two species differing in the orientation of the axial ligand has been suggested.[451] E.s.r. spectra with $g_{\parallel}$ = 2.00 and $g_{\perp}$ = 5.7 have been interpreted assuming that the ground state is an admixture of 85% of S = 5/2 and 15% of S = 3/2 states in the (triflato)aquoiron(III) picket-fence porphyrin derivative [Fe(TP$_{piv}$P)(OSO$_2$CF$_3$)(H$_2$O)].[452] Increasing the percentage of S = 3/2 state causes a shift of $g_\perp$ towards the value 4.0. Other ferric porphyrins have been reviewed in ref. 453 The spin-crossover transitions S = 5/2 → S = 1/2 in octahedral complexes, which determine either a gradual variation over a wide temperature range or an abrupt change near the critical temperature of the e.s.r. spectra, have been studied in bis(7-amino-4-methyl-5-aza-3-hepten-2-onato(1-))iron(III) tetraphenylborate,[454] in bis(pyridoxal-4-phenyl-thiosemicarbazonato)iron(III) chloride[455] and in iron(III) thioseleno- and diselenocarbamates.[456,457]

Detailed e.s.r. investigations on ruthenium(III) compounds are still rare, but the technique has been used in the characterization of structure and reactivity of the three geometrical isomers of [RuL$_2$Cl$_2$]ClO$_4$ · H$_2$O (L = 2-(phenylazo)pyridine),[458] of the cis- and trans-isomers of [Ru(ROCS$_2$)$_2$(PPh$_3$)$_2$]$^+$, where the metal ion has a S$_4$P$_2$ coordination environment,[459] of the compounds [Ru(PPh$_3$)$_2$(m-ClC$_6$H$_4$CO$_2$)Cl$_2$]^{460} and trans-[Ru(NH$_3$)$_4$P(OEt)$_3$(H$_2$O)] (CF$_3$SO$_3$)$_3$.[461] The synthesis of a number of stable ruthenium(III) carbonyl chelates of Schiff bases (L) complexes of the type [RuLX(Cl)]$^{n+}$ and [RuLX(CO)]$^{n+}$, with axial ligands (X) chloro, imidazole and 2-methyl imidazole has been reported. The e.s.r. spectra have been analyzed using the approach developed by Hill and have suggested that the replacement of Cl$^-$ with CO in the axial position of the octahedral environment of the metal ion resulted in a d_{xy} ground state irrespective of the nature of L.[462,463] The oxygenation and carbonylation of the chloro complexes have been also followed, and e.s.r. has been found to be useful in individuating the formation of a ruthenium(IV) superoxo complex.[464]

Detailed spectroscopic studies, based on IR, XPS, e.s.r. and ESEEM spectroscopies, have been reported on the ruthenium species formed in Ru(NH$_3$)$_6^{3+}$ exchanged X-zeolites. The existence of alkali-metal ion cocation effects on the formation and adsorbate geometry of the paramagnetic ion in X-zeolites has been established[465] and the two major ruthenium(III) species formed in the exchanged hydrogen form of the X-zeolite as well as the interactions of the paramagnetic centres with adsorbate carbon monoxide, nitric oxide, oxygen an water have been described and discussed.[466]

d^7 **Configuration.** - *Bivalent Cobalt and Rhodium, and Univalent Iron.* The ground state of low-spin *d*^{7} is not generally a pure state but an admixture of doublet states, with possible contribution of low-lying quartet states through spin-orbit coupling. This gives a large variability in the observed *g* values, which can be hardly related to the symmetry of the chromophores. This effects show up in a series of adducts of alkylcobaloximes as in the penta co-ordinated [Co(Hdmg)$_2$L] and six co-ordinated [Co(Hdmg)$_2$L$_2$] compounds (H$_2$dmg = dimethylglyoxime, L = phosphine, phosphite);[467] in [Co(Hdmg)$_2$(py)L] complexes (L = tetrahydrothiophene, γ-butyrothiolactone);[468] in the transient [RCo(dmgBF$_2$)$_2$A]$^-$, obtained by one-electron reduction of [RCo(dmgBF$_2$)$_2$A] (R = 4-BrC$_6$H$_4$CH$_2$ and A = H$_2$O, py);[469] and in a series of [N^1,N^4-bis(salicylidene)-S-alkyl-isothio-semicarbazonato] cobalt(II) complexes, doped into the analogue zinc(II) compounds.[470]

The interest towards the adducts of cobalt(II) complexes with the oxygen molecule is still large. One of the most oxidation-resistant, reversible oxygen carriers, the complex Co(CN)$_5$$^{3-}$, has been obtained by reaction of cesium exchanged Co-zeolite Y with methanolic NaCN solutions, and characterized through electron spin resonance.[471] The electrochemical behaviour[472] and the reversibility of the formation of dioxygen adducts[473] have been studied for the complex (*meso*-tetrakis-(1-methylpyridinium-4-yl)-porphyrinato)cobalt(II), [(TMpyP)Co]$^{4+}$ and for polymer supported cobalt(II) tetraaza macrocyclic complexes of the lacunar cyclidene type.[474] E.s.r., IR and Mössbauer spectroscopic measurements have suggested that the 1-methylimidazole adducts of the cobalt(II) and iron(II) complexes of the highly symmetric bis-fenced 5,10,15,20-tetra(2,6-dipivaloyl-oxyphenyl)porphyrin are able to form dioxygen compounds. The possibility of an electrostatic interaction between the bound dioxygen and the ester fences is investigated.[475]

E.s.r. spectroscopy has been used to characterize the cobalt(II) ion environment in the five co-ordinate compounds Co(NO$_2$)$_2$R$_3$, with R = PMe$_3$ and PMe$_2$Ph,[476] in the square planar bis[(cyclooctane-1,5-diyl)bis(pyrazol-1-yl)-borate]cobalt(II);[477] in the hydride complexes [P(CH$_2$CH$_2$PPh$_2$)$_3$CoH]BF$_4$ and [N(CH$_2$CH$_2$PPh$_2$)$_3$CoH]ClO$_4$;[478]

and in [Co([10]-aneS$_3$)$_2$]$^{2+}$ and [Ni([10]-aneS$_3$)$_2$]$^{3+}$ low spin d^7 species ([10]-aneS$_3$ = 1,4,7-trithiacyclododecane.[479]

A very interesting S = 1/2 neutral square-planar cobalt complex with the macrocycle ligand H$_4$[L] = 5,6-(4,5-dichlorobenzo)-3,8,11,13-tetraoxo-2,2,9,9-tetramethyl-12,12-diethyl-1,4,7,10-tetraaza-cyclotridecane, Co(η^4-[L]) has been reported. The metal ion has a formal oxidation state (IV), but X-ray and e.s.r. data have suggested that ligand-centred oxidation is important in the neutral species giving rise to limit resonance structures of the type show in Figure 5.[480]

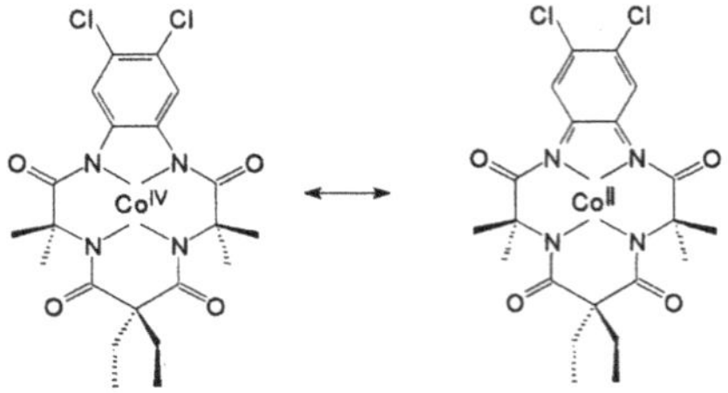

Figure 5. Two resonance contributors to the structure of Co(η^4-[L])

The temperature dependence of the spin hamiltonian parameters for the [Rh(CN)$_6$]$^{4-}$ complex in the KCl and NaCl host lattices has been measured by e.s.r. in the temperature range 12-300 K.[481] The dynamic behaviour has been attributed to the effect of localized anharmonic vibration modes and interpreted using the theory proposed by Vugman and Rothier.[482]

Electron spin resonance has been used to analyze the effect of channel zeolite structures of the types RhCa-L zeolite and RhCa-M(mordenite) on the formation of paramagnetic rhodium intermediates during ethylene dimerization. The signals coming from different rhodium(II) sites have been recognized and a comparison with the behaviour of cage type zeolites has been performed.[483] A rhombic e.s.r. spectrum, showing evidence of hyperfine splitting in the parallel region, has been also observed for [Rh(η^3-TMPP)$_2$](BF$_4$)$_2$ (TMPP = tris(2,4,6-trimethoxyphenyl)phosphine) in a CH$_2$Cl/Me-THF glass.[484]

Tervalent Nickel, Palladium and Platinum. The less-common oxidation states of nickel, namely nickel(III) and nickel(I), play a crucial role in the activity of a number of enzymes, so that much effort has been expended in order to investigate the factors leading to the stabilization of d^7 and d^9 electronic configurations. A characteristic feature of hydrogenases is the low value of Ni^{III}/Ni^{II} redox potential and an increasing interest has been devoted to the synthesis of nickel(II) complexes exhibiting such a behaviour. Nickel(III), as well as nickel(I) species are often generated through electrochemical one-electron oxidation or reduction of the corresponding nickel(II) compounds and electron spin resonance has been widely employed to characterize their stability and their electronic structure.[485] The electrochemical oxidation of [Ni(3,5-Cl$_2$saloph)] in strong donating solvents yields [Ni(3,5-Cl$_2$saloph)(sol)$_2$]$^+$ (saloph = bis(salicylaldehyde)-o-phenylenediimine; sol = solvent DMF or (CH$_3$)$_2$SO)), with a N$_2$O$_2$ donor set. The e.s.r. frozen solution spectra of this complex and of the substituted pyridines bis-adducts are indicative of a $^2A_1$ [a (d_z2) + b $(d_{x^2-y^2})$] ground state, with a » b; in the pyridine adducts a superhyperfine splitting due to the two axially co-ordinated nitrogen atoms has been observed.[486] A mainly d_z2 ground state, with $g_\perp > g_{\parallel}$, has been also proposed for the pentaco-ordinated electrogenerated [Ni(MC1)Et$_2$Dtc]$^{2+}$ (MC1 = 2,4,4-trimethyl-1,5,9-triazacyclodec-1-ene, Et$_2$Dtc = diethyldithiocarbamate) having a N$_3$S$_2$ co-ordination environment,[487] for [NiR$_x$S]$^+$ (H$_2$R$_x$S = MeC(=NOH)C(Me)=N(CH$_2$)$_2$S(CH$_2$)$_x$S(CH$_2$)$_2$N=C(Me)C(=NOH)Me; x = 2, 3) with a NiN$_2$S$_2$ chromophore,[488] for the nickel(III) complexes of the low-potential Ni^{II}/Ni^{III} couples [Ni(ema)]$^-$, [Ni(phma)]$^-$ and [Ni(emi)]$^-$ (H$_4$ema = N,N'-ethylenebis(2-mercaptoacetamide); H$_4$phma = N,N'-1,2-phenylenebis(2-mercaptoacetamide); H$_4$emi = N,N'-ethylenebis(2-mercaptoisobutyramide)) having a N$_2$S$_2$ donor set,[489] and for [NiCl$_2$(H$_2$L$_3$)]$^{3+}$ (L$_3$ = 2,5,9,12,16,19-hexaaza-7-spiro[6.13]eicosane) having a N$_4$Cl$_2$ co-ordination sphere around the nickel(III) ion.[490]

The chemistry of nickel(III) co-ordinated to sulphur donor ligands has received some impulse in recent years. The tris-chelate xanthates Ni(RX)$_3$ complexes (X = ROC(S)S$^-$; R = alkyl radical) can be obtained by one-electron oxidation of the nickel(II) analogues[491] and their e.s.r. spectra have been recorded both in frozen acetonitrile/toluene solution and in nickel(III) doped Co^{III}(RX)$_3$ polycrystalline samples. The NiS$_6$ chromophores have $g_\perp > g_{\parallel}$, with a small rhombic splitting in the perpendicular region, and $g_{\parallel}$ in the range 2.03-2.04. These values fall in the range of values observed in other hexaco-ordinated compounds.[492] The oxidation of [Ni^{II}(bdt)$_2$]$^{2-}$ (bdt = butane-2,3-dithiolate) in DMF has been reported to afford the nickel(III) analogue, whose e.s.r. spectrum, with $g_\perp$ = 2.042 and $g_\parallel$ = 2.187, was assigned to a square planar NiS$_4$ chromophore.[493]

For the complex Ni(η^4-[L])$^-$ (H$_4$[L] = 3,6,9,12,14-pentaoxo-2,2,5,5,7,7,10,10-octamethyl- 13,13-diethyl- 1,4,8,11-tetraaza-cyclotetradecane) the X-ray structure and frozen solution e.s.r. spectra in co-ordinating and non- co-ordinating solvent have been reported. The authors have attributed a spectrum with g_1 = 2.366, g_2 = 2.303 and $g_3 = g_\parallel$ = 1.994 to a four co-ordinated distorted square planar nickel(III) species.[494]

Nickel(III) spin hamiltonian parameters have been used to clarify the local distortion of the NiO$_6$[495] and CoO$_6$[496] octahedra in the K$_2$NiF$_4$-type oxides with the general formula A$_{0.50}$La$_{1.50}$Mg$_{0.50}$Co$_x$(Ni$_y$)O$_4$ (A = Ca, Sr, Ba; in the pure compounds x= 0 and y = 0.50 and x = 0.50 and y = 0 respectively; in the doped samples x = 0.45 and y = 0.05).

Accurate spectroscopic studies, comprehensive of e.s.r., ESEM and X-ray photoelectron spectroscopy (XPS) measurements, have been performed on [Pd(NH$_3$)$_4$]$^{2+}$-doped Al$_{13}$-pillared laponite clay and the oxidation states of palladium, which depend on the activation conditions, have been established.[497] Different oxidized palladium(III) species have been evidenced in samples activated in flowing oxygen from 250 to 500 °C and the dependence of their migration from an Al$_{13}$-pillar to the surface of the three-sheet laponite layer on the temperature of activation investigated.[498]

The temperature dependence of $g_\perp$ in the complexes [Pd(CN)$_4$Cl$_2$]$^{3-}$ and [Ni(CN)$_4$Cl$_2$]$^{3-}$ obtained by X-ray irradiation of the corresponding M(CN)$_4^{2-}$ doped in KCl and NaCl have been proposed as mainly determined by the presence of localized anharmonic vibration modes of the host lattices,[499] according to the interpretation given for the analogous rhodium(II) compound.[481,482]

d^9 **configuration.** - *Univalent Nickel.* Two new e.s.r. active centres NIRIM-1 and NIRIM-2 have been studied using synthetic diamond crystals grown from Ni solvent and attributed to interstitial nickel(I) associated with a vacancy and with local charge compensation, respectively. The temperature dependence of the spectra has been investigated and electron spin echo experiments have been also performed.[500]

As previously observed the importance of nickel containing biological systems is well stated and the role of the metal in its +1 oxidation state is actively investigated, especially through electrochemical experiments. A nickel(I) species has been obtained by one-electron reduction of the [Ni^{III}L$_2$]Cl$_2$ (L = o-C$_6$H$_4$(OH)CH=N-NHCSNH$_2$), a suggested model compound for hydrogenase, where the paramagnetic ion is in a distorted N$_2$O$_2$S$_2$ co-ordination environment. The e.s.r. spectrum gives rhombic **g** tensor with g_1 = 2.25, g_2 = 2.12 and g_3 = 2.06.[501] It is well known that the one-electron reduction of nickel(II) porphyrins, chlorins and isobacteriochlorins yields species that can be described as a nickel(II) porphyrin π-anion radical or as a nickel(I) porphyrin anion or as a mixture of both. The effects of temperature, porphyrin macrocycle, solvent and degree of axial co-ordination on the site of electron reduction have been studied using e.s.r and UV-visible spectroscopies for the (T(p-Et$_2$N)PP)Ni^{II} and (T(p-Me$_2$N)F$_4$PP)Ni^{II} metalloporphyrins, where (T(p-Et$_2$N)PP) and (T(p-Me$_2$N)F$_4$PP) are the dianions of meso- tetrakis(p-(diethylamino)phenyl)porphyrin and of meso-tetrakis(o,o,m,m-tetrafluoro-p-(diethylamino)phenyl)porphyrin.[502] In order to gain some insights into the chemistry of Factor 430, the reductive behaviour of a series of nickel(II) porphyrin complexes has been examined. Only isobacteriochlorins unambiguously yield a nickel(I) species, whose reactivity with metal halides has been studied and compared with that of the π radicals obtained from the other starting nickel(II) compounds.[503] In order to mimic the active site of several iron-nickel hydrogenases the short living [Ni(terpy)(SR)$_2$]$^-$ (terpy = 2,2':6',2''-terpyridine; R = C$_6$F$_5$, 2,4,6-(^{i}Pr)$_3$C$_6$H$_2$) compounds, having the metal ion in a N$_3$S$_2$ co-ordination environment, has been obtained by reduction of the corresponding nickel(II) complexes and the binding of CO and H$^-$ at the sixth co-ordinative site of the paramagnetic centres has been investigated through e.s.r. spectroscopy.[504]

Bivalent Copper, Silver and Gold. Electron paramagnetic resonance has been widely used as a routine technique, together with X-ray structure determination, optical and magnetic measurements, in order to characterize a number of copper(II) complexes useful to investigate the co-ordination ability of new ligands, to clarify catalytical activity of different substrates, to find new models for molecules of biological or technological interest. In many cases only room temperature powder or solution e.s.r. spectra have been described, generally without discussion. The listing of so large a bibliography appears to be too boring for the readers and we decided to report only those papers in which the e.s.r. technique is the dominant one and in which the e.s.r. data are determinant for the conclusions.

A detailed single crystal e.s.r. study has been performed on copper(II) doped in the monoclinic lattice of $KHCO_3$. The hyperfine parameters have been measured for both ^{63}Cu and ^{65}Cu isotopes, and the forbidden $\Delta m = \pm 1$ transitions, allowed by the nuclear electric quadrupole interaction, have been detected in some crystal orientations. The spin hamiltonian parameters ($g_{//} = 2.234$, $g_{\perp} = 2.045$, $A^{63Cu}{}_{//} = 192.3 \cdot 10^{-4}$ cm^{-1}, $A^{63Cu}{}_{\perp} = 28.6 \cdot 10^{-4}$ cm^{-1}, $A^{65Cu}{}_{//} = 206.0 \cdot 10^{-4}$ cm^{-1}, $A^{65Cu}{}_{\perp} = 30.6 \cdot 10^{-4}$ cm^{-1}) agree with a square planar co-ordination for the metal ion and interstitial sites have been suggested for the guest copper ions.[505] Single crystal e.s.r. experiments have employed in studying the effects of thermal treatment on copper(I) doped NaF lattice, where a tetragonal copper(II)-NNN Skotty vacancy complex has been characterized.[506] E.s.r. single crystal measurements have been also reported for copper(II) ions doped in $NaNH_4SO_4 \cdot 2H_2O$[507] and KNH_4SO_4,[508] where the paramagnetic ions enter at the alkali metal sites, in $Tl_2Ni(SO_4)_2 \cdot 6H_2O$,[509] where the copper(II) guest substitutes nickel(II), and in calcium di(hydrogenmaleate) pentahydrate.[510]

The location, stoichiometry, migrations and reactivity of copper(II) species in catalysts have been studied using e.s.r. and ESEM spectroscopies during the copper(II)-catalyzed propylene oxidation over Cu(II)-exchanged X[511] and Y[512] zeolites; in a series of silicoaluminophosphate (SAPO) molecular sieves like CuH-SAPO-5,[513] CuH-SAPO-34[514] and CuH-SAPO-11;[515,516] in metal substituted fluorohectorite;[517] and in single crystals of β"- alumina with nominal composition $Na_{1.67}Mg_{0.67}Al_{10.33}O_{17}$, where sodium has been entirely substituted with Cu^+/Cu^{2+}.[518] A wide series of copper-thorium oxide catalysts, with different copper contents, has been investigated[519]: the copper(II) species obtained for different values of the Cu/Th atomic ratio and for different calcination temperatures were identified through their e.s.r. spectra[520-522] and their surface redox reactions with H_2 and O_2[523] investigated. The analogous Cu/Ce oxide with an atomic ratio of 0.01 was also studied.[524]

E.s.r. has been widely used to study metal transition ions in glasses[525-527]. The X- and Q-band e.s.r. and ESEM spectra of a novel 43PbO-56PbCl$_2$-CuCl$_2$ ternary glass have suggested a distribution of different geometries for the copper(II) centres.[528,529]

The ionomer morphology have been studied for nafion perfluorinated membranes[530,531], partially neutralized by copper(II) and swollen by acetonitrile, CH_3CN and CD_3CN. The gradual replacement of oxygen ligands around copper(II) with nitrogen ligands has been also followed using L- (1.2 GHz), S- (2.8 GHz) and C (4.7 GHz) e.s.r. frequencies[532].

Complexes of copper(II) with amino-acids and copper(II) compounds exhibiting antitumoral and biological activity have been actively studied using e.s.r. and ENDOR[533,534] spectroscopies, which are very useful in obtaining information on their electronic and molecular structure. Selected examples of these investigations include copper(II) aminoanthracyclines;[535] γ- and α-cyclodextrins 1:2 and 1:1 inclusion complexes with bis(2-pyridilcarbinolato) copper(II);[536] bis(amidinourea) copper(II) and bis(O-alkyl-1-amidinourea)copper(II) complexes;[537] copper(II)-substituted biguanidine compounds;[538,539] substituted 1,10-phenanthroline adducts of copper(II) dipeptide;[540] copper(II) cinnamate;[541] copper(II) ternary complexes of cyclo-L-histidyl-L-histidine and L- or D-amino acids.[542,543] Three new cupric pteridine complexes have been found to mimic the behaviour of the pterin cofactor of metalloenzymes,[544] [HB(^{t}Bupz)$_3$]Cu(NO$_2$) (HB(^{t}Bupz)$_3$ = Hydrotris(3-tert-butylpyrazol-1-yl)borate) was used as model for the A. Cycloclastes nitrite reductase,[545] and a number of benzimidazolate[546,547], substituted imidazolate[548] and tripodal imidazolate donor ligands[549] copper(II) complexes, models for copper-containing proteins, have been studied through solid and solution e.s.r. spectra. A number of copper(II) compounds possessing cytotoxic and antitumoral activity contain sulphur in the metal co-ordination sphere. Two bis(dithiophosphinato) copper(II) complexes, having a CuS$_4$ chromophore, have been studied by single crystal spectra of copper doped nickel(II) analogues.[550] Solution e.s.r. spectra have been reported also for the stable Lewis-base adducts of CuL (HL = 2-

formylpyridinethiosemicarbazone) with nitrogen and sulphur donor ligands[551] and for the {[2,2'-[1-(1-ethoxy-ethyl)-1,2-ethane-diylidene]-bis-[N,N-dimethyl-hydrazine-carbothio-amidato]](2-)$N^2,N^{2'},S,S'$} copper(II) complex, (CuKTSM$_2$).[552]

ENDOR induced e.s.r. spectroscopy (EI-EPR) has been applied to measure hyperfine and superhyperfine tensors in copper glutamate complexes. A brief survey of the theory has been also presented.[553]

The interest in structural and magnetic properties of copper(II) halides continues to be large because of variety of their stereochemical features and the formation of extended magnetic systems. X-ray and magnetic measurements clearly indicate that in the complex (3-chloroanilinium)$_8$[CuCl$_6$]Cl$_4$ an antiferromagnetic exchange pathways with J = -11.2 cm^{-1} is operative between the six-co-ordinate CuCl$_6^{4-}$ units. The g values, obtained by single crystal measurements, $g_\perp$ = 2.19 and $g_\parallel$ = 2.06, are in agreement with a tetragonally compressed octahedral co-ordination environment[554] of the metal ion. E.s.r. spectra with $g_\parallel > g_\perp$ have been observed for (H$_2$Me$_2$pipz)[CuCl$_3$(H$_2$O)]$_2$ and (HAmpym)[CuCl$_3$(H$_2$O)] (H$_2$Me$_2$pipz = N,N'-Dimethylpiperazinium, HAmpym = 2-aminopyrimidinium), both containing planar [CuCl$_3$(H$_2$O)]$^-$ anions linked together by longer semico-ordinate Cu --- Cl bonds to form stack.[555] A dipolar interaction between copper(II) centres has been invoked to justify the angular and temperature dependences of the linewidth in the single crystal e.s.r. spectra of (NMe$_4$)$_2$CuCl$_4$.[556] Solution e.s.r. has been reported for the monomeric and dimeric complexes [Cu(TMPA)F]$_n^{n+}$ (n = 1 and 2, TMPA = Tris[(2-pyridyl)-methyl]amine)[557] and for dichloro(1,2,4-triazole)copper(II) and dichlorobis(1,2,4-triazole)copper(II) in non aqueous media.[508]

Bis(N-alkilsalicylaldiminato)copper(II) complexes, Cu(R-sal)$_2$, have been investigated by spectroscopic and kinetic methods both in solid state (single crystal) and in solution, and the effect of the alkyl substituents in controlling the degree of nonfluxional distortion discussed. It should be noted that the reported angular dependence of the g values for the complex with R = ^{i}Pr does not follow the symmetry requirements for an orthorhombic crystal, probably due to a misalignment of 10-15°.[558] Single crystal e.s.r. measurements on the pentaco-ordinated [Cu(terpy)X$_2$] (X = I, NCS; terpy = 2,2':6',2''-terpyridine) are in agreement with structural results, excluding the possibility of a dynamically averaged geometry.[559] Solution spectra for copper(II) complexes with N-(aminoalkyl)salicylaldimines have been reported in refs. 560,561.

As previously stated the metallomesogens form a wide class of liquid crystals for which synthetic procedure and physical properties are actively investigated. The synthesis of a series of rod-like liquid crystals bis[5-(4-alkoxybenzoyloxy)-salicylaldiminato] copper(II) complexes and the temperature dependence of the e.s.r. spectra of one of them (Figure 6) have been reported.[562] The compounds undergo a relatively low temperature phase transition to a smectic S_C phase, whose e.s.r. spectrum can be interpreted within the same orientational model previously used for vanadyl analogues.[396]

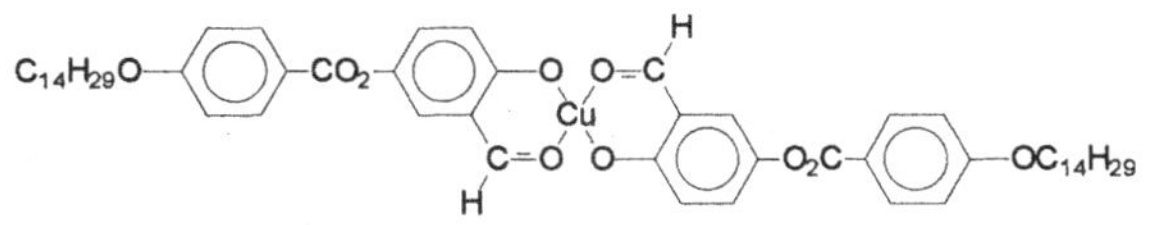

Figure 6. - Schematic view of the structure of bis[5-(4-alkoxybenzoyloxy)-salicylaldiminato] copper(II) complexes

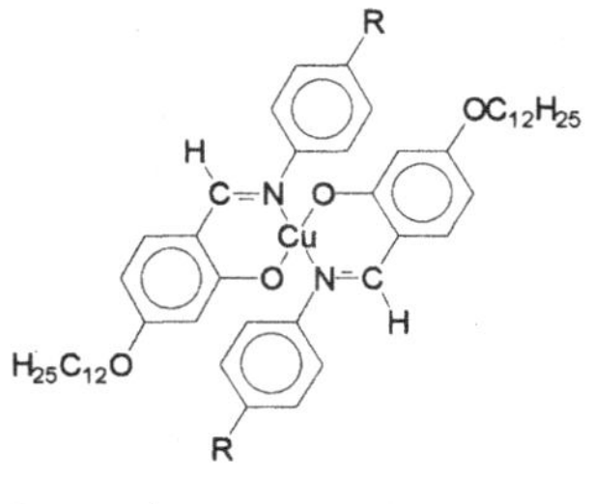

Figure 7. - Schematic view of the structure of copper(II) complexes with N(-4-(dodecyloxy)-salicylidene)-4'-alkylanilines

The synthesis, mesomorphic properties and e.s.r. characterization of an homologous series of copper(II) complexes with N(-4-(dodecyloxy)-salicylidene)-4'-alkylanilines of the type shown in Figure 7 (R = C_nH_{2n+1} ; n = 1-4,6,8) have been described. All the compounds had a complex temperature dependent behaviour, and different phases, with characteristic e.s.r. spectra, were achieved. In particular the smectic S_A phase has given rise to a peculiar e.s.r. line shape that has appeared to be independent of the compound, indicating that the formation of such a phase is accompanied by a disordering in the structure with an increase in the misalignment in the principal directions of the molecular **g** tensor[563].

The chemistry of bismetal complexes of 1,3-dithiole-2-thione-4,5-dithiolate (dmit) and related ligands continues to attract the attention of a number of researchers[564] because of the peculiar magnetic and conductive properties of this kind of solid. The interest in chalcogen-rich metal complexes is now extending to the selenium analogues. Single crystal e.s.r. study has been performed on tetra-*n*-butylammonium-bis(2-thioxo-1,3-dithiole-4,5-diselenolato)-cuprate(II), (*n*-Bu$_4$N)$_2$[Cu(dsit)$_2$], doped in the nickel(II) analogue. The rhombic anisotropy of the **g** tensor, larger than that observed for the corresponding "all sulphur" complex, has been rationalized in terms of large ligand orbital contributions of different weight to the corresponding *g* components.[565]

The partial iodine oxidation of metalloorganic complexes with aromatic ligands provides a class of molecular conductor and, as a part of a systematic study on the properties of this kind of compound, the characterization of a number of copper(II)-aryltetraazaannulene complexes has been performed with the aid of electrochemical and spectroscopic methods[566,567].

Only few papers have been published regarding silver(II) and gold(II) species, due to the difficulty in stabilizing such oxidation state for both the elements. The chloroform solution and mixed chloroform/toluene frozen solution e.s.r. spectra of highly soluble silver(II) tetraneopentoxyphthalocyanine, (AgTNPc), have been interpreted with spin hamiltonian parameters (g_{iso} = 2.049, $A_{iso}{}^{Ag}$ = 52 G, $A_{iso}{}^{N}$ = 23.6 G and $g_{//}$ = 2.087, $g_{\perp}$ = 2.030, $A_{//}{}^{Ag}$ = 40 G, $A_{\perp}{}^{Ag}$ = 58 G, $A_{//}{}^{N}$ = 20 G, $A_{\perp}{}^{N}$ = 25.4 G) which are in agreement with previously reported data on silver phthalocyanine[568].

The +2 oxidation state of gold can be stabilized with good σ-donor and π-acceptor ligands. E.s.r. and magnetic measurements results for monomeric and dimeric gold(II) complexes [Au(mta)$_2$Cl]Cl, [Au(mta)$_2$X]$_2$X$_2$ and [Au(dae)$_2$Y]Y (X = Cl, Br, SCN; Y = Cl, Br; Hmta = o-(methylthio)aniline; dae =1,2-bis((o-amino-phenyl)-thio)-ethane) have been found to be consistent with a gold(II)-stabilized radical structure, where the delocalization of the unpaired electron onto the ligands is indicated by the low value of the gold hyperfine coupling constants, $A_{iso}{}^{Au}$ = 5.5 - 6.5 G and $A_{iso}{}^{Au}$ = 4.0 G for the monomeric and dimeric species respectively. For the dimeric complexes the ferromagnetic coupling between the two gold(II) ion leads to a ground triplet state.[569] Two methods have been presented in order to generate gold(II) defects in gold(III) fluorosulphate or solvated gold(II) ions in strongly acidic HSO_3F and e.s.r. techniques have been used to identify and characterize the Au^{2+} species[570].

d^{10} **configuration.** - *Zerovalent Silver and Gold.* The five doublets observed in the e.s.r. spectra of ^{107}Ag + SiO systems in adamantane matrix at 77 K have been assigned to isolated Ag(0) atoms, and to the Ag(SiO), $Ag(Si_2O_2)$, $Ag(Si_3O_3)$ and $Ag(Si_nO_n)$ complexes.[571]. The $Ag(0)PF_3$ species, produced in a solid adamantane matrix, has been suggested to be a silver centred radical with a $^2A_1(C_{3v})$ ground state[572]. E.s.r. and ESEM spectroscopies have been used to characterize two Ag(0) species generated by γ-irradiation in different silver(I)-exchanged samples of synthetic smectite clay minerals[573] and sodium montmorillonite and its Al_{13} pillared derivative[574].

4 S = 1

d^2 **configuration.** - *Quadrivalent Chromium and Molybdenum.* Chromium(IV) is an unstable species which has been revealed using X-ray photoelectron and e.s.r. spectroscopies in $Bi_4Ge_3O_{12}$ where chromium substitutes a Ge(IV) ion at a tetrahedral site[575]. Chromium doped forsterite, a solid having good qualities for laser applications, has been studied through e.s.r. and fluorescence-line-narrowing techniques. The near infrared lasing centre has a C_s site symmetry, with an effective S = 1 spin, supporting a previously reported attribution to a tetrahedral chromium(IV) ion substituting silicon, even if conclusive identification cannot be raised[576].

A Mo(IV) species has been electrochemically generated as the dication complex $[Mo^{IV}Cl(NCH_3)(Ph_2PCH_2CH_2PPh_2)_2]^{2+}$, and its e.s.r. and ENDOR spectra have been reported.[577]

d^4 **configuration.** - *Quadrivalent iron.* Some complexes containing the ferryl ion have been reported with porphyrins, and have been discussed in the section dealing with Paramagnetic Ligands. Spin intermediate S = 1 iron(IV) species are still rare. Synthesis and crystal structure of the complexes [FeL$_2$(PMe$_3$)$_x$]^n (n = 0 x = 1,2; n = -1 x = 2) (L = 1,2-benzenedithiolate) have been reported. The anionic compound is a low spin iron(III) d^5 system and the neutral compounds can be described as iron(IV) d^4 systems with a S = 1 spin state (μ_{eff} = 2.75 μ_B at room temperature). The e.s.r. spectrum shows a very broad low intensity signal at $g \approx 2$.[578]

d^8 **configuration.** *Bivalent Palladium.* The e.s.r. spectra of a triplet state with $g_{//}$ = 2.18(3), $g_\perp$ = 2.16(3) and D = 0.25(2) cm^{-1} have been observed for a single crystal of CsMgCl$_3$ doped with palladium(II) and attributed to a trigonal palladium(II) site, substituting magnesium(II)[579].

5 S = 3/2

d^3 **configurations.** - *Tervalent Chromium and Molybdenum, Quadrivalent Manganese and Quinquevalent iron.* In the last two years the analysis of the e.s.r. spectra of chromium(III) doped in a large variety of host lattices has been greatly improved since chromium(III) has been largely used as paramagnetic impurity in solids in order to obtain information, from the spin hamiltonian parameters, on the local symmetry and distortions of the host lattices. Many efforts have been also directed to the solution of theoretical and computational problems and the chromium(III) ion doped into diamagnetic host has been often used to test models and methods of calculation. Among the studied systems, we cite chromium(III) doped RbCdF$_3$ (fluoroperovskite);[580] Cs$_2$NaAlF$_6$ (hexagonal perovskite);[581] K(H$_3$O)$_2$AlF$_6$, where also iron(III) has been used as paramagnetic probe;[582] SnO$_2$[583] and α-TeO$_2$.[584] The influence of paramagnetic doping on the ferroelectric properties of LiNbO$_3$ have been investigated through e.s.r. and ENDOR of host chromium(III) species: a Cr^{3+} centre with an unusually low D value substituting niobium(V) has been evidenced in highly magnesium codoped LiNbO$_3$[585,586] and signals assigned to the S = 3 state of a chromium(III) pair have been recognized in the X-band e.s.r. spectrum of single crystal of LiNbO$_3$ doped with 0.8 weight percent of the paramagnetic ion.[587] Crystal-field models have been applied to correlate local structure and spin hamiltonian parameters of three chromium(III) centres in the K$_2$NiF$_4$-type layer fluorides A$_2$MF$_4$ (A = K, Rb, Cs; M = Zn, Cd, Mg) crystals,[588] of axial chromium(III) centres in the piezoelectric PbTiO$_3$ ceramic;[589] and of chromium(III) doped in ZnGa$_2$O$_4$ spinel.[590] The linewidth of the e.s.r. lines have been correlated to internal stress and strain distribution in chromium(III) doped in yttrium aluminum garnets,[591,592] in Ca$_3$Ga$_2$Ge$_3$O$_{12}$ garnet.[593] The molecular structure and the dynamics of the interactions between chromium and the surface sites of hydrous α-Al$_2$O$_3$ have also been studied.[594]

Spin hamiltonian parameters have been obtained from simulation of polycrystalline powder spectra for a number of iminodiacetate, 6-methylpicolinate and L-histidinate chromium(III) complexes[595] and for [Cr(L-asp)(phen)(H$_2$O)]NO$_3$·2H$_2$O, [Cr(terpy)(pydca)] [Cr(pydca)$_2$]·4H$_2$O (L-asp = L-aspartate(2-), phen = 1,10-phenanthroline, terpy = 2,2':6',2''-terpyridyl, pydca = pyridine-2,6-dicarboxylate).[596] The preparation and characterization of ternary chromium(III)-nucleotide and aminoacid complexes (nucleotide = 5'AMP and 5'CMP, amino-acid = L-serine, L-methionine, glycine) have been presented with preliminary e.s.r. results[597]. The spin hamiltonian parameters of the [Cr(diammac)](ClO$_4$)$_3$ (diammac = 6,13-diamino-6,13-dimethyl-1,4,8,11-tetraazacyclotetradecane), where the ligand acts a as a sexidentate one, have been also reported[598].

The X-band polycrystalline powder e.s.r. spectrum of hexaaquomolybdenum(III) ions diluted in the cesium indium β-alum has been interpreted using a S = 3/2, I = 5/2 spin hamiltonian with $g_{//}$ = 1.954, $g_{\perp}$ = 1.950, |D| = 1.15 cm^{-1}, $|A_{//}^{95,97Mo}|$ = 50.5·10^{-4} cm^{-1} and $|A_{\perp}^{95,97Mo}|$ = 47.3·10^{-4} cm^{-1}.[599] The e.s.r. spectra of Mo(H$_2$O)$_6$$^{3+}$ diluted in the ammonium aluminum α-alum have been reinterpreted.

A series of mononuclear manganese(IV) complexes with MnO$_4$N$_2$ chromophores have been prepared and their frozen solution e.s.r. spectra have been recorded.[600,601] The spectra showed resonances at $g_{\perp} \approx 4$ and $g_{//} \approx 2$ characteristic of a octahedral S = 3/2 system split by large zero field splitting (2|D|>hν). Similar spectra have also been observed for the up to now unique manganese(IV) complex with a MnO$_6$ chromophore,

Na$_2$[Mn(HIB)$_3$]·2MeOH (H$_2$HIB = 2-hydroxyisobutyric acid).[602] Small zero field splitting is observed in two complexes of manganese(IV) with the hexadentate ligands 1,8-bis(2-hydroxybenzamido)-3,6-diazaoctane and 1,10-bis(2-hydroxybenzamido)-4,7-diazadecane, having a N$_4$O$_2$ co-ordination environment of the paramagnetic ion[603].

Iron(V) centre has been observed in ordered cubic perovskite La$_2$LiV$_{0.9}$Fe$_{0.1}$O$_6$: the room temperature e.s.r spectrum consists of a perfectly Lorentzian line centred at g = 2.013(9), with a linewidth of 148 G. The positive Δg shift has been ascribed to strong covalence in the iron-oxygen bond.[604]

d^7 **configuration.** - *Bivalent Cobalt.* High spin cobalt(II) centres have been clearly recognized in dried and heated silica gels prepared with the sol gel process[605]. The single crystal e.s.r. spectra of cobalt(II) ions substituting cadmium(II) into Cd$_2$P$_2$S$_6$ and Cd$_2$P$_2$Se$_6$ have been interpreted with an axial S = 1/2 effective spin hamiltonian with $g_{//}$ = 4.86, $g_\perp$ = 4.01 and $g_{//}$ = 4.98, $g_\perp$ = 3.85 respectively, in agreement with a trigonal CoX$_6$ (X = S and Se) chromophore. The zero field splitting, D, was estimated to be larger than 50 cm^{-1}.[606] Single crystal e.s.r., magnetic susceptibility and polarized absorption data have been used to characterize (Ph$_4$P)$_2$[Co(SPh)$_4$], a complex possessing a pseudo-tetrahedral CoS$_4$ chromophore in an approximate D_{2d} symmetry. The spin hamiltonian parameters of the ground S = 3/2 state are $g_{//}$ = 2.60(5), $g_\perp$ = 2.2(1), D = -70(10) cm^{-1} and |E/D| = 0.09. The large value of $g_{//}$ has been justified putting a $^4T_2$ state ~1000 cm^{-1} above the ground $^4A_2$ state.[607] Angular overlap calculations on the pentaco-ordinated cobalt(II) complex [Co(TMC)(CH$_3$CN)](ClO$_4$)$_2$ (TMC = 1,4,8,11-tetramethyl-1,4,8,11-tetraazatetradecane) have reproduced the highly anisotropic **g** tensor, obtained by single crystal e.s.r. measurements, showing that the observed pattern is mainly determined by the in-plane σ interaction of the macrocyclic ligand[608].

6 S = 2

d^6 **configuration.** - *Bivalent iron.* In order to improve the understanding of the photo refractive effect in iron doped LiNbO$_3$, thermally detected single crystal e.s.r. spectra of a reduced sample have been presented. The method used has been briefly described and the results have been discussed on the basis of the electronic structures of this system[609].

7 S = 5/2

d^5 **configuration.** - *Univalent Chromium*. E.s.r. spectra of pairs of group III-acceptors in a substitutional site of silicon and transition metal ions in nearest neighbour (NN) or next nearest neighbour (NNN) interstitial positions have shown that the paramagnetic centre lies in a trigonal or in an orthorhombic site for the NN and the NNN configuration, respectively. E.s.r. spectra of silicon doped with chromium and indium, obtained after illumination, have been assigned on the basis of the fine and hyperfine structure to a Cr$^+$-In$^-$ pair, *i.e.* to transitions within the $^6S_{5/2}$ ground manifold of the chromium(I) ion in a tetrahedral crystal field trigonally distorted by the associated In$^-$ ion[610].

Bivalent Manganese. V.K. Jain and G. Lehmann have reviewed the e.s.r. spectra of manganese(II) in orthorhombic and higher symmetry crystals, giving a survey of the theoretical models which can be used for the interpretation of the spectra. The form of the S = 5/2 spin-hamiltonian to be used in cubic, hexagonal, trigonal tetragonal, and orthorhombic symmetries is given. The zero field splitting parameters are expressed as a sum of contributions from nearest neighbours according to the Newman superposition model (SPM) and the effect of phase transition on the e.s.r. spectra as well as the host spin-lattice relaxation narrowing effect has been discussed.[611] The SPM has been also applied to interpret the zero field splitting of manganese(II) in zinc oxalate dihydrate.[612] The zero field splitting observed in manganese(II) doped CaCO$_3$ is axial (D_{3d} symmetry) and has been calculated in a ligand field framework using eighth-order perturbation theory .[613]

Manganese(II) has been used as a paramagnetic probe in a number of solids, among which we cite: Bi$_4$Ge$_3$O$_{12}$, a compound which has applications as scintillator and for non-linear optical devices in which manganese(II) substitutes bismuth(III) in a O$_6$ co-ordination environment[614] and a method to distinguish between inequivalent centres is given;[615] ferro- and para-elastic crystals of bismuth vanadate[616] and LiNbO$_3$.[617] Manganese(II) has been used as e.s.r. active probe to analyze the mechanism of intercalation of solvated cationic species in Cd$_2$P$_2$S$_6$ lattice[618] and to estimate the spin-lattice relaxation times of the host in single crystals of [Co(H$_2$O)$_6$](ClO$_4$)$_2$[619] and of cobalt hydrogenmaleate tetrahydrate.[620] Manganese-indium and manganese-gallium pairs in silicon have been evidenced through the measurement of the hyperfine splitting[610] and a dependence of the magnitude of the zero field splitting on the size of the acceptor has been proposed.[621,622]

Manganese(II) has been used as a probe in a number of semiconducting solids which we have not explicitly mentioned in the Semiconductors section. They include: PbTe,[623] zinc selenide,[624] neutron irradiated GaP and GaAs.[625]

The dependence of the quenching of the electroluminescence emission in manganese(II) doped (Zn,Cd)S phosphors on the magnetic ion concentration have been analyzed,[626] as well as the deformation of the crystal field surrounding manganese(II) ions as a function of the film processing conditions[627] and of the thicknesses in epitaxially grown ZnS:Mn films on a GaAs(100) substrate.[628]

Optical and e.s.r. studies have been performed on manganese(II) doped NaBr, in order to analyze the paramagnetic ion environment in the formed Suzuki phase.[629] X- and Q-band liquid helium temperature single crystal e.s.r. and ENDOR spectra of manganese(II) ions at the orthorhombic substitutional site in the rutile like salt MgF$_2$ have indicated that the ligand octahedra of the substitutional ions is axially compressed with a configuration close to that observed in MnF$_2$.[630] The variable temperature single crystal e.s.r. spectra of manganese(II) doped in Sr(CH$_3$COO)$_2$·0.5H$_2$O have been reinterpreted and a significant temperature dependence of the local symmetry of the manganese(II) ions was found.[631] Single crystal e.s.r. spectra have been reported for manganese(II) doped in cadmium maleate dihydrate,[632] in magnesium maleate pentahydrate,[633] in Sm$_2$Zn$_3$(NO$_3$)$_{12}$·24H$_2$O,[634] in ZnTl$_2$(SeO$_4$)$_2$·6H$_2$O[635] and in calcium strontium fluoride.[636]

X- and Q-band powder e.s.r. spectra have been reported of manganese(II) doped in [M(DA)$_2$(H$_2$O)$_2$]X$_2$, M(DA)$_2$Cl$_2$ (M = Mg, Co, Ni; X = Cl, Br), [Mg(DA)$_2$(H$_2$O)$_2$](NO$_3$)$_2$, [Co(DA)$_4$(H$_2$O)$_2$]Br$_2$ and M(DA)$_2$(NO$_3$)$_2$ (M = Co, Zn) where DA is diacetamide. The

1: Transition Metal Ions

zero-field splitting parameters D and E have been measured and the positive sign of D has been deduced from the hyperfine spacing at 35 GHz.[637]

As a part of an extensive investigation on paramagnetic ions/ SAPO molecular sieves systems,[513-516] Kevan *et al.* have reported and discussed TGA (thermogravimetric analysis), IR, e.s.r. and ESEM spectroscopic comparison between the co-ordination geometry of the manganese(II) site in a framework position of synthesized MnSAPO-11 and the extra framework manganese(II) centre in ion-exchanged MnH-SAPO-11.[638] Both X- and Q-band e.s.r and ESEM spectroscopies have been used to get information on the immediate environment and the spatial distribution of manganese(II) in aluminophosphate molecular sieves (AlPO).[639,640] The importance of the method of the incorporation and of the concentration of the paramagnetic ions as well as the effects of the dehydration have been discussed for MnAlPO-5,[639] for MnAlPO-11[640] and for manganese impregnated AlPO-11.[640]

Tervalent iron. The electronic structure of iron(III) has been actively investigated in the past two years and a lot of work has been devoted to the analysis of the local structure of this S = 5/2 ion in host lattices as yttrium aluminium garnet;[641] in ferroelectric crystals of $Cd_2Nb_2O_7$[642] and of $KTaO_3$;[643,644] in single crystal of Bi_2WO_6;[645] in $KTiOPO_4$;[646] in nominally pure Li_2O as grown and after x-ray irradiation and annealing,[647] and as naturally occurring defects in plumbo-jarosite mineral.[648]

The fine structure parameters $b_n{}^m$ for the tetragonal uncompensated iron(III), for the substitutional nearest V_{Cd} -associated orthorhombic iron(III)[649] and for the nearest V_{Rb} -associated monoclinic iron(III)[650] centres in the K_2NiF_4-type layer perovskite fluoride Rb_2CdF_4 crystals have been analyzed to obtain information on the local environments around substitutional magnetic ions through the zero field splitting parameters. The SPM has revealed to be not applicable to reproduce the spin hamiltonian parameters of iron(III) doped in single crystals of aluminium and gallium acetylacetonates, probably depending on the delocalization of the conjugated π system of the ligand[651]. The spin hamiltonian tensors **g** and **D**, as well as parameter of fourth order S terms, have been obtained for alkali- metal ion compensated iron(III) centres in synthetic α - quartz. The low C_1 symmetry observed for the $[FeO_4/Na]^0$ site at 20 K has been associated with the off-axis location of the interstitial sodium(I) compensator,[652] and the characterization of three lithium(I) compensated iron(III) centres[653] has allowed the authors to establish their substantial similar symmetry with a twofold axis bearing both Fe^{3+} and Li^+. The effects of temperature annealing and ionizing radiation on the mobility of the interstitial charge compensating alkali ions have also been investigated using e.s.r. spectroscopy.[654]

It is known that the photo refractive effect of $LiNbO_3$ is due to transition metal ion impurities present even in pure samples of the compound and its photoconductivity has been demonstrated to be strongly dependent on the magnesium(II) doping. The dependence of the e.s.r. spectra of iron doped $LiNbO_3$ on magnesium doping and Li/Nb ratio has been investigated[655] as well as the effect of the neodymium(III) doping on the photoconductivity of pure and iron-magnesium doped $LiNbO_3$.[656,657] Two iron(III) sites in highly magnesium doped $LiNbO_3$ have been studied and the analysis of their zero field splitting parameters through the SPM assigned them as substitutional iron centres for lithium(I) and niobium(V) respectively.[658] The oxidation state of iron, the thermal stability and the catalytic properties have been studied in the molecular sieves VPI-5[659] and template-free synthesized iron-modified ZMS-5 zeolites with different Fe_2O_3/Al_2O_3 ratios.[660]

Only few studies have been performed on heme and non-heme iron(III) complexes which could be used as models for biologically relevant molecules. The interest of the authors seems to be mainly directed to the synthetic procedures allowing the isolation of solid compounds, in order to obtain first hand information on the co-ordination environment of the metal ion through X-ray crystal structure determination, to the pH dependent solution equilibria and to the electrochemical properties of the compounds. E.s.r. techniques, as well as other spectroscopic methods, are used only to characterize complexes of new synthesis, such as the σ-bonded axial tetrazolato and triazolato iron(III) porphyrin adducts,[661] or to follow the reactions of iron(III)-EDTA (ethylenediaminetetraacetic acid) with H_2O_2, t-BuOOH and n-BuOOH,[662] and for the characterization of the solution equilibria of the iron(III)-DTPA (Diethylenetriaminepentaacetic acid) system.[663]

References

1. A. Bencini, and C. Zanchini, *Electron Spin Resonance - Specialist Periodical Reports*, Royal Society of Chemistry, 1991, **12B**, 1.
2. J.R. Pilbrow, Transition Ion Electron Paramagnetic Resonance, Clarendon, Press, Oxford 1990.
3. A.X. Trautwein, E. Bill, E.L. Bominaar, and H. Winkler, *Structure and Bonding (Berlin)*, 1991, **78**, 1.
4. M. Moreno, *J. Phys. Chem. Solids*, 1990, **51**, 835.
5. B. Bleaney, *Proc. R. Soc. London, Ser. A*, 1991, **433**, 461.
6. M. Che, and E. Giamello, *Stud. Surf. Sci. Catal.*, 1990, **57**, B265.
7. G.R. Eaton, and S.S. Eaton, *Biol. Magn. Reson.*, 1989, **8**, 339.
8. C.M. Srivastava, *Ferrite Mater.*, 1990, 85.
9. H. Benner, and J.P. Boucher, *Phys. Chem. Mater. Low-Dimens. Struct.*, 1990, **9**, 323.
10. M.W. Evans, *J. Mol. Spectrosc.*, 1991, **150**, 120.
11. J.M. Spaeth, and F. Lohse, *J. Phys. Chem. Solids*, 1990, **51**, 861.
12. A. Schweiger, *Angew. Chem., Int. Ed. Engl.*, 1991, **30**, 265.
13. H. Thomann, and M. Bernardo, *Spectroscopy (Ottawa)*, 1990, **8**, 119.
14. C. Gemperle and A. Schweiger, *Chem. Rev.*, 1991, **91**, 1481.
15. W.T. Wenckebach, *Hyperfine Interact.*, 1991, **65**, 751.
16. J.H. Freed, *J. Chem. Soc., Faraday Trans.*, 1990, **86**, 3173
17. L.R. Dalton, A. Bain, and C.L. Young, *Ann. Rev. Chem. Phys.*, 1990, **41**, 389.
18. F. Herlach, and L. Van Bockstal, *Infrared Phys.*, 1991, **32**, 145.
19. D. Duret, M. Beranger, M. Moussavi, P. Turek, and J.J. Andre, *Rev. Sci. Instrum.*, 1991, **62**, 685.
20. J.P. Hornack, M. Spacher, and R.G. Briant, *Meas. Sci. Technol.*, 1991, **2**, 520.
21. C.R. Bowers, S.K. Buratto, P.J. Carson, H.M. Cho, J.Y. Hwang, L.J. Mueller, P.J. Pizzarro, D.N. Shykind, and D.P. Weitekamp, *Proc. SPIE-Int. Soc. Opt. Eng.*, 1991, **1436**, 36.
22. J.S. Hyde, M.E. Newton, R.A. Strangeway, T.G. Camenish, and W. Froncisz, *Rev. Sci. Instrum.*, 1991, **62**, 2969.
23. M.D. Forbes, J. Peterson, and C.S. Breivogel, *Rev. Sci. Instrum.*, 1991, **62**, 2662.
24. P. Ludowise, S.S. Eaton, and G.R. Eaton, *J. Magn. Reson.*, 1991, **93**, 410.
25. F.P. Auteri, A.H. Beth, and B.H. Robinson, *J. Magn. Reson.*, 1988, **80**, 493.
26. M. Bonori, C. Franconi, P. Galuppi, and M. Guerrisi, *Meas. Sci. Technol.*, 1991, **2**, 1046.
27. P.B. Sczaniecki, J.S. Hyde, and W. Fronsicz, *J. Chem. Phys.*, 1990, **93**, 3891.
28. G. Lassmann, R. Wessel, P. Grasshoff, C. Jakobs, and D. Schwarz, *Exp. Tech. Phys.*, 1990, **38**, 249.

29 C. Gemperle, G. Aebli, A. Schweiger, and R.R. Ernst, *J. Magn. Reson.*, 1990, **88**, 241.
30 T. Wacker, and A. Schweiger, *Chem. Phys. Lett.*, 1991, **186**, 27.
31 H. Cho, S. Pfenninger, J. Forrer, and A. Schweiger, *Chem. Phys. Lett.*, 1991, **180**, 198.
32 G.A. Sierra, A. Schweiger, and R.R. Ernst, *Chem. Phys. Lett.*, 1991, **184**, 363.
33 A.V. Astashkin and A. Schweiger, *Chem. Phys. Letters*, 1990, **174**, 595.
34 M. Furusawa, and M. Ikeya, *J. Phys. Soc. Jpn.*, 1990, **59**, 2340.
35 M. Furusawa, and M. Ikeya, *Jpn. J. Appl. Phys.*, Part 2, 1991, **30**, L1682.
36 A.I. Smirnov, O.G. Poluektov, and Y.S. Lebedev, *J. Magn. Reson.*, 1992, **97**, 1.
37 W. Jantsch, and G. Hendorfer, *J. Cryst. Growth*, 1990, **101**, 404.
38 C.J.M. Coremans, E.J.J. Groenen, and J.H. van der Waals, *J. Chem. Phys.*, 1990, **93**, 3101.
39 Y.C. Zhong, and J.R. Pilbrow, *J. Magn. Reson.*, 1992, **97**, 111.
40 S. Taleb, T.W. Hijmans, and W.G. Clark, *Rev. Sci. Instrum.*, 1990, **61**, 2354.
41 K. Dyrek, A. Madej, E. Mazur, and A. Rokosz, *Colloids Surf.*, 1990, **45**, 135.
42 P.N. Keizer, J.R. Morton, and K.F. Preston, *J. Chem. Soc., Faraday Trans.*, 1991, **87**, 3147.
43 K.W.H. Stevens, *Proc. Roy. Soc. London*, 1952, **65**, 209.
44 Y. Wan-Lu, and C. Rudowicz, *Phys. Rev. B*, 1992, **45**, 9736.
45 Z. Kang-Wei, X. Jun-Kai, N. You-Ming, Z. Sang-Bo, and W. Ping-Feng, *Phys. Rev. B*, 1991, **44**, 7499.
46 Z. Sang-Bo, W. Hui-Su, and W. Ping-Feng, *J. Phys.: Condens. Matter*, 1990, **2**, 687.
47 Z. Kang-Wei, Z. Sang-Bo, W. Ping-Feng, and X. Jun-Kai, *Phys. Stat. Sol. (b)*, 1990, **162**, 193.
48 A.G. Breñosa, M. Moreno, and J.A. Aramburu, *J. Phys.: Condens. Matter*, 1991, **3**, 7743.
49 S.R. Nogueira, N.V. Vugman, and D. Guenzburger, *Hyperfine Interact.*, 1990, **60**, 631.
50 G. Scholz, R. Lueck, R. Stoesser, H.J. Lunk, and F. Ritschl, *J. Chem. Soc., Faraday Trans.*, 1991, **87**, 717.
51 P.A. Reynolds, and B.N. Figgis, *Inorg. Chem.*, 1991, **30**, 2294.
52 A. Grand, P. Rey, R. Subra, V. Barone, and C. Minichino, *J. Phys. Chem.*, 1991, **95**, 9238.
53 G. Shen, C. Xu, and G. Bai, *Phys. Stat. Sol. B*, 1990, **158**, K185.
54 C.P. Keijzers, *Specialist Periodical Reports: Electron Spin Resonance*, 1987, **10B**, 1.
55 A. Bencini, and D. Gatteschi, *Electron Paramagnetic Resonance of Exchange Coupled Systems*, Springer-Verlag, Heidelberg, 1990.
56 Y.C. Zhong and J.R. Pilbrow, *J. Magn. Reson.*, 1991, **93**, 447.
57 R. Aasa and T. Vanngard, *J. Magn. Reson.*, 1975, **19**, 308.

58 G. van Veen, *J. Magn. Reson.*, 1978, **30**, 91.
59 C.P. Keijzers, E.J. Reijerse, P. Stam, M.F. Dumont, and M.C.M. Gribnau, *J. Chem. Soc., Farady Trans. I*, 1987, **83**, 3493.
60 M.C.M. Gribnau, J.L.C. van Tits, and E.J. Reijerse, *J. Magn. Reson.*, 1990, **90**, 474.
61 A. Kreiter, and J. Hüttermann, *J. Magn. Reson.*, 1991,**93**, 12.
62 M.D. Pace, and M.S. Masumura, *Reports*, 1989, NRL-MR-6417; order no. AD-A206441.
63 R. Berger, and M. Haddad, *Phys. Status Solidi B*, 1991, **163**, 463.
64 D. Collison, and A.K. Powell, *Inorg. Chem.*, 1990, **29**, 4735.
65 M. Valko, M. Melník, P. Pelikán, F. Valach, and M. Mazúr, *Chem. Phys. Letters*, 1990, **174**, 591.
66 J.S. Hyde, M. Pasenkiewicz-Gierula, A. Jesmanowicz, and W.E. Antholine, *Appl. Magn. Reson.*, 1991, **1**, 483.
67 A. Cabral-Prieto, H. Jiménez-Domínguez, L. González-Tovany, S Galindo, H. Flores-Llamas, and M. Torres-Valderrama, *J. Magn. Reson.*, 1990, **89**, 568.
68 H. Flores-Llamas, A. Cabral-Prieto, H. Jiménez-Domínquez, and M. Torres-Valderrama, *Nucl. Instrum. Methods Phys. Res., Sect. A*, 1991, **A300**, 159.
69 A. Cabral-Prieto, H. Jiménez-Domínguez, L. González-Tovany, S Galindo, and M. Torres-Valderrama, *J. Magnet. Reson.*, 1992, **96**, 473.
70 V. Bertrán-López, and J. Castro-Tello, *J. Magn. Reson.*, 1980, **39**, 437.
71 V. Bertrán-López, and J. Castro-Tello, *J. Megn. Reson.*, 1982, **47**, 19.
72 L. González-Tovany, and V. Beltrán-López, *J. Magn. Reson.*, 1990, **89**, 227.
73 N.P. Benetis, L. Sjöqvist, A. Lund, and J. Maruani, *J. Magn. Reson.*, 1991, **95**, 523,
74 J. Köhler, and P. Reineker, *Chem. Phys.*, 1990, **146**, 415.
75 Z. Zimpel, and S.K. Hoffmann, *Physica B (Amsterdam)*, 1991, **172**, 499.
76 D. Marsh, and L.I. Horvath, *J. Magn. Reson.*, 1992, 97, 13.
77 W. Zheng, *Philos. Mag. Lett.*, 1990, **62**, 115.
78 R. Hrabanski, and J. Lech, *Phys. Stat. Sol. B*, 1990, **162**, 275.
79 V. Cerny, *Philosof. Mag. B*, 1992, **65**, 401.
80 K.A. Müller, and J.C. Fayet, *Top. Curr. Phys.*, 1991, **45**, 1.
81 G. Zwanenburg, J.J.M. Michiels, and E. de Boer, *Phys. Rev. B*, 1990, **42**, 7783.
82 W. Zapart, and M.B. Zapart, *Ferroelectrics*, 1990, **107**, 343.
83 W. Zapart, *Solid. State Commun.*, 1990, **76**, 959.
84 W. Zapart, and M.B. Zapart, *Phys. Stat. Solidi A*, 1990, **121**, K43.
85 W. Zapart, *Phase Transitions*, 1991, **34**, 161.
86 W. Zapart, Phys. Stat. Sol. A, 1990, 118, 447.
87 W. Zapart, *Ferroelectrics*, 1990, **107**, 337.
88 R. Hrabanski, *Ferroelectrics*, 1991, **124**, 333.
89 M. Suhara, T. Bandoh, T. Kitai, T. Kobayashi, and H. Katsuda, *Phase Transitions*, 1992, **37**, 111.

90 P.J. Alonso, R. Alcala, and J.M. Spaeth, *Phys. Rev. B: Condens. Matter*, 1990, **41**, 10902.
91 M. Kahziri, S.K. Misra, J. Kotlinski, M.O. Steiniz, and T. Smith-Palmer, *Solid State Commun.*, 1991, **79**, 167.
92 C.J. Wu, J.T. Yu, M.N. Tsai, and S.H. Lou, *J. Phys.: Condens. Matter*, 1991, **3**, 3795.
93 M. Das, and A.K. Pal, *Indian J. Cryog.*, 1990, **15**, 537.
94 M. Das, and A.K. Pal, *Indian J. Cryog.*, 1990, **15**, 547.
95 N.N. Kolpakova, A. Pietraszko, S. Waplak, and L. Szczepanska, *Solid State Commun.*, 1991, **79**, 707.
96 V.K. Jain, and V. Kapoor, *Phys. Rev. B: Condens. Matter*, 1992, **45**, 3126.
97 M. Shimokoshi, and K. Ohi, *J. Phys. Soc. Jpn.*, 1990, **59**, 3629.
98 K.A. Müller, W. Berlinger, and E. Tosatti, *Z. Phys. B: Condens. Matter*, 1991, **84**, 277.
99 L. Sadlowski, J. Kuriata, B. Bojanowski, J. Walczak, M. Kurzawa, M.L. Falin, and V.V. Izotov, *Phys. Status Solidi A*, 1990, **121**, K95.
100 H. Tanaka, *J. Magn. Magn. Mater.*, 1990, **90-91**, 251.
101 A.S. Lagutin, and A.V. Dmtriev, *J. Magn. Magn. Mater.*, 1990, **90-91**, 83.
102 G.I. Bersuker, L.F. Chibotaru, V.Z. Polinger, and A.O. Solonenko, *Mol. Phys.*, 1990, **70**, 1031.
103 G.I. Bersuker, L.F. Chibotaru, V.Z. Polinger, and A.O. Solonenko, *Mol. Phys.*, 1990, **70**, 1045.
104 J.V. Folgado, W. Henke, R. Allmann, H. Stratemeier, D. Beltran-Porter, T. Rojo, and D. Reinen, *Inorg. Chem.*, 1990, **29**, 2035.
105 S.K. Misra, X. Li, and C. Wang, *J. Phys.: Condens. Matter*, 1991, **3**, 8479.
106 S.K. Misra, and J. Sun, *Phys. Rev. B: Condens. Matter*, 1991, **44**, 11862.
107 L.W. Parker, C.A. Bates, J.L. Dunn, A. Vasson, and A.M. Vasson, *J. Phys.: Condens. Matter*, 1990, **2**, 2841.
108 T.W. Kool, and M. Glasbeek, *J. Phys.: Condens. Matter*, 1991, **3**, 9747.
109 A. Monnier, A. Gerber, and H. Bill, *J. Chem. Phys.*, 1991, **94**, 5891.
110 Yu.A. Koksharov, V.V. Moshchalkov, V.I. Voronkova, and A.A. Gippius, *Physica C (Amsterdam)*, 1991, **185-189**, 1149.
111 M. Höhne, and W. Gehlhoff, *Phys. Status Solidi B*, 1991, **164**, 503.
112 M. Hoehne, and W. Gehlhoff, *Phys. Status Solidi B*, 1991, **165**, 189.
113 C. Benelli, A. Dei, D. Gatteschi, and L. Pardi, *Inorg. Chem.*, 1990, **29**, 3409.
114 A. Dei, D. Gatteschi, L. Pardi, A.L. Barra, and L. Brunel, *Chem. Phys. Lett.*, 1990, **175**, 589.
115 A. Dei, D. Gatteschi, L. Pardi, and U. Russo, *Inorg. Chem.*, 1991, **30**, 2589
116 A. Dei, D. Gatteschi, and L. Pardi, *Inorg. Chim. Acta*, 1991, **189**, 125.
117 G. Speier, S. Tisza, A. Rockenbauer, S.R. Boone, and C.G. Pierpont, *Inorg. Chem.*, 1992, **31**, 1017.
118 B.G. Maiya, Y. Deng, and K.M. Kadish, *J. Chem. Soc., Dalton Trans.*, 1990, 3571.

119 M.A. El-Sayed, A. El-Toukhy, K.Z. Ismael, and A.A. El-Maradne, *Inorg. Chim. Acta*, 1990, **177**, 155.
120 M.A. El-Sayed, A. El-Thouky, K.Z. Ismael, A.A. El-Maradne, and G. Davies, *Inorg. Chim. Acta*, 1991, **182**, 213.
121 G. Ricciardi, A. Rosa, G. Morelli, and F. Lelj, *Polyhedron*, 1991, **10**, 955.
122 S.C. Shoner, and P.P. Power, *Inorg. Chem.*, 1992, **31**, 1001.
123 C. Bianchini, P. Frediani, F. Laschi, A. Meli, F. Vizza, and P. Zanello, *Inorg. Chem.*, 1990, **29**, 3402.
124 M. Haga, K. Isobe, S.R. Boone, and C.G. Pierpont, *Inorg. Chem.*, 1990, **29**, 3795.
125 H. Masui, A.B.P. Lever, and P.R. Auburn, *Inorg. Chem.*, 1991, **30**, 2402.
126 S.B. Kumar, and M. Chaudhury, *J. Chem. Soc., Dalton Trans.*, 1991, 2169.
127 N. Bag, A. Pramanik, G.K. Lahiri, and A. Chakravorty, *Inorg. Chem.*, 1992, **31**, 40.
128 N. Bag, G.K. Lahiri, P. Basu, and A. Chakravorty, *J. Chem. Soc., Dalton Tran.*, 1992, 113.
129 P.R. Auburn, E.S. Dodsworth, M. Haga, W. Liu, W.A. Nevin, and A.B.P. Lever, *Inorg. Chem.*, 1991, **30**, 3502.
130 E.L. Bomonaar, X.-Q. Ding, A. Gismelseed, E. Bill, H. Winkler, A.X. Trautwein, H. Nasri, J. Fischer, and R. Weiss, *Inorg. Chem.*, 1992, **31**, 1845.
131 L. Salmon, C. Bied-Charreton, A. Gaudemer, P. Moisy, F. Bedioui, and J. Devynck, *Inorg. Chem.*, 1990, **29**, 2734.
132 P.O. Sandusky, W.A. Oertling, C.K. Chang, and G.T. Babcock, *J. Phys. Chem.*, 1991, **95**, 4300.
133 M. Satoh, Y. Ohba, S. Yamauchi, and M. Iwaizumi, *Inorg. Chem.*, 1992, **31**, 298.
134 E. Bill, E.L. Bominaar, X.Q. Ding, A.X. Trautwein, H. Winkler, D. Mandon, R. Weiss, A. Gold, K. Jayaraj, and G.E. Toney, *Hyperfine Interact.*, 1990, **58**, 2343.
135 R.A. Leising, B.A. Brennan, and L. Que, Jr., *J. Am. Chem. Soc.*, 1991, **113**, 3988.
136 M.D. Assis, O.A. Serra, Y. Iamamoto, and O.R. Nascimento, *Inorg. Chim. Acta*, 1991, **187**, 107.
137 A. Caneschi, F. Ferraro, D. Gatteschi, P. Rey, and R. Sessoli, *Inorg. Chem.*, 1991, **30**, 3162.
138 C. Benelli, A. Caneschi, D. Gatteschi, M.C. Melandri, and P. Rey, *Inorg. Chim. Acta*, 1990, **172**, 137.
139 P. Lacroix, O. Kahn, L. Valade, P. Cassoux, and L.K. Thompson, *Synth. Met.*, 1990, **39**, 81.
140 W.W. Wang, and S.S. Wang, *Synth. Met.*, 1991, **42**, 1729.
141 S.L. Bartley, and K.R. Dunbar, *Angew. Chem., Int. Ed. Engl.*, 1991, **30**, 448.
142 A. Bencini, and C. Zanchini, *Inorg. Chem.*, 1991, **30**, 4245.
143 Y.I. Kim, and W.E. Hatfield, *Inorg. Chim. Acta*, 1991, **188**, 15.
144 D. Reefman, J.P. Cornelissen, R.A.G. De Graaff, J.G. Haasnoot, and J. Reedijk, *Inorg. Chem.*, 1991, **30**, 4928.
145 A.A. Eagle, M.F. Mackay, and C.G. Young, *Inorg. Chem.*, 1991, **30**, 1425.
146 C. Harding, V. McKee, and J. Nelson, *J. Am. Chem. Soc.*, 1991, **113**, 9684.

147 L.P. He, C.L. Yao, M. Naris, J.C. Lee, J.D. Korp, and J.L. Bear, *Inorg. Chem.*, 1992, **31**, 620 - 625.
148 H. So, and C.W. Lee, *Bull. Korean Chem. Soc.*, 1990, **11**, 115.
149 D.L. Hughes, M.Y. Mohammed, and C.J. Pickett, *J. Chem. Soc., Dalton Trans.*, 1990, 2013.
150 J.M. Ball, P.M. Boorman, J.F. Fait, H.B. Kraatz, B. Heinz, J.F. Richardson, D. Collison, and F.E. Mabbs, *Inorg. Chem.*, 1990, **29**, 3290.
151 A.R. Oki, J. Glerup, and D.J. Hodgson, *Inorg. Chem.*, 1990, **29**, 2435.
152 P.A. Goodson, and D.J. Hodgson, *Inorg. Chim. Acta*, 1990, **172**, 49.
153 P.A. Goodson, J. Glerup, D.J. Hodgson, K. Michelsen, and H. Weihe, *Inorg. Chem.*, 1991, **30**, 4909.
154 R. Bhula, S. Collier, W.T. Robinson, and D.C. Weatherburn, *Inorg. Chem.*, 1990, **29**, 4027.
155 G. Swarnabala, and M.V. Rajasekharan, *Proc. Indian Acad. Sci., Chem. Sci.*, 1990, **102**, 87.
156 H.H. Thorp, *Chemtracts: Anal., Phys., Inorg. Chem.*, 1990, **2**, 405.
157 J.E. Sarneski, M. Didiuk, H.H. Thorp, R.H. Crabtree, G.W. Brudvig, J.W. Faller, G.K. Schulte, *Inorg. Chem.*, 1991, **30**, 2833.
158 A.R. Schake, E.A. Schmitt, A.J. Conti, W.E. Streib, J.C. Huffman, D.N. Hendrickson, and G. Christou, *Inorg. Chem.*, 1991, **30**, 3192.
159 R. Bhula, and D.C. Weatherburn, *Angew. Chem. Int. Ed. Engl.*, 1991, **30**, 688.
160 P. Barbaro, A. Bencini, I. Bertini, F. Briganti, and S. Midollini, *J. Am. Chem. Soc.*, 1990, **112**, 7238.
161 S. Ciurli, S. Yu, R.H. Holm, K.K.P. Srivastava, and E. Münck, *J. Am. Chem. Soc.*, 1990, **112**, 8169.
162 E.K.H. Roth, J.M. Greneche, and J. Jordanov, *J. Chem. Soc., Chem. Commun.*, 1991, 105.
163 H.L. Blonk, J. Mesman, J.G.M. Van der Linden, J.J. Steggerda, J.M.M. Smits, G. Beurskens, P.T. Beurskens, C. Tonon, and J. Jordanov, *Inorg. Chem.*, 1992, **31**, 962.
164 C. Chen, J. Cai, Q. Liu, D. Wu, X. Lei, K. Zhao, B. Kang, and J. Lu, *Inorg. Chem.*, 1990, **29**, 4878.
165 N. Kobayashi, H. Lam, W.A. Nevin, P. Janda, C.C. Leznoff, and A.B.P. Lever, *Inorg. Chem.*, 1990, **29**, 3415.
166 H. Kitagawa, N. Kojima, N. Matsushida, T. Ban, and I. Tsujikawa, *J. Chem. Soc., Dalton Trans.*, 1991, 3115.
167 N. Sreehari, B. Varghese, and P.T. Manoharan, *Inorg. Chem.*, 1990, **29**, 4011.
168 J. Ribas, A. Garcia, R. Costa, M. Monfort, S. Alvarez, C. Zanchini, X. Solans, and M.V. Domenech, *Inorg. Chem.*, 1991, **30**, 841.
169 L. Soto Tuero, J. Garcia-Lozano, E. Escriva Monto, M. Beneto Borja, F. Dahan, J.P. Tuchagues, and J.P. Legros, *J. Chem. Soc., Dalton Trans.*, 1991, 2619.

170 L.P. Battaglia, A. Corradi-Bonamartini, S. Ianelli, M.A. Zoroddu, and G. Sanna, *J. Chem. Soc., Faraday Trans.*, 1991, **87**, 3863.
171 J.P. Costes, F. Dahan, and J.P. Laurent, *Inorg. Chem.*, 1992, **31**, 284.
172 A. Tosik, W. Maniukiewicz, M. Bukowska-Strzyzewska, J. Mrozinski, M.P. Sigalas, and C.A. Tsipis, *Inorg. Chim. Acta*, 1991, **190**, 193.
173 I. Castro, J. Faus, M. Julve, C. Bois, J.A. Real, and F. Lloret, *J. Chem. Soc., Dalton Trans.*, 1992, 47.
174 M. Valko, M. Melník, H. Morris, R.F. Bilton, and P. Pelikán, *Chem. Phys. Letters*, 1991, **183**, 372.
175 S. Kawata, H. Yokoi, and M. Iwaizumi, *Bull. Chem. Soc. Japan*, 1990, **63**, 3414.
176 K.J. Oberhausen, J.F. Richardson, R.M. Buchanan, J.K. McCusker, D.N. Hendrickson, and J.M. Latour, *Inorg. Chem.*, 1991, **30**, 1357.
177 S.K. Mandal, L.K. Thompson, M.J. Newlands, E.J. Gabe, and F.L. Lee, *Inorg. Chem.*, 1990, **29**, 3556.
178 H. Yokoi, A. Takeuchi, and S. Yamada, *Bull. Chem. Soc. Japan*, 1990, **63**, 1462.
179 T.M. Donlevy, L.R. Gahan, T.W. Hambley, G.R. Hanson, A. Markiewicz, K.S. Murray, I.L. Swann, and S.R. Pickering, *Aust. J. Chem.*, 1990, **43**, 1407.
180 P.V. Bernhardt, P. Comba, L.R. Gahan, and G.A. Lawrance, *Aust. J. Chem.*, 1990, **43**, 2035.
181 A. McAuley, K. Beveridge, S. Subramanian, and T.W. Whitcombe, *J. Chem. Soc., Dalton Trans.*, 1991, 1821.
182 M. Hoshino, Y. Iimura, and S. Konishi, *J. Phys. Chem.*, 1992, **96**, 179.
183 J.J. Borrás-Almenar, E. Coronado, D. Gatteschi, and C. Zanchini, *Inorg. Chem.*, 1992, **31**, 294.
184 V.K. Voronkova, L.V. Mosina, A.E. Usachev, and Yu.V. Yablokov, *Phys. Status Solidi B*, **158**, 337.
185 J.L.E. Burn, C. Woods, and D.P. Rillema, *Inorg. Chim. Acta*, 1991, **188**, 113. See also ref. 127 in ref. 1
186 S.J. Chen, and K.R. Dunbar, *Inorg. Chem.*, **30**, 2018.
187 F.A. Cotton, and R.C. Torralba, *Inorg. Chem.*, 1991, **30**, 2196.
188 F.A. Cotton, and R.C. Torralba, *Inorg. Chem.*, 1991, **30**, 4392.
189 F.A. Cotton, and R.C. Torralba, *Inorg. Chem.*, 1991, **30**, 3293.
190 F.A. Cotton, and R.C. Torralba, *Inorg. Chem.*, 1991, **30**, 4386.
191 G. Swarnabala, and M.V. Rajasekharan, *Inorg. Chim. Acta*, 1990, **168**, 167.
192 Y. Nishida, S. Haga, and T. Tokoi, *Chem. Lett.*, 1989, 321.
193 T.R. Holman, M.P. Hendrich, and L. Jr. Que, *Inorg. Chem.*, 1992, **31**, 937.
194 C. Flassbeck, K. Wieghardt, E. Bill, C. Butzaff, A.X. Trautwein, B. Nuber, and J. Weiss, *Inorg. Chem.*, 1992, **31**, 21.
195 A.S. Borovik, M.P. Hendrich, T.R. Holman, E. Münck, V. Papaefthymiou, and L. Que Jr., *J. Am. Chem. Soc.*, 1990, **112**, 6031.
196 W.B. Tolman, S.C. Liu, J.G. Bentsen, and S.J. Lippard, *J. Am. Chem. Soc.*, 1990, **112**, 152.

197 M.P. Hendrich, E. Münck, B.G. Fox, and J.D. Lipscomb, *J. Am. Chem. Soc.*, 1990, **112**, 5861.
198 T.R. Holman, C. Juarez-Garcia, M.P. Hendrich, L. Que Jr., and E. Münck, *J. Am. Chem. Soc.*, 1990, **112**, 7611.
199 C. Juarez-Garcia, M.P. Hendrich, T.R. Holman, L. Que Jr., and E. Münck, *J. Am. Chem. Soc.*, 1991, **113**, 518.
200 X.L. Tan, Y. Gultneh, J.E. Sarneski, and C.P. Scholes, *J. Am. Chem. Soc.*, 1991, **113**, 7853.
201 J.E. Sarneski, H.H. Thorp, G.W. Brudvig, R.H. Crabtree, and G.K. Schulte, *J.Am. Chem. Soc.*, 1990, **112**, 7255.
202 H.H. Thorp, J.E. Sarneski, R.J. Kulawiec, G.W. Brudvig, R.H. Crabtree, and G.C. Papaefthymiou, *Inorg. Chem.*, 1991, **30**, 1153.
203 D.N. Hendrickson, G. Christou, E.A. Schmitt, E. libby, J.S. Bashkin, S. Wang, H.L. Tsai, J.B. Vincent, P.D.W. Boyd, J.C. Huffman, K. Folting, Q. Li, and W.E. Streib, *J. Am. Chem. Soc.*, 1992, **114**, 2455.
204 Y. Journaux, F. Lloret, and O. Kahn, *Inorg. Chem.*, 1990, **29**, 3048.
205 C.J. Gómez-Garcia, E. Coronado, and J.J. Borrás-Almenar, *Inorg. Chem.*, 1992, **31**, 1667.
206 E. Bouwman, W.L. Driessen, J. Reedijk, C. Smykalla, J.M.M. Smits, P.T. Beurskens, F. Laschi, and P. Zanello, *Inorg. Chem.*, 1990, **29**, 4881.
207 M.A. El-Sayed, and G. Davies, *Inorg. Chem.*, 1990, **29**, 4891.
208 F. Lloret, Y. Journaux, and M. Julve, *Inorg. Chem.*, 1990, **29**, 3967.
209 P. Chaudhuri, M. Winter, B.P.C. Della Vedova, E. Bill, A. Trautwein, S. Gehring, P. Fleischhauer, B. Nuber, and J. Weiss, *Inorg. Chem.*, 1991, **30**, 2148.
210 B. Xu, and L. Kevan, *J. Phys. Chem.*, 1991, **95**, 1147.
211 M. Ebihara, K. Isobe, Y. Sasaki, and K. Saito, *Inorg. Chem.*, **31**, 1644.
212 A. Bencini, S. Midollini, and C. Zanchini, *Inorg. Chem.*, 1992, **31**, 2132.
213 A. Caneschi, D. Gatteschi, R. Sessoli, A.L. Barra, L.C. Brunel, and M. Guillot, *J. Am. Chem. Soc.*, 1991, **113**, 5873.
214 F.D.M. Haldane, *Phys. rev. Lett.*, 1983, **50**, 1153.
215 J.C. Bonner, *J. Appl. Phys.*, 1987, **61**, 3941.
216 J.P. Renard, M. Verdaguer, L.P. Regnault, W.A.C. Erkelens, J. Rossat-Mignod, J. Ribas, W.G. Stirling, and C. Vittier, *J. Appl. Phys.*, 1988, **63**, 3538.
217 a) K. Katsumata, H. Hori, T. Takeuchi, M. Date, A. Yamagishi, and J.P. Renard, *Phys. Rev. Lett.*, 1989, **63**, 86; b) Y. Ajiro, T. Goto, H. Kikuchi, T. Sakakibara, and T. Inami, *Phys. Rev. Lett.*, 1989, **63**, 1424.
218 K. Kindo, T. Yoshida, and M. Date, *J. Magn. Magn. Mat..*, 1990, **90-91**, 227.
219 M. Datę, and K. Kindo, *Phys. Rev. Lett.*, 1990, **65**, 1659.
220 M. Hagiwara, K. Katsumata, I. Affleck, B.I. Halperin, and J.P. Renard, *Phys. Rev. Lett. B*, 1990, **65**, 3181.
221 M. Hagiwara, K. Katsumata, and J.P. Renard, *J. Magn. Magn. Mat.*, 1990, **91-92**, 225.

222 M. Hagiwara, and K. Katsumata, *Mod. Phys. Lett. B*, 1991, **5**, 741.
223 S.H. Glarum, S. Geschwind, K.M. Lee, M.L. Kaplan, and J. Michel, *Phys. Rev. Lett.*, 1991, **67**, 1614.
224 D.H. Reich, L.P. Levy, and T. Giamarchi, *Phys. Rev. Lett.*, 1991, **67**, 2207.
225 R. Calvo, P.R. Levstein, E.E. Castellano, S.M. Fabiane, O.E. Piro, and S.B. Oseroff, *Inorg. Chem.*, 1991, **30**, 216.
226 P.R. Levstein, R. Calvo, E.E. Castellano, O.E. Piro, and B.E. Rivero, *Inorg. Chem.*, 1990, **29**, 3918.
227 R. Calvo, and M.C.G. Passeggi, *J. Phys.: Condens. Matter*, 1990, **2**, 9113.
228 C. Zanchini, and R.G. Willett, *Inorg. Chem.*, 1990, **29**, 3027.
229 C. Balagopalakrishna, and M.V. Rajasekharan, *Phys. Rev. B: Condens. Matter*, 1990, **42**, 7794.
230 R. Calco, M.C.G. Passeggi, M.A. Novak, O.G. Symko, S.B. Oseroff, O.R. Nascimento, and M.C. Terrile, *Phys. Rev. B: Condens. Matter*, 1991, **43**, 1074.
231 D.M. Martino, C.A. Staren, R. Calvo, and O.E. Piro, *J. Solid. State Chem.*, 1991, **90**, 211.
232 P.R. Levstein, H.M. Pastawski, and R. Calvo, *J. Phys.: Condens. Matter*, 1991, **3**, 1877.
233 C.A. Steren, R. Calvo, E.E. Castellano, M.S. Fabiane, and O.E. Piro, *Physica B (Amsterdam)*, 1990, **164**, 323.
234 R. Calvo, and M.C.G. Passeggi, *Phys. Rev. B: Condens. Matter*, 1991, **44**, 5111.
235 M.R. Bond, R.D. Willett, R.S. Rubins, P. Zhou, C.E. Zaspel, S.L. Hutton, and J.E. Drumheller, *Phys. Rev. B: Condens. Matter*, 1990, **42**, 10280.
236 C.E. Zaspel, *J. Appl. Phys.*, 1990, **67**, 6011.
237 T. Ishii, and L. Yamada, *J. Phys.: Condens. Matter*, 1990, **2**, 5771.
238 M. Beneto, L. Soto, J. Garcia-Lozano, E. Escriva, J.P. Legros, and F. Dahan, *J. Chem. Soc., Dalton Trans.*, 1991, 1057.
239 M.B. Inoue, Q. Fernando, and M. Inoue, *Synth. Met.*, 1991, **42**, 2069.
240 J. Goslar, W. Hilczer, S.K. Hoffmann, and M. Krupski, *Phys. Status Solidi B*, 1991, **167**, 291.
241 S.K. Hoffmann, and W. Hilczer, *Inorg. Chem.*, 1991, **30**, 2210.
242 J.J. Borrás-Almenar, E. Coronado, C.J. Gomez-Garcia, D. Gatteschi, and C. Zanchini, *J. Appl. Phys.*, 1990, **67**, 6006.
243 J.J. Borrás-Almenar, R. Burriel, E. Coronado, D. Gatteschi, C.J. Gomez-Garcia, and C. Zanchini, *Inorg. Chem.*, 1991, **30**, 947.
244 D. Skrzypek, *Phys. Status Solidi B*, 1990, **157**, 695.
245 T. Grieb, T. Pabst, A. Kieslich, J. Lindner, H. Rauschmann, K. Strobel, W. Treutmann, and R. Geick, *J. Phys.: Condens. Matter*, 1991, **3**, 9757.
246 D. Deguenon, G. Berardinelli, J.P. Tuchagues, and P. Castan, *Inorg. Chem.*, 1990, **29**, 3031.
247 A. Nakanishi, and M. Date, *J. Phys. Soc. Jpn.*, 1991, **60**, 3818.
248 J. Tuchendler, and K. Katsumata, *Solid State Commun.*, 1990, **74**, 1159.

249 P.F. McDonald, A. Parasiris, R.K. Pandey, B.L. Gries, and W.P. Kirk, *J. Appl. Phys.*, 1991, **69**, 1104.
250 R.W. Saalfrank, C.J. Lurz, K. Schobert, O. Struck, E. Bill, and A. Trautwein, *Angew. Chem. Int. Ed. Engl.*, 1991, **30**, 1494.
251 H. Okamoto, K. Toriumi, T. Mitani, and M. Yamashita, *Phys. Rev. B: Condens. Matter*, 1990, **42**, 10381.
252 K.H. Lii, C.H. Li, C.Y. Cheng, and S.L. Wang, *J. Solid State Chem.*, 1991, **95**, 352.
253 S. Chehab, P. Biensan, J. Amiell, and S. Flandrois, *J. Phys. I*, 1991, **1**, 537.
254 S. Chehab, P. Biensan, S. Flandrois, and J. Amiell, *Phys. Rev. B: Condens. Matter*, 1992, **45**, 2844.
255 S. Chehab, J. Amiell, P. Biensan, and S. Flandrois, *Physica B (Amsterdam)*, 1991, **173**, 211.
256 J. Kuriata, L. Sadlowski, B. Bojanowski, M. Wabia, M.L. Falin, V.V. Izotov, and Zh.S. Yakovleva, *J. Phys. Chem. Solids*, 1992, **53**, 23.
257 M. Handa, H. Sono, K. Kasamatsu, K. Kasuga, M. Mikuriya, and S. Ikenoue, *Chem. Lett.*, 1992, 453.
258 Y.K. Shin, and D.G. Nocera, *J. Am. Chem. Soc.*, 1992, **114**, 1264.
259 C.I. Cabello, A. Caneschi, R.L. Carlin, D. Gatteschi, P. Rey, and R. Sessoli, *Inorg. Chem.*, 1990, **29**, 2582
260 A. Caneschi, D. Gatteschi, P. Rey, and R. Sessoli, *Inorg. Chem.*, 1991, **30**, 3936.
261 K. Nakatani, J. Sletten, S. Halut-Desporte, S. Jeannin, Y. Jeannin, and O. Kahn, *Inorg. Chem.*, 1991, **30**, 164.
262 N. Kuroda, M. Sakai, M. Suezawa, Y. Nishina, and K. Sumino, *J. Phys. Soc. Jpn.*, 1990, **59**, 3049.
263 G. Matsubayashi, and A. Yokozawa, *J. Chem. Soc., Dalton Trans.*, 1990, 3535.
264 A. Yokozawa, and G. Matsubayashi, *Inorg. Chim. Acta*, 1991, **186**, 165.
265 H. Strzelecka, R. Vicente, J. Ribas, J.-P. Legros, P. Cassoux, P. Petit, and J.-J. Andre, *Polyhedron*, 1991, **10**, 687.
266 C. Bourbonnais, R.T. Henriques, P. Wzietek, D. Köngeter, J. Voiron, and D. Jérome, *Phys. Rev. B*, 1991, **44**, 641.
267 L.G. Caron, and C. Bourbonnais, *Synth. Met.*, 1991, **43**, 3941.
268 T. Sugano, H. Takenouchi, D. Shiomi, and M. Kinoshida, *Synth. Met.*, 1991, **42**, 2217.
269 M. Kurmoo, D. Kanazawa, and P. Day, *Synth. Met.*, 1991, **42**, 2123.
270 S. Kutsumitzu, N. Kojima, and I. Tsujikawa, *J. Chem. Soc., Dalton Trans.*, 1991, 169.
271 T. Story, C.H.W. Swuste, H.J.M. Svagten, R.J.T. Van Kempen, and W.J.M. De Jonge, *J. Appl. Phys.*, 1991, **69**, 6037.
272 E.A. Eivazov, A.F. Safarov, S.M. Atakishiev, and Ya.M. Abasov, *Phys. Status Solidi A*, 1990, **117**, K147.
273 J.T. Yu, S.S. Lin, and Y.S. Huang, *J. Appl. Phys.*, 1990, **68**, 1796.
274 D.M. Ioshpe, *Phys. Lett. A*, 1991, **160**, 579.

275 K.A. Müller, M. Aguilar, W. Berlinger, and K.W. Blazey, *J. Phys.: Condens. Matter*, 1990, **2**, 2735.
276 Z. Wilamowski, W. Jantsch, and G. Hendorfer, *Semicond. Sci. Technol.*, 1990, **5**, S266.
277 J.T. Yu, Y.H. Huang, and S.S. Lin, *J. Phys: Condens. Matter*, 1990, **2**, 5587.
278 A. Goerger, and J.M. Spaeth, *Semicond. Sci. Technol.*, 1991, **6**, 800.
279 T. Nakamura, Y. Takeuchi, H. Muramatsu, H. Fujiyasu, and Y. Nakanishi, *Appl. Surf. Sci.*, 1991, **48-49**, 478.
280 W. Gehlhoff, P. Emanuelsson, P. Omling, and H.G. Grimmeiss, *Phys. Rev. B: Condens. Matter*, 1990, **41**, 8560.
281 S. Dannefaer, *Radiat. Eff. Defects Solids*, 1989, **111-112**, 65.
282 N.S. Dalal, and P.K. Kahol, *Prog. High Temp. Supercond.*, 1988, **7**, 196.
283 K.N. Shrivastava, *Phys. Rep.*, 1991, **200**, 51.
284 D.A. Huse, M.P.A. Fisher, and D.S. Fisher, *Nature*, 1992, **358**, 553.
285 K. Moorjani, B.F. Kim, F.J. Adrian, and J. Bohandy, *Prog. High Temp. Supercond.*, 1989, **17**, 225.
286 B. Czyzak, T. Zuk, and J. Stankowski, *Acta Phys. Pol. A*, 1990, **A78**, 769.
287 B. Czyzak, and J. Stankowski, *Acta Phys. Pol. A*, 1991, **80**, 99.
288 A.T. Mahal, J.P. DeLooze, P.K. Kahol, and N.S. Dalal, *Prog. High Temp. Supercond.*, 1989, **17**, 239.
289 D. Shaltiel, H. Bill, A. Grayevski, A. Junod, D. Lowy, S. Sadowski, and E. Walker, *Supercond. Sci. Technol.*, 1991, **4**, S85.
290 B. Ravkin, M. Pozek, and A. Dulcic, *Solid State Commun.*, 1989, **72**, 199.
291 B. Ravkin, M. Pozek, and A. Dulcic, *Physica C*, 1990, **170**, 166.
292 Yu.N. Shvachko, A.A. Koshta, A.A. Romanyukha, V.V. Ustinov, and A.I. Akimov, *Physica C (Amsterdam)*, 1991, **174**, 447.
293 K. Sugawara, D.J. Baar, Y. Shiohara, and S. Tanaka, *Mod. Phys. Lett. B*, 1991, **5**, 779.
294 K. Sugawara, D.J. Baar, M. Murakami, A. Kondoh, K. Yamaguchi, Y. Shiohara, and S. Tanaka, *Mod. Phys. Lett. B*, 1991, **5**, 1001.
295 M. Wahlen, M. Ruebsam, O. Dobbert, and K.P. Dinse, *Physica C (Amsterdam)*, 1991, **185-189**, 1797.
296 N. Bontemps, D. Davidov, P. Monod, and R. Even, *Phys. Rev. B*, 1991, **43**, 11512.
297 B. Ravkin, T.A. Mahl, A.S. Bhalla, Z.Z. Sheng, and N.S. Dalal, *Ferroelectrics*, 1991, **117**, 313.
298 N. Bontemps, P.Y. Bertin, D. Davidov, P. Monod, C. Lacour, R. Even, and V.V. Moshchalkov, *Physica C (Amsterdam)*, 1991, **185-189**, 1809.
299 J.T. Masiakowski, M. Puri, and L. Kevan, *J. Phys. Chem.*, 1991, **95**, 8968.
300 S.N. Ekbote, *Bull. Mater. Sci.*, 1991, **14**, 631.
301 K. Kindo, M. Honda, T. Kohashi, and M. Date, *J. Phys. Soc. Jpn.*, 1990, **59**, 2332.
302 K. Muraleedhran, and T.K. Gundu Rao, *J. Magn. Magn. Mater*, 1990, **89**, L277.

303 C.B. Azzoni, A. Paleari, and G.B. Pallavicini, *J. Phys.: Condens. Matter*, 1992, **4**, 1359.
304 E.W. Ong, G.H. Kwei, R.A. Robinson, B.L. Ramakrishna, and R.B. Von Dreele, *Phys. Rev. B: Condens. Matter*, 1990, **42**, 4255.
305 H. Masuda, F. Mizuno, I. Hirabayashi, and S. Tanaka, *Phys. Rev. B: Condens. Matter*, 1991, **43**, 7871.
306 J. Stankowski, W. Hilczer, J. Baszynski, B. Czyzak, and L. Szczepanska, *Solid State Commun.*, 1991, **77**, 125.
307 D.C. Vier, S. Schultz, C. Rettori, D. Rao, S.B. Oseroff, M. Thova, Z. Fisk, and S.W. Cheong, *J. Appl. Phys.*, 1991, **69**, 4872.
308 A.V. Lazuta, *Physica C (Amsterdam)*, 1991, **181**, 127.
309 R. Janes, K.K. Singh, S.D. Burnside, and P.P. Edwards, *Solid State Commun.*, 1991, **79**, 241.
310 T. Ishida, K. Kanoda, T. Takahashi, and K. Koga, *Physica C (Amsterdam)*, 1991, **178**, 231.
311 K. Sugawara, D.J. Baar, Y. Shiohara, and S. Tanaka, *Mod. Phys. Lett. B*, 1991, **5**, 895.
312 R.J. Barham, and D.C. Doetschman, *J. Mater. Res.*, 1992, **7**, 565.
313 E. Buluggiu, A. Vera, and G. Amoretti, *Physica C (Amsterdam)*, 1990, **171**, 271.
314 J.T. Lue, and C.C. Wu, *Solid State Commun.*, 1990, **75**, 643.
315 J.T. Masiakowski, M. Puri, S. Cuvier, M. Romanelli, R.N. Schwartz, H. Kimura, and L. Kevan, *Physica C (Amsterdam)*, 1990, **166**, 140.
316 J.T. Masiakowski, M. Puri, and L. Kevan, *J. Phys. Chem.*, 1991, **95**, 1393.
317 R.N. Bagchi, and D. Byrt, *Indian J. Cryog.*, 1990, **15**, 106.
318 S. Cuvier, M. Puri, J. Bear, and L. Kevan, *Chem. Mater.*, 1991, **3**, 115.
319 M. Puri, S. Marrelli, L. Li, S. Cuvier, J. Bear, and L. Kevan, *J. Chem. Soc. Faraday Trans.*, 1991, **87**, 167.
320 R. Janes, R.-S. Liu, P.P. Edwards, A.D. Stevens, and M.C.R. Symons, *J. Chem. Soc. Faraday Trans.*, 1991, **87**, 1209.
321 R. Karim, H. How, C. Vittoria, and A. Widom, *J. Appl. Phys.*, 1991, **69**, 4143.
322 M. Puri, S. Marrelli, J. Bear, and L. Kevan, *Appl. Magn. Reson.*, 1991, **1**, 509.
323 S.B. Oseroff, D. Rao, F. Wright, D.C. Vier, S. Schultz, J.D. Thompson, Z. Fisk, S.W. Cheong, M.F. Hundley, and M. Tovar, *Phys. Rev. B: Condens. Matter*, 1990, **41**, 1934.
324 M.D. Sastry, K.S. Ajayakumar, R.M. Kadam, G.M. Phatak, and R.M. Iyer, *Physica C (Amsterdam)*, 1990, **170**, 41.
325 R. Janes, R.S. Liu, P.P. Edwards, and J.L. Tallon, *Physica C (Amsterdam)*, 1990, **167**, 520.
326 R. Janes, R.S. Liu, A.D. Stevens, and M.C.R. Symons, *Bull. Mater. Sci.*, 1991, **14**, 811.
327 F.J. Owens, and Z. Iqbal, *Physica C (Amsterdam)*, 1990, **171**, 238.

328 M. Puri, J. Masiakowski, J. Bear, and L. Kevan, *J. Chem. Soc. Faraday Trans.*, 1991, **87**, 3687.
329 F.J. Owens, *Phys. Status Solidi B*, 1990, **162**, 565.
330 F.J. Owens, Z. Iqbal, and E. Wolf, *J. Phys.: Condens. Matter*, 1991, **4**, 205.
331 Y. Hayashi, M. Fukui, T. Fujita, H. Shibayama, K. Iwahashi, and S. Sako, *Supercond. Sci. Technol.*, 1991, **4**, S322.
332 D.J. Keeble, D.S. Ginley, E.H. Pointdexter, and G.J. Gerardi, *NATO ASI Ser., Ser. E*, 1990, **181**, 439.
333 K.N. Shrivastava, *Solid State Commun.*, 1991, **78**, 403.
334 K.N. Shrivastava, and S. Koka, *Solid State Commun.*, 1991, **78**, 735.
335 G.B. Teitelbaum, E.F. Kukovitski, S.G. L'vov, Yu.I. Talanov, and R.I. Khasanov, *Physica C (Amsterdam)*, 1991, **185-189**, 2369.
336 R. Karim, R. Seed, H. How, A. Widom, C. Vittoria, G. Balestrino, and P. Paroli, *J. Appl. Phys.*, 1990, **67**, 5064.
337 H. Maruyama, and I. Shiozaki, *Jpn. J. Appl. Phys. Part 2*, 1991, **30**, L694.
338 T. Singh, and E. Zacharias, *Bull. Mater. Sci.*, 1991, **14**, 343.
339 N. Guskos, S.K. Patapis, and C. Politis, *Mod. Phys. Lett. B*, 1991, **5**, 701.
340 S.K. Misra, M. Kahrizi, J. Kotlinski, M.O. Steinitz, and F.S. Razavi, *Solid State Commun.*, 1992, **81**, 503.
341 N. Guskos, G.P. Triberis, M. Calamiotou, C. Trikalinos, A. Koufoudakis, C. Mitros, H. Gamari-Sale, and D. Niarchos, *Phys. Status Solidi B*, 1990, **162**, 243.
342 K. Nagata, K. Saito, Y. Egawa, R. Liang, and T. Nakamura, *J. Magn. Magn. Mater.*, 1990, **90**, 649.
343 Z. Zuotao, *J. Phys. Chem. Solids*, 1991, **52**, 659.
344 N. Guskos, C.A. Lodros, C. Trikalinos, S.M. Paraskevas, A. Koufoudakis, C. Mitros, H. Gamari-Sale, and D. Niarchos, *Phys. Status Solidi B*, 1991, **165**, 249.
345 S. Yuan, X. Zhou, S. Jin, Y. Wang, S. Wang, Z. Chen, *Mod. Phys. Lett. B*, 1991, **5**, 651.
346 P. Wang, S. Bandow, J.W. Zhan, and D.B. Zhu, *Synth. Met.*, 1991, **43**, 4077.
347 R. Rai, *Bull. Mater. Sci.*, 1991, **14**, 1083.
348 J. Stankowski, W. Kempinski, and Z. Trybula, *Acta Phys. Pol. A*, 1991, **80**, 571.
349 E. Buluggiu, A. Vera, D.C. Giori, G. Amoretti, and F. Licci, *Supercond. Sci. Technol.*, 1991, **4**, 595.
350 C.N.R. Rao, G.N. Subbanna, R. Nagarajan, A.K. Ganguli, L. Ganapathi, R. Vijayaraghavan, S.V. Bhat, and A.R. Raju, *J. Solid State Chem.*, 1990, **88**, 163.
351 B.H. Hou, Y.X. Wang, R.Y. Liang, and S.Z. Jin, *Mod. Phys. Lett. B*, 1991, **5**, 1745.
352 M. Kaise, H. Kondoh, M. Mizuno, C. Nishihara, H. Nozoye, and H. Shindo, *Physica C (Amsterdam)*, 1991, **185-189**, 1207.
353 B.V. Rozentuller, S.V. Stepanov, M.M. Shabarchina, and A.I. Tsapin, *Phys. Lett. A*, 1990, **148**, 119.
354 A.M. Finkel'stein, V.E. Kataev, E.F. Kukovitskii, and G.B. Teitel'baum, *Physica C (Amsterdam)*, 1990, **168**, 370.

355 K. Sugawara, *Progr. High Temp. Supercond.*, 1989, **12**, 217.
356 K. Sugawara, K. Okibayashi, S. Ehara, N. Hashizume, and S. Kataoka, *Progr. High Temp. Supercond.*, 1989, **12**, 221.
357 N. Brnicevic, L. Tusek-Bozic, P. Planinic, S. Popovic, B. Rakvin, M. Pozek, A. Dulcic, G. Leising et al., *J. Less Common Met.*, 1990, **166**, 63.
358 A.A. Romanyukha, Yu.N. Shvachko, V.Yu. Irkhin, M.I. Katsnelson, A.A. Koshta, and V.V. Ustinov, *Physica C (Amsterdam)*, 1990, **171**, 276.
359 A. Moto, A. Morimoto, M. Kumeda, and T. Shimizu, *Supercond. Sci. Technol.*, 1990, **3**, 579.
360 H.C. Kim, H. So, and H.K. Lee, *Bull. Korean Chem. Soc.*, 1991, **12**, 499.
361 M. Coldea, *Physica C (Amsterdam)*, 1991, **185-189**, 1197.
362 G. Whichard, H. Hosono, R.A. Weeks, R.A. Zuhr, and R.H. III Magruder, *J. Appl. Phys.*, 1990, **67**, 7526.
363 A. Tuel, J. Diab, P. Gelin, M. Dufaux, J.F. Dutel, and Y. Ben Taarit, *J. Mol. Catal.*, 1990, **63**, 95.
364 K. Akagi, D. Hashimoto, H. Shirakawa, and J. Isoya, *Polym. Commun.*, 1990, **31**, 411.
365 K. Akagi, D. Hashimoto, H. Shirakawa, and J. Isoya, *Synth. Met.*, 1991, **40**, 197.
366 M. P. Scripsick, G. J. Edwards, L.E. Halliburton, and L.F. Belt, *J. Appl. Phys.*, 1991, **70**, 2991.
367 P. Bailey, and J. A. Weil, *J. Phys. Chem. Solids*, 1992, **53**, 309.
368 D. Gourier, D. Vivien, and C. Wyon, *J. Solid State Chem.*, 1990, **93**, 378.
369 D. Gourier, A.M. Lejus, and D. Vivien, *Ann. Phys. (Paris)*, 1991, **16**, 215.
370 F.W. Breitbarth, A. Feltz, and M. Steinbrück, *Phys. Status Solidi A*, 1991, **127**, 253.
371 C.B. Azzoni, and A. Paleari, *Phys. Rev. B*, 1991, **44**, 6858.
372 M. Venkateswarlu, T.B. Rao, and A. Hussain, *Solid State Commun.*, 1991, **78**, 1073.
373 G.L. Narendra, J.L. Rao, and S.V.J. Lakshman *Solid State Commun.*, 1991, **77**, 235.
374 A.S. Rao, J.L. Rao, and S.V.J. Lakshman, *Phys. Chem. Glasses*, 1992, **33**, 1.
375 S.K. Misra, and C. Wang, *Solid State Commun.*, 1990, **76**, 751.
376 R.S. Bansal, V.P. Seth, and P. Cham, *Spectrochim. Acta*, 1990, **46A**, 1139.
377 D. Collison, F.E. Mabbs, and J. Temperley, *Spectrochim. Acta*, 1991, **47A**, 691.
378 S.K. Misra, and J. Sun, *Phys. Rev. B: Condens. Matter*, 1990, **42**, 8601.
379 S.K. Misra, and J. Sun, *Physica B (Amsterdam)*, 1990, **162**, 331.
380 S.K. Misra, J. Sun, and U. Orhun, *Phys. Status Solidi B*, 1990, **162**, 585.
381 S.K. Misra, J. Sun, and X. Li, *Physica B (Amsterdam)*, 1991, **168**, 170.
382 S.K. Misra, and J. Sun, *Phys. Rev. B: Condens. Matter*, 1991, **44**, 10333.
383 A.S. Mangrich, L.F.S. Viana, R.R. Sobral, and N.V. Vugman, *Chem. Phys.*, 1990, **46**, 257.
384 H. Utsumi, T. Tatebe, and A. Hamada, *Chem. Lett.*, 1992, 277.
385 R.P. Ferrari, E. Laurenti, S. Poli, and L. Casella, *J. Inorg. Biochem.*, 1992, **45**, 21.

386 B.I. Whittington, and J.R. Anderson, *J.Phys. Chem.*, 1991, **95**, 3306.
387 K.L. Walther, A. Wokaun, and A. Baiker, *J. Chem. Soc., Faraday Trans.*, 1991, **87**, 1217.
388 S. Ristori, M.F. Ottaviani, and G. Martini, *Langmuir*, 1991, **7**, 755.
389 K.V.R. Chary, B.R. Rao and V.S. Subrahmanyam, *Appl. Catal.*, 1991, **74**, 1.
390 G. Busca, and E. Giamello, *Mater. Chem. Phys.*, 1990, **25**, 475.
391 G. Centi, E. Giamello, D. Pinelli, and F. Trifiro, *J. Catal.*, 1991, **130**, 220.
392 M. Sanati, L.R. Wallenberg, A. Andersson, S. Jansen, and Y. Tu, *J. Catal.*, 1991, **132**, 128.
393 M. Morris, J. M. Adams, and A. Dyer, *J. Mater. Chem.*, 1991, **1**, 43.
394 M. Ghedini, S. Morrone, R. Bartolino, V. Formoso, O. Francangeli, B. Yang, D. Gatteschi, and C. Zanchini, *Chem. Mater.*, in press
395 N. Hoshino, A. Kodama, T. Shibuya, Y. Matsunaga, and S. Miyajima, *Inorg. Chem.*, 1991, **30**, 3091.
396 P.J. Alonso, M.L. Sanjuán, P. Romero, M. Marcos, and J.L. Serrano, *J. Phys.: Condens. Matter*, 1990, **2**, 9173.
397 R. Guilard, N. Senglet, Y.H. Liu, D. Sazou, E. Findsen, D. Faure, T. Des Courieres, and K.M. Kadish, *Inorg. Chem.*, 1991, **30**, 1898.
398 M. Rietzel, H.W. Roesky, K.V. Katti, M. Noltemeyer, M.C.R. Symons, and A. Abu-Raqabah, *J. Chem. Soc., Dalton Trans.*, 1991, 1285.
399 A. Watterich, G.J. Edwards, O.R. Gilliam, L.A. Kappers, R. Capelletti, and B. Zelei *Phys. Lett. A*, 1991, **160**, 477.
400 C.J. Wu, and J.T. Yu, *J. Solid State Chem.*, 1991, **93**, 549.
401 L.A. Bottomley, and F.L. Neely, *Inorg. Chem.*, 1990, **29**, 1860.
402 V. Indovina, D. Cordischi, S. De Rossi, G. Ferraris, G. Ghiotti, and A. Chiorino, *J. Mol. Catal.*, 1991, **68**, 53.
403 A. Cimino, D. Cordischi, S. De Rossi, G. Ferraris, D. Gazzoli, V. Indovina, and M. Occhiuzzi, *J. Catal.*, 1991, **127**, 761.
404 T.J. Collins, C. Slebodnick, and E.S. Uffelman, *Inorg. Chem.*, 1990, **29**, 3433.
405 R. Bramley, J.Y. Ji, R.J. Hudd, and P.A. Lay, *Inorg. Chem.*, 1990, **29**, 3089.
406 R. Bramley, J.Y. Ji, and P.A. Lay, *Inorg. Chem.*, 1991, **30**, 1557.
407 R. Bramley, R.D. Farrell, J.Y. Ji, and P.A. Lay, *Aust. J. Chem.*, 1990, **43**, 263.
408 S.L. Brauer, and K.E. Wetterhahn, *J. Am. Chem. Soc.*, 1991, **113**, 3001.
409 M. Quiros, and D.M.L. Goodgame, *Polyhedron*, 1992, **11**, 1.
410 S.B. Kumar, and M. Chaudhury, *J. Chem. Soc. Dalton Trans.*, 1991, 1149.
411 N. Ueyama, N. Yoshinaga, A. Kajiwara, A. Nakamura, and M. Kusunoki, *Bull. Chem. Soc. Japan*, 1991, **64**, 2458.
412 M. Minelli, M.R. Carson, D.W. Whisenhunt Jr., W. Imhof, and G. Huttner, *Inorg. Chem.*, 1990, **29**, 4801.
413 V. Sanz, T. Picher, P. Palanca, P. Gómez-Romero, E. Llopis, J.A. Ramirez, D. Beltrán, and A. Cervilla, *Inorg. Chem.*, 1991, **30**, 3113.

414 G.L. Wilson, R.J. Greenwood, J.R. Pilbrow, J.T. Spence, and A.G. Wedd, *J. Am. Chem. Soc.*, 1991, **113**, 6803.
415 C.S.J. Chang, D. Collison, F.E. Mabbs, and J.H. Enemark, *Inorg. Chem.*, 1990, **29**, 2261.
416 C.S.J. Chang, and J.H. Enemark, *Inorg. Chem.*, 1991, **30**, 683.
417 M. Rietzel, H.W. Roesky, K.V. Katti, H.G. Schmidt, R. Herbst-Irmer, M. Noltemeyer, G.M. Sheldrick, M.C.R. Symons, and A. Abu-Raqabah, *J. Chem. Soc. Dalton Trans.*, 1990, 2387.
418 J.M. Giraudon, J. Sala-Pala, J.E. Guerchais, and L.Toupet, *Inorg. Chem.*, 1991, **30**, 891.
419 B. Deroide, Y. Bensimon, P. Belougne, and J.V. Zanchetta, *J. Phys. Chem. Solids*, 1991, **52**, 853.
420 M. Haddad, R. Berger, Y. Servant, A. Levasseur, and S. Gharbage, *J. Chim. Phys. Phys. Chim. Biol.*, 1991, **88**, 2013.
421 K. Dyrek, and M. Labanowska, *J. Chem. Soc., Faraday Trans.*, 1991, **87**, 1003.
422 Y. Xu, M. Furusawa, M. Ikeya, Y. Kera, and K. Kuwata, *Chem. Letters*, 1991, 293.
423 Z. Sojka, K. Dyrek, P.C. Roberge, and M. Che, *Pol. J. Chem.*, 1991, **65**, 637.
424 U. Abram, K. Köhler, and R. Kirmse, *Zentalinst. Kernforsch., Rossendorf Dresden, [Ber.] ZfK*, 1990, **ZfK-711**, 58.
425 U. Abram, K. Köhler, R. Kirmse, N.B. Kalinichenko, and I.N. Marov, *Zentalinst. Kernforsch., Rossendorf Dresden, [Ber.] ZfK*, 1990, **ZfK-711**, 60.
426 U. Abram, K. Köhler, R. Kirmse, N.B. Kalinichenko, and I.N. Marov, *Zentalinst. Kernforsch., Rossendorf Dresden, [Ber.] ZfK*, 1990, **ZfK-711**, 62.
427 U. Abram, R. Münze, R. Kirmse, K. Köhler, W. Dietzsch, and L. Golic, *Inorg. Chim. Acta*, 1990, **169**, 49.
428 J. Baldas, J.F. Boas, S.F. Colmanet, and G.A. Williams, *J. Chem. Soc., Dalton Trans.*, 1991, 2441.
429 U. Abram, K. Köhler, R. Kirmse, N.B. Kalinichenko, and I.N. Marov, *Inorg. Chim. Acta*, 1990, **176**, 139.
430 R. Kirmse, K. Köhler, U. Abram, R. Böttcher, L. Golic, and E. De Boer, *Chem. Phys.*, 1990, **143**, 75.
431 R. Kirmse, K. Köhler, R. Böttcher, U. Abram, M.C.M. Gribnau, C.P. Keijzers, and E. De Boer, *Chem. Phys.*, 1990, **143**, 83.
432 K. Köhler, R. Kirmse, R. Böttcher, and U. Abram, *Chem. Phys.*, 1992, **160**, 281.
433 U. Abram, and R. Kirmse, *Proc. 4th Meet. Nucl. Anal. Methods*, 1987, **2**, 324.
434 J.C. Brewer, H.H. Thorp, K.M. Slagle, G.W. Brudvig, and H.B. Gray, *J. Am. Chem. Soc.*, 1991, **113**, 3171.
435 A.C. Dengel, and W.P. Griffith, *Inorg. Chem.*, 1991, **30**, 869.
436 R.C. Maurya, and D.D. Mishra, *Synth. React. Inorg. Met.-Org. Chem.*, 1991, **21**, 1627 and references therein.
437 G. Stopa, Z. Stasicka, and D. Rehorek, *Pol. J. Chem.*, 1991, **65**, 649.

438 R.N. Bagchi, A.M. Bond, R. Colton, I. Creece, K. McGregor, and T. Whyte, *Organometallics*, 1991, **10**, 2611.
439 D.M. Wang, S.M. Meijers, and E. de Boer, *Mol. Phys.*, 1990, **70**, 1135.
440 D.M. Wang, and E. de Boer, *J. Chem. Phys.*, 1990, **92**, 4698.
441 P.V. Bernhardt, P. Comba, T.W. Hambley, and G.A. Lawrance, *Inorg. Chem.*, 1991, **30**, 942.
442 H. Stratemeier, M.A. Hitchman, P. Comba, P.V. Bernhardt, and M.J. Riley, *Inorg. Chem.*, 1991, **30**, 4088.
443 D.V. Stynes, H. Noglik, and D.W. Thompson, *Inorg. Chem.*, 1991, **30**, 4567.
444 H. Nasri, J.A. Goodwin, and W.R. Scheidt, *Inorg. Chem.*, 1990, **29**, 185.
445 H. Nasri, Y. Wang, B.H. Huynh, F.A. Walker, and W.R. Scheidt, *Inorg. Chem.*, 1991, **30**, 1483.
446 Y.O. Kim, and H.M. Goff, *Inorg. Chem.*, 1990, **29**, 3907.
447 T. Yoshimura, H. Toi, S. Inaba, and H. Ogoshi, *Inorg. Chem.*, 1991, **30**, 4315.
448 K. Tajima, M. Shigematsu, J. Jinno, K. Ishizu, and H. Ohya-Nishiguchi, *J. Chem. Soc., Chem. Comm.*, 1990, 144.
449 M.K. Safo, W.R. Scheidt, G.P. Gupta, R.D. Orosz, and C.A. Reed, *Inorg. Chim. Acta*, 1991, **184**, 251.
450 M.K. Safo, G.P. Gupta, F.A. Walker, and W.R. Scheidt, *J. Am. Chem. Soc.*, 1991, **113**, 5497.
451 K. Hatano, M.K. Safo, F.A. Walker, and W.R. Scheidt, *Inorg. Chem.*, 1991, **30**, 1643.
452 A. Gismelseed, E.L. Bominaar, E. Bill, A.X. Trautwein, H. Winkler, H. Nasri, P. Doppelt, D. Mandon, J. Fisher, and R. Weiss, *Inorg. Chem.*, 1990, **29**, 2741.
453 E.T. Kintner, and J.H. Dawson, *Inorg. Chem.*, 1991, **30**, 4892.
454 J.P. Costes, F. Dahan, and J.P. Laurent, *Inorg. Chem.*, 1990, **29**, 2448.
455 N.S. Gupta, M. Mohan, N.K. Jha, and W.E. Antholine, *Inorg. Chim. Acta*, 1991, **184**, 13.
456 E. Gelerinter, N.V. Duffy, W. Dietzsch, T. Thanyasiri, and E. Sinn, *Inorg. Chim. Acta*, 1990, **177**, 185.
457 E. Gelerinter, N.V. Duffy, S.S. Yarish, W. Dietzsch, and R. Kirmse, *Chem. Phys. Letters*, 1991, **184**, 375.
458 G.K. Lahiri, S. Bhattacharya, S. Goswami, and A. Chakravorty, *J. Chem. Soc., Dalton Trans.*, 1990, 561.
459 S. Chattopadhyay, N. Bag, P. Basu, G.K. Lahiri, and A. Chakravorty, *J. Chem. Soc., Dalton Trans.*, 1990, 3389.
460 N. Bag, G.K. Lahiri, and A. Chakravorty, *J. Chem. Soc., Dalton Trans.*, 1990, 1557.
461 N.M.S. de Rezende, S. de CastroMartins, L. Alves Marinho, J.A. Viana dos Santos, M. Tabak, J. Rodrigues Perussi, and D.W. Franco, *Inorg. Chim. Acta*, 1991, **182**, 87.
462 M.M.T. Khan, D. Srinivas, R.I. Kureshy, and N.H. Khan, *Inorg. Chem.*, 1990, **29**, 2320.

463 M.M. Taqui Khan, D. Srinivas, R.I. Kureshy, and N.H. Khan, *Polyhedron*, 1991, **10**, 2559.
464 M.M.T. Khan, N.H. Khan, R.I. Kureshy, and A.B. Boricha, *Inorg. Chim. Acta*, 1990, **174**, 175.
465 G.D. Lei, and L. Kevan, *J. Phys. Chem.*, 1990, **94**, 6384.
466 G.D. Lei, and L. Kevan, *J. Phys. Chem.*, 1991, **95**, 4506.
467 B. de Castro, M. Rangel, and J.B. Raynor, *J. Chem. Soc., Dalton Trans.*, 1990, 3311.
468 M. Tadu, and R. Shino, *J. Inorg. Biochem.*, 1991, **44**, 89.
469 S. Shi, A. Bakak, and J.H. Espenson, *Inorg. Chem.*, 1991, **30**, 3410.
470 K. Drabent, J.A. Wolny, M.F. Rudolf, and P.J. Chmielewski, *Polyhedron*, 1992, **11**, 271.
471 R.J. Taylor, R.S. Drago, and J.P. Hage, *Inorg. Chem.*, 1992, **31**, 253 - 258.
472 C. Araullo-McAdams, and K.M. Kadish, *Inorg. Chem.*, 1990, **29**, 2749.
473 D. Sazou, C. Araullo-McAdams, B.C. Han, M.M. Franzen, and K.M. Kadish, *J. Am. Chem. Soc.*, 1990, **112**, 7879.
474 J. Cameron, and S. Graham, *J. Chem. Soc., Dalton Trans.*, 1992, 385.
475 E. Tsuchida, T. Komatsu, T. Nagata, E. Hasegawa, H. Nishide and H. Inoue, *J. Chem. Soc., Dalton Trans.*, 1991, 3285.
476 O. Alnaji, M. Dartiguenave, Y. Dartiguenave, M. Simard, and A.L. Beauchamp, *Inorg. Chim. Acta*, 1991, **187**, 31.
477 S. Trofimenko, F.B. Hulsbergen, and J. Reedijk, *Inorg. Chim. Acta*, 1991, **183**, 203.
478 C. Bianchini, F. Laschi, M. Peruzzini, M.F. Ottaviani, A. Vacca, and P. Zanello, *Inorg. Chem.*, 1990, **29**, 3394.
479 S. Chandrasekhar, and A. McAuley, *Inorg. Chem.*, 1992, **31**, 480.
480 T.J. Collins, R.D.Powell, C.Slebodnick, and E.S. Uffelman, *J. Am. Chem. Soc.*, 1991, **113**, 8419.
481 N.V. Vugman, and W.O. Franco, *Physic Letters A*, 1991, **155**, 516.
482 N.V. Vugman, and M. Rothier Amaral Jr., *Phys. Rev. B*, 1990, **42**, 9837.
483 J.S. Bass, and L. Kevan, *J. Phys. Chem.*, 1990, **94**, 4640.
484 S.C. Haefner, K.R. Dunbar, and C. Bender, *J. Am. Chem. Soc.*, 1991, **113**, 9540.
485 A. McAuley, and S. Subramanian, *Inorg. Chem.*, 1990, **29**, 2830.
486 B. de Castro, and C. Freire, *Inorg. Chem.*, 1990, **29**, 5113.
487 A. Escuer, J. Ribas, R. Vicente, C. Faulmann, D. De Montauzon, and P. Cassoux, *Polyhedron*, 1991, **10**, 2025.
488 S. Bhanja Choudhury, D. Ray, and A. Chakravorty, *Inorg. Chem.*, 1991, **30**, 4354.
489 H.-J. Krüger, G. Peng, and R.H. Holm, *Inorg. Chem.*, 1991, **30**, 734 - 742.
490 A. McAuley, and S. Subramanian, *Inorg. Chem.*, 1991, **30**, 371 - 378.
491 S. Bhanja Choudhury, and A. Chakravorty, *Inorg. Chem.*, 1992, **31**, 1055 - 1058.
492 S. Bhanja Choudhury, D. Ray, and A. Chakravorty, *Inorg. Chem.*, 1990, **29**, 4603.
493 N. Baidya, P.K. Mascharak, D.W. Stephan, and C.F. Campagna, *Inorg. Chim. Acta*, 1990, **177**, 233 - 238.

494 T.J. Collins, T.R. Nichols, and E.S. Uffelman, *J. Am. Chem. Soc.*, 1991, **113**, 4708.
495 S.H. Byeon, G. Demazeau, J.M. Dance, and J.H. Choy, *Eur. J. Solid State Inorg. Chem.*, 1991, **28**, 643.
496 S.H. Byeon, G. Demazeau, L. Fournes, J.M. Dance, and J.H. Choy, *Solid State Commun.*, 1991, **80**, 457.
497 V. Luca, R.K. Kukkadapu, and L. Kevan, *J. Chem. Soc., Faraday Trans.*, 1991, **87**, 3083.
498 R.K. Kukkadapu, V. Luca, and L. Kevan, *J. Phys. Chem.*, 1992, **96**, 415.
499 N.V. Vugman, M.L. Netto Grillo, and V.K. Jain, *Chem. Phys. Letters*, 1992, **188**, 419.
500 J. Isoya, H. Kanda, and Y. Uchida, *Phys. Rew. B*, 1990, **42**, 9843.
501 M. Zimmer, G. Schulte, X.L. Luo, and R.H. Crabtree, *Angew. Chem., Int. Ed. Engl.*, 1991, **30**, 193.
502 K.M. Kadish, M.M. Franzen, B.C. Han, C. Araullo-McAdams, and D. Sazou, *J. Am. Chem. Soc.*, 1991, **113**, 512.
503 M. W. Renner, L.R. Furenlid, K.M. Barkigia, A. Forman, H.K. Shim, D.J. Simpson, K.M. Smith, and J. Fajer, *J. Am. Chem. Soc.*, 1991, **113**, 6891.
504 N. Baida, M. Olmstead, and P.K. Mascharak, *Inorg. Chem.*, 1991, **30**, 929.
505 T.S. Gau, and J.T. Yu, *Solid State Commun.*, 1991, **78**, 123.
506 E. Minner and H. Bill, *Chem. Phys. Letters*, 1990, **175**, 231.
507 J.L. Rao, R.M. Krishna, and S.V.J. Lakshman, *J. Phys. Chem. Solids*, 1990, **51**, 323.
508 Y.N. Naidu, J.L. Rao, and S.V.J. Lakshman, *Phys. Status Solidi B*, 1992, **169**, K109.
509 R.M. Krishna, J.L. Rao, and S.V.J. Lakshman, *Spectrochim. Acta, Part A*, 1992, **48A**, 245.
510 V.C. Mouli, R.S. Sunder, and G.S. Sastry, *Indian J. Pure Appl. Phys.*, 1990, **28**, 478.
511 J.S. Yu, and L. Kevan, *J. Phys. Chem.*, 1990, **94**, 7620.
512 J.S. Yu, and L. Kevan, *J. Phys. Chem.*, 1990, **94**, 7612.
513 X. Chen, and L. Kevan, *J. Am. Chem. Soc.*, 1991, **113**, 2861.
514 M. Zamadics, X. Chen, and L. Kevan, *J. Phys. Chem.*, 1992, **96**, 2652.
515 C.W. Lee, X. Chen, and L. Kevan, *J. Phys. Chem.*, 1991, **95**, 8626.
516 C.W. Lee, X. Chen, and L. Kevan, *J. Phys. Chem.*, 1992, **96**, 357.
517 V. Luca, X. Chen, and L. Kevan, *Chem. Mater.*, 1991, **3**, 1073.
518 D. Gourier, D. Vivien, B. Dunn, and L. Salmon, *J. Mater. Chem.*, 1991, **1**, 265.
519 R. Bechara, A. Abou-Kais, M. Guelton, A. D'Huysser, J. Grimblot, and J.P. Bonnelle, *Spectrosc. Letters*, 1990, **23**, 1237.
520 R. Bechara, G. Wrobel, C.F. Aissi, M. Guelton, J.P. Bonnelle, and A. Abou-Kais, *Chem. Mater.*, 1990, **2**, 518.

521 R. Bechara, A. D'Huysser, C.F. Aissi, M. Guelton, J.P. Bonnelle, and A. Abou-Kais, *Chem. Mater.*, 1990, **2**, 522.
522 A. Abou-Kais, R. Bechara, C.F. Aissi, and M. Guelton, *Chem. Mater.*, 1991, **3**, 557.
523 A. Abou-Kais, R. Bechara, D. Ghoussoub, C.F. Aissi, M. Guelton, and J.P. Bonnelle, *J. Chem. Soc., Faraday Trans.*, 1991, **87**, 631.
524 A. Abou-Kais, A. Bennani, C.F. Aissi, G. Wrobel, and M. Guelton, *J. Chem. Soc., Faraday Trans.*, 1992, **88**, 615.
525 L.D. Bogomolova, V.A. Yachkin, N.A. Krasil'nikova, V.L. Bogdanov, E.B. Fedorushkova, and V.D. Khalilev, *J. Non-Cryst. Solids*, 1990, **125**, 32.
526 B. Sreedhar, J.L. Rao, and S.V.J. Lakshman, *J. Non-Cryst. Solids*, 1990, **124**, 216.
527 G.L. Narendra, B. Sreedhar, J.L. Rao, and S.V.J. Lakshman, *J. Mater. Sci.*, 1991, **26**, 5342.
528 P. Raghunathan, and S. Sivasubramanian, *J. Phys. Chem.*, 1991, **95**, 6346.
529 P. Raghunathan, and S. Sivasubramanian, *J. Non-Cryst. Solids*, 1991, **136**, 14.
530 J. Bednarek, and S. Schlick, *J. Am. Chem. Soc.*, 1990, **112**, 5019.
531 J. Bednarek, and S. Schlick, *Langmuir*, 1992, **8**, 249.
532 J. Bednarek, and S. Schlick, *J. Am. Chem. Soc.*, 1991, **113**, 3303.
533 C.A. McDowell, A. Naito, D.L. Sastry, Y. Cui, and K. Sha, *J. Phys. Chem.*, 1990, **94**, 8113.
534 R. Miyamoto, Y. Ohba, and M. Iwaizumi, *Inorg. Chem.*, 1990, **29**, 3234.
535 L. Costa, F. Morazzoni, R. Scotti, F. Angelucci, A. Suarato, and V. Malatesta, *Inorg. Chim. Acta*, 1990, **168**, 123.
536 H. Yokoi, M. Satoh, and M. Iwaizumi, *J. Am. Chem. Soc.*, 1991, **113**, 1530.
537 R.K. Ray, and G.B. Kauffman, *Inorg. Chim. Acta*, 1990, **173**, 207.
538 R.K. Ray, and G.B. Kauffman, *Inorg. Chim. Acta*, 1990, **174**, 237.
539 R.K. Ray, and G.B. Kauffman, *Inorg. Chim. Acta*, 1990, **174**, 257.
540 R.G. Bhirud, and T.S. Srivastava, *Inorg. Chim. Acta*, 1991, **179**, 125.
541 M.A. Zoroddu, M.I. Pilo, R. Seeber, R. Pogni, and R. Basosi, *Inorg. Chim. Acta*, 1991, **184**, 185.
542 G. Arena, R. Bonomo, L. Casella, M. Gullotti, G. Impellizzeri, G. Marracone, and E. Rizzarelli, *J. Chem. Soc., Dalton Trans.*, 1991, 3203.
543 E. Farkas, and B. Kurzak, *J. Coord. Chem.*, 1990, **22**, 145.
544 J. Perkinson, S. Brodie, K. Yoon, K. Mosny, P.J. Carroll, T. V. Morgan, and S.J. Nieter Burgmayer, *Inorg. Chem.*, 1991, **30**, 719.
545 W.B. Tolman, *Inorg. Chem.*, 1991, **30**, 4877.
546 R. Bonomo, E. Rizzarelli, M. Bressan, and A. Morvillo, *Inorg. Chim. Acta*, 1991, **186**, 21.
547 G. Batra, and P. Mathur, *Inorg. Chem.*, 1992, **31**, 1575.
548 H.G. Hang, W.L. Kwik, G.R. Hanson, J.A. Crowther, M. McPartlin, and N. Choi, *J. Chem. Soc., Dalton Trans.*, 1991, 3193.
549 K.J. Oberhausen, R.J. O'Brian, J.F. Richardson, and R.M. Buchanan, *Inorg. Chim. Acta*, 1990, **173**, 145.

550 N.D. Yordanov, G. Gochev, O. Angelova and J. Macicek, *Polyhedron*, 1990, **9**, 2597.
551 E.W. Ainscough, A.M. Brodie, J.D. Ranford, and J.M. Waters, *J. Chem. Soc., Dalton Trans.*, 1991, 1737.
552 N. Marchettini, R. Pogni, A. Diaz, and R. Basosi, *J. Chem. Soc., Faraday Trans.*, 1990, **86**, 2575.
553 S.P. Greiner, R.W. Kreilick, and K.A. Kraft, *J. Am. Chem. Soc.*, 1992, **114**, 391.
554 D.A. Tucker, P.S. White, K.L. Trojan, M.L. Kirk, and W.E. Hatfield, *Inorg. Chem.*, 1991, **30**, 823.
555 T. Manfredini, G.C. Pellacani, A. Bonamartini-Corradi, L.P. Battaglia, G.G.T. Guarini, J. Gelsomini Giusti, G. Pon, R.D. Willett, and D.X. West, *Inorg. Chem.*, 1990, **29**, 2221.
556 M. Suhara, and T. Kobayashi, *J. Phys. Chem. Solids*, 1992, **53**, 27.
557 R.J. Jacobson, Z. Tyeklár, K.D. Karlin, and J. Zubieta, *Inorg. Chem.*, 1991, **30**, 2035.
558 R. Knoch, A. Wilk, K.J. Wannowius, D. Reinen, and H. Elias, *Inorg. Chem.*, 1990, **29**, 3799.
559 A. Kutoglu, R. Allmann, J.V. Folgado, M. Atanasov, and D. Reinen, *Z. Naturforsch., B: Chem. Sci.*, 1991, **46**, 1193.
560 E. Casassas, A. Izquierdo-Ridorsa, and R. Tauler, *J. Chem. Soc., Dalton Trans.*, 1990, 2341.
561 M. Leluk, B. Jezowska-Trzebiatowska, and J. Jezierska, *Polyhedron*, 1991, **10**, 1653.
562 E. Campillos, M. Marcos, J.L. Serrano, and P.J. Alonso, *J. Mater. Chem.*, 1991, **1**, 197.
563 M. Ghedini, S. Morrone, D. Gatteschi, and C. Zanchini, *Chem. Mater.*, 1991, **3**, 752.
564 W. Dietzsch, S. Rauer, R.M. Olk, R. Kirmse, K. Köhler, L. Golic, and B. Olk, *Inorg. Chim. Acta*, 1990, **169**, 55.
565 R. Kirmse, K. Köhler, R.M. Olk, W. Dietzsch, E. Hoyer, and G. Zwanenburg, *Inorg. Chem.*, 1990, **29**, 4073.
566 F. Lelj, G. Morelli, G. Ricciardi, and A. Rosa, *Polyhedron*, 1991, **10**, 523.
567 F. Lelj, G. Morelli, G. Ricciardi, A. Rosa, and M.F. Ottaviani, *Polyhedron*, 1991, **10**, 1911.
568 G. Fu, Y. Fu, K. Jayaraj, and A.B.P. Lever, *Inorg. Chem.*, 1990, **29**, 4090.
569 A.P. Koley, S. Purohit, L.S. Prasad, S. Ghosh, and P.T. Manoharan, *Inorg. Chem.*, 1992, **31**, 305.
570 F.G. Herring, G. Hwang, K.C. Lee, F. Mistry, P.S. Phillips, H. Willner, and F. Aubke, *J. Am. Chem. Soc.*, 1992, **114**, 1271.
571 J.H.B. Chenier, J.A. Howard, H.A. Joly, B. Mile, and P.L. Timms, *J. Chem. Soc., Chem. Commun.*, 1990, 581.

572 M. Histed, J.A. Howard, R. Jones, M. Tomietto, and H.A. Joly, *J. Phys. Chem.*, 1992, **96**, 1144.
573 V. Luca, D.R. Brown, and L. Kevan, *J. Phys. Chem.*, 1991, **95**, 10065.
574 D.R. Brown, V. Luca, and L. Kevan, *J. Chem. Soc., Faraday Trans.*, 1991, **87**, 2479.
575 F.J. Lopez, E. Moya, and C. Zaldo, *Solid State Commun.*, 1990, **76**, 1169.
576 K.R. Hoffman, J. Casas-Gonzales, S.M. Jacobsen, and W.M. Yen, *Phys. Rev. B: Condens. Matter*, 1991, **44**, 12589.
577 D.L. Hughes, D.J. Lowe, M.Y. Mohammed, C.J. Pickett, and N.M. Pinhal, *J. Chem. Soc., Dalton Trans.*, 1990, 2021.
578 D. Sellmann, M. Geck, F. Knoch, G. Ritter and J. Dengler, *J. Am. Chem. Soc.*, 1991, **113**, 3819.
579 T.C. Vanoy, and G.L. McPherson, *Solid State Commun.*, 1990, **74**, 5.
580 B.Villacampa, J. Casas Gonzales, R. Alcalá, and P.J. Alonso, *J. Phys.: Condens. Matter*, 1991, **3**, 8281.
581 E. Fargin, B. Lestienne, and J.M. Dance, *Solid State Commun.*, 1990, **75**, 769.
582 J.J. Rousseau, B. Boulard, H. Duroy, and J.L. Fourquet, *Eur. J. Solid State Inorg. Chem.*, 1990, **27**, 913.
583 H. Hikita, K. Takeda, and Y. Kimura, *Phys. Status Solidi A*, 1990, **119**, 251.
584 A. Watterich, K. Raksanyi, O.R. Gilliam, R. H. Bartram, L.A. Kappers, H. Soethe, and J.M. Spaeth, *J. Phys. Chem. Solids*, 1992, **53**, 189.
585 G. Corradi, H. Söthe, J-M Spaeth, and K. Polgár, *J. Phys.: Condens. Matter*, 1991, **3**, 1901.
586 A. Martin, F.J. lopez, and F. Agullo-Lopez, *J. Phys.: Condens. Matter*, 1992, **4**, 847.
587 G.G. Siu, and M.G. Zhao, *Phys. Rev. B: Condens. Matter*, 1991, **43**, 13575.
588 M.L. Du, and M.G. Zhao, *Solid State Commun.*, 1990, **76**, 565.
589 R. Heidler, W. Windsch, R. Böttcher, and C. Klimm, *Chem. Phys. Letters*, 1990, **175**, 55.
590 W.C. Zheng, *Solid State Commun.*, 1991, **80**, 213.
591 W.C. Zheng, *Solid State Commun.*, 1990, **76**, 859.
592 W.B. Lewis, *Appl. Phys. A*, 1992, **A54**, 31.
593 W.C. Zheng, Phys. Lett. A, 1990, 151, 77.
594 R.Karthein, H. Motschi, A. Schweiger, S. Ibric, B. Sulzberger, and W. Stumm, *Inorg. Chem.*, 1991, **30**, 1606.
595 R.P. Bonomo, A.J. Di Bilio, and F. Riggi, *Chem. Phys.*, 1991, **151**, 323.
596 U. Casellato, R. Graziani, R.P. Bonomo, and A.J. Di Bilio, *J. Chem. Soc., Dalton Trans.*, 1991, 23.
597 A.M. Calafat, J.J. Fiol, A. Terron, D.M.L. Goodgame, and I. Hussain, *Inorg. Chim. Acta*, 1990, **169**, 133.
598 P.V. Bernhardt, P. Comba, N.F. Curtis, T.W. Hambley. G.A. Lawrence, M. Maeder, and A. Siriwardena, *Inorg. Chem.*, 1990, **29**, 3208.

599 J.H. Jacobsen, and E. Pedersen, *Inorg. Chem.*, 1991, **30**, 4477.
600 S.K. Chandra, P. Basu, D. Ray, S. Pal, and A. Chakravorty, *Inorg. Chem.*, 1990, **29**, 2423.
601 S. Dutta, P. Basu, and A. Chakravorty, *Inorg. Chem.*, 1991, **30**, 4031.
602 S.M. Saadeh, M.S. Lah, and V.L. Pecoraro, *Inorg. Chem.*, 1991, **30**, 8.
603 S.K. Chandra, and A. Chakravorty, *Inorg. Chem.*, 1992, **31**, 760.
604 J.H. Choy, G. Demazeau, and S.H. Byeon, *Solid State Commun.*, 1991, **77**, 647.
605 K. Kojima, H. Taguchi, and J. Matauda, *J. Phys. Chem.*, 1991, **95**, 7595.
606 G.T. Long, and D.A. Cleary, *J. Phys.: Condens. Matter*, 1990, **2**, 4747.
607 K. Fukui, N. Kojima, H. Ohya-Nishiguchi, and N. Hirota, *Inorg. Chem.*, 1992, **31**, 1338.
608 M. Wicholas, and C. Zanchini, *Chem. Phys. Lett.*, 1990, **171**, 114.
609 S. Juppe, and O.F. Schirmer, *Solid State Commun.*, 1990, **3**, 299.
610 P. Emanuelsson, P. Omling, H.G. Grimmeiss, J. Kreissl, and W. Gehlhoff, *Phys. Rev. B: Condens. Matter*, 1991, **44**, 6125.
611 V.K. Jain and G. Lehmann, *Phys. Status Solidi B*, 1990, **159**, 495.
612 I. Sledzinska, K. Maletka, and L. Tykarski, *Phys. Status Solidi B*, 1991, **167**, K107.
613 F.Z. Li, and Z.M. Li, *Phys. Status Solidi B*, 1990, **161**, K29.
614 D. Bravo, L. Arizmendi, M. Aguilar, and F.J. López, *J. Phys.: Condens. Matter*, 1990, **2**, 10123.
615 D. Bravo, and F.J. López, *J. Phys.: Condens. Matter*, 1991, **3**, 7691.
616 T.H. Yeom, S.H. Choh, and M.S. Jang, *J. Phys.: Condens. Matter*, 1992, **4**, 587.
617 G. Corradi, H. Soethe, J.M. Spaeth, and K. Polgar, *J. Phys.: Condens. Matter*, 1990, **2**, 6603.
618 E. Lifshitz, and A.H. Francis, *Synth. Met.*, 1989, **34**, 653.
619 J. Lech, A. Slezak, and R. Hrabanski, *J. Phys. Chem. Solids*, 1991, **52**, 685.
620 R.S. Bansal, V.P. Seth, and P. Chand, *Phys. Scr.*, 1991, **44**, 493.
621 J. Kreissl, W. Gehlhoff, P. Omling, and P. Emanuelsson, *Phys. Rev. B: Condens. Matter*, 1990, **42**, 1731.
622 J. Kreissl, K. Irmscher, W. Gehlhoff, P. Omling, and P. Emanuelsson, *Phys. Rev. B: Condens. Matter*, 1991, **44**, 3678.
623 Yu.S. Gromovoj, V.S. Plyatsko, F.F. Sizov, and L.A. Korovina, *J. Phys.: Condensed Matter*, 1990, **2**, 10391.
624 S.A. Marshall, D.R. Yoder-Short, C. Yu, Y.N. Zhang, and J.K. Furdyna, *Phys. Status Solidi B*, 1990, **160**, 591.
625 S.J.C.H.M. van Gisbergen, M. Godlewski, T. Gregorkiewicz, and C.A.J. Ammerlaan, *Phys. Rev. B: Condens. Matter*, 1991, **44**, 3012.
626 J.K. Nandagawe, P.K. Patil, P.R. Bote, and R.D. Lawangar-Pawar, *Solid State Commun.*, 1991, **77**, 513.
627 Y.H. Lee, M.H. Oh, and S.H. Choh, *J. Appl. Phys*, 1991, **70**, 7042.
628 T. Nakamura, H. Muramatsu, Y. Takeuchi, H. Fujiyasu, and Y. Nakanishi, *J. Mater. Chem.*, 1991, **1**, 357.

629 C. Marco de Lucas, F. Rodriguez, and M. Moreno, *J. Phys.: Condens. Matter*, 1990, **2**, 1891.
630 T. Yoshida, H. Aoki, H. Takeuchi, M. Arakawa, and K. Horai, *J. Phys. Soc. Japan*, 1991, **60**, 625.
631 S.K. Misra, and S.Z. Korczak, *Physica B (Amsterdam)*, 1991, **168**, 143.
632 R.S. Bansal, and V.P. Seth, *J. Phys. Chem. Solids*, 1990, **51**, 329.
633 Y.N. Naidu, J.L. Rao, and S.V.J. Lakshman, *Solid State Commun.*, 1992, **81**, 437.
634 J.S. Phor, *Spectrochim. Acta, Part A*, 1990, **46A**, 1535.
635 J.S. Phor, *Radiat. Eff. Defects Solids*, 1991, **116**, 149.
636 R. Nakata, T. Nakajima, K. Kawano, and M. Sumita, *J. Phys. Chem. Solids*, 1991, **52**, 1011
637 M. Goodgame, and I. Hussain, *Inorg. Chim. Acta*, 1990, **174**, 245.
638 C.W. Lee, X. Chen, G. Brouet, and L. Kevan, *J. Phys. Chem.*, 1992, **96**, 3110.
639 Z. Levi, A.M. Raitsimring, and D. Goldfarb, *J. Phys. Chem.*, 1991, **95**, 7830.
640 G. Brouet, X. Chen, C.W. Lee, and L. Kevan, *J. Am. Chem. Soc.*, 1992, **114**, 3720.
641 W.C. Zheng, *J. Phys.: Condens. Matter*, 1991, **3**, 177.
642 Y.Y. Zhou, *Phys. Lett. A*, 1990, **150**, 53.
643 Y.Y. Zhou, *Phys. Rev. B*, 1990, **42**, 917.
644 A.P. Pechenyi, M.D. Glinchuk, T.V. Antimirova, and W. Kleeman, *Phys. Status Solidi B*, 1991, **168**, K27.
645 M. Arakawa, T. Hirose, and H. Takeuchi, *J. Phys. Soc. Japan*, 1991, **60**, 4319.
646 J.M. Gaite, J.F. Stenger, Y. Dusausoy, G. Marnier, and H. Rager, *J. Phys.: Condens. Matter*, 1991, **3**, 7877.
647 J.M. Baker, A.A. Jenkins, and R.C.C. Ward, *J. Phys.: Condens. Matter*, 1991, **3**, 8467.
648 S. Lakshmi Reddy, P.S. Rao, and B.J. Reddy, *Phys. Lett. A*, 1991, **161**, 74.
649 H. Takeuchi, H. Ebisu, and M. Arakawa, *J. Phys. Soc. Japan*, 1991, **60**, 304.
650 H. Takeuchi, M. Arakawa, and H. Ebisu, *J. Phys.: Condens. Matter*, 1991, **3**, 4405.
651 S. Hansen, and G. Lehmann, *Appl. Magn. Reson.*, 1990, **1**, 47.
652 J. Minge, M.J. Mombourquette, and J.A. Weil, *Phys. Rev. B: Condens. Matter*, 1990, **42**, 33.
653 D. Choi, and J.A. Weil, *Phys. Rev. B: Condens. Matter*, 1990, **42**, 9759.
654 M.R. Hantehzadeh, C.S. Han, and L.E. Halliburton, *J. Phys. Chem. Solids*, 1990, **51**, 425.
655 H. Feng, J. Wen, H. Wang, S. Han and Y. Xu, *J. Phys. Chem. Solids*, 1990, **51**, 397.
656 X. Feng, L. Tang, L. Xu, and X. Jiang, *Phys. Status Solidi A*, 1990, **119**, 561.
657 X. Feng, L. Tang, and J. Ying, *Ferroelectrics*, 1990, **107**, 21.
658 A. Böker, H. Donnerberg, O.F. Schirmer, and X. Feng, *J. Phys.: Condens. Matter*, 1990, **2**, 6865.
659 R.F. Shinde, and I. Balakrishnan, *J. Phys. D: Appl. Phys.*, 1991, **24**, 1496.

660 M. Ricter, R. Fricke, B. Parlitz, U. Haedicke, and G. Oehlmann, Z. Phys. Chem. (Munich), 1991, **171**, 97.
661 R. Guilard, I. Perrot, A. Tabard, P. Richard, C. Lecomte, Y.H. Liu, and K.M. Kadish, Inorg. Chem., 1991, **30**, 27.
662 S. Fujii, H. Ohya-Nishiguchi, and N. Hirota, Inorg. Chim. Acta, 1990, **175**, 27.
663 D.C. Finnen, A.A. Pinkerton, W.R. Dunham, R.H. Sands, and M.O. Funk Jr., Inorg. Chem., 1991, **30**, 3960.

2
Metalloproteins

BY G. R. HANSON

1 Introduction

In the previous two years there has been a considerable expansion of interest in the characterisation of metalloproteins, especially with electron paramagnetic resonance (EPR), electron nuclear double resonance (ENDOR) and electron spin echo envelope modulation (ESEEM) spectroscopy. This is not only reflected in the number of papers reviewed herein but in the number of symposia held during the last two years. These include Copper in Biological Systems[1], Copper Coordination Chemistry: Bioinorganic Perspectives[2], International Society of Magnetic Resonance Workshop on ESEEM[3], the 26th. Congress Ampere and the International Conference on Bioinorganic Chemistry[4] and associated satellite meetings on Iron and Iron Containing Proteins, the Molecular Biology of Hydrogenases, Metal Gene Communication and the Chemistry and Biochemistry of Nitrogen Fixation. The nomenclature committee of the International Union of Biochemistry has seen fit to update and extend the enzyme nomenclature for electron transfer proteins.[5]

2 Copper Proteins

Copper plays a vital role in the chemistry of organisms ranging from bacteria to mammals. Copper is an essential active site cofactor in a large number of metalloproteins whose functions[6] range from electron transfer, dioxygen transport, substrate oxygenation, oxidation and reduction. The structural factors which control long range electron transfer in metalloproteins and in particular azurin[7] have been described.

2.1 Type I Copper Proteins.- Q-band ^{1}H and ^{14}N ENDOR spectra of the type I copper centres in plastocyanin, azurin, stellacyanin and type II reduced fungal and tree laccase revealed ^{1}H resonances (16-31 MHz) attributed to strongly coupled methylene protons of cysteine. A single ^{14}N resonance from the two magnetically equivalent nitrogens (histidine) in plastocyanin and two resonances from two inequivalent histidine ligands in the other proteins was also observed.[8] Pulsed ENDOR and a new two dimensional pulsed electron nuclear triple resonance method for recording hyperfine selective ENDOR spectra have been used to characterise the copper(II) ions coordination sphere at pH 5.4 and 11.0. At high pH evidence for a [Cu(N-His)$_2$ S-Cys N-amide] coordination sphere is presented.[9]

^{14}N ENDOR measurements on a series of N$_4$, cis- and trans-N$_2$O$_2$, NO$_3$ and N$_2$S$_2$ copper(II) complexes have shown that the magnitude of the ^{14}N hyperfine coupling constants is related to the extent of deformation of the orbital containing the unpaired electron by the ligand field of the donor atoms. In copper N$_2$S$_2$ complexes delocalisation of the electron spin onto the sulfur atoms can also affect the magnitude of the nitrogen coupling constants.[10]

Alteration of the hydrogen bond network around the copper site in azurin by site directed mutagenesis (Asn-47 -> Leu-47) was shown to have little effect on the structure of the copper site although the stability and midpoint potential were altered.[11] EPR, optical absorption and fluorescence spectroscopy demonstrated that the interaction of short chain n- and iso-alcohols with azurin did not involve the Cu(II)N$_2$SS* active site.[12]

The factors effecting the reconstitution of apoplantocyanin with copper-thionein have been examined, together with the oxidation of plantocyanin by cytochrome c oxidase and its autooxidation in the presence of cardiolipin.[13] Three small type I copper proteins designated auracyanin-A, -B1 and -B2 have been isolated from the thermophilic green gliding photosynthetic bacterium *Chloroflexus aurantiacus* and show similar optical, circular dichroism (CD) and resonance raman spectra but have quite different EPR signals.[14]

Insulin stabilised CuN$_3$L chromophores in which L represents an exchangeable aromatic thiolate ligand have been characterised by EPR, CD and electronic absorption spectroscopy. Type I copper centres are formed with L equal to benzenethiolate (BzT) and 4-MeBzT.[15]

2.2 Type II Copper Proteins.- Galactose oxidase contains a mononuclear type II copper(II) ion exchange coupled to a free radical. EPR and ENDOR studies of the copper depleted enzyme with cysteine substituted for tyrosine (Tyr-272) show that this amino acid is the source of the free radical in the native enzyme.[16] X-ray absorption near edge structure (XANES) and EPR spectra of the oxidatively activated and reductively inactivated forms of galactose oxidase together with a substrate complex support a catalytic cycle involving [Cu(II)

+ enzyme radical] ⇌ [Cu(I)].[17] Oxidation of apogalactose oxidase with ferricyanide produces a stable free radical associated with a tyrosine residue. Reconstitution of the apoenzyme in the oxidised form with copper restores the spectroscopic and catalytic properties of the enzyme. However, the free radical signal is abolished indicating that this radical is associated with the radical-coupled copper active site.[18]

A combination of site directed mutagenesis (Thr-137 -> Ser-137, Ala-137[19]; His-46 -> Cys-46; His-80 -> Cys-80[20]) and spectroscopy has allowed the characterisation of the copper binding site in superoxide dismutase. The ability of Arg-141 to stabilise the copper centre in superoxide dismutase was examined through the chemical modification of Arg-141 with phenylglyoxal or butanedione and subsequent characterisation with nuclear magnetic resonance (NMR) and EPR spectroscopy.[21] Reconstitution of superoxide dismutase (100% active) from the apoprotein with copper(I) glutathione was monitored by optical, EPR and NMR spectroscopy. A copper(I) glutathione protein intermediate was identified during the reaction.[22] A copper(II) complex of desferrithiocin from *Streptomyces antibioticus* has been isolated and structurally characterised. The complex has a square planar pyramidal structure and exhibits superoxide dismutase activity.[23]

ESEEM studies of the nuclear quadrupole interactions of the remote ^{14}N in copper(II) diethyltriamine-substituted imidazole complexes suggest that the variation of quadrupole coupling constants found in copper(II) proteins may arise from variations in hydrogen bonding at the remote nitrogen of the coordinated histidine.[24] A single crystal ^{14}N ESEEM study of copper(II) doped into zinc *bis*-(1,2-dimethylimidazole) chloride was performed to aid the interpretation of spectra from distorted tetrahedral copper-histidine centres in type I copper proteins.[25]

Evidence for low-density lipoprotein-lipid derived radicals during copper ion- and lipoxygenase-catalysed oxidation of the low-density lipoprotein was obtained using spin trapping methods.[26] Hydantoins (hyd), a class of antiepileptic drugs have been shown to coordinate to copper(II) ions [Cu(hyd)$_2$(pyridine)$_2$] in a square planar geometry.[27] Bleaching of eumelanin has been studied in model systems consisting of synthetic dopa-melanin and various concentrations of hydrogen peroxide, dioxygen and copper(I) ions at neutral or alkaline pH.[28] Copper, iron, zinc and manganese were found to accumulate in organs of rats which spontaneously develop jaundice with hereditary hepatitis, the copper being associated with metallothionein.[29] Three different orientations of type II copper(II) complexes (coppper bleomycin, the tridentate 2-formylpyridine monothiosemicarbazonato copper(II) complex and the tripeptide (Gly-His-Lys, Gly-His-Gly) copper(II) complexes) bind to DNA with their equatorial planes orientated at 65°, perpendicular and parallel to the DNA fibre axis.[30]

2.3 Multicentred Copper Proteins.- A brief review of the structure and function of the copper-pyrroloquinoline amine oxidases has been presented.[31] Reduction of copper amine oxidase with substrate produces a free radical which has been identified as a 6-hydroxydopamine radical from EPR and optical studies of the enzyme substrate complex and model systems.[32] The radical intermediates formed upon reduction of porcine kidney diamine oxidase and *Arthrobacter* P1 methylamine oxidase with substrates in the presence of cyanide under anaerobic
conditions were characterised using ESEEM spectroscopy. The results support a mechanism entailing the binding of the substrate to the covalently linked quinone-cofactor rather than tyrosine or cyanide radicals.[33] Lysyl oxidase was shown to have a single type II copper(II) centre *per*. 32 kDa monomer which was catalytically essential. The EPR spectrum of the enzyme was used to identify three histidine residues in the copper ions equatorial coordination sphere.[34]

EPR and CD studies of the binding of azide to the type II-copper depleted zucchini ascorbate oxidase reveal that one azide ion binds to each of the two binuclear type III copper(II) centres with weak affinity.[35] A combination of electronic spectroscopy, low temperature magnetic circular dichroism (MCD), EPR and exogenous ligand (azide) perturbation has shown that one molecule of azide bridges the type II and type III copper(II) centres while the second and third molecules bind terminally to the type II and III centres respectively.[36] The removal of the type II copper from ascorbate oxidase and *Rhus vernicifera* laccase with nitriloacetate was monitored with EPR spectroscopy.[37] Anaerobic reduction of either ascorbate oxidase[38] or *Rhus vernicifera* laccase[39] and their type II copper depleted derivatives produced axially symmetric EPR spectra whose origin was attributed to an adduct involving the type III copper centres and hexacyanoferrate(II).

Reconstitution of the type II copper depleted form of isotopically enriched laccase with different copper isotopes reveals that this copper(II) ion is not responsible for the type II EPR signal at low temperature, but is EPR active at room temperature. This has been interpreted as a temperature dependent structural reorganisation and a change in the antiferromagnetic coupling in the type II / III cluster.[40] A combination of electronic absorption and EPR spectroscopy, magnetic susceptibility and exogenous ligand perturbations was used to probe the electronic structure of the trinuclear site (type II centre and the coupled binuclear type III centre) in laccase.[41] Approximately 10% of copper in tree laccase was found to be transferred to type I copper apoproteins. EPR spectra and thiol titrations showed that the source of the transferrable copper was the type II copper and a small amount of type III copper.[42] Appreciable autoreduction of the type I copper centre in type II copper-depleted *Rhus vernicifera* laccase at pH 9.5 produced the stable type III copper EPR spectrum.[43]

ESEEM spectra of the type II copper binding site in the mercury derivative of laccase identified an equatorial coordination sphere consisting of two imidazolate nitrogens and a water molecule which can be replaced by azide.[44] Temperature dependent EPR spectra of this derivative indicate the presence of multiple species. Similar studies on the binding of anions reveal preferential binding to the type III centres which in turn provides an insight into the flexibilty of the type II and III copper centres.[45]

EPR spectra of *Coliolus consors* laccase was shown to contain one type I, one type II and two type III copper centres.[46] Extracellular laccase from the lignin-degrading fungus *Phlebia radiata* was found to contain type I and type II copper centres and a pyrroloquinoline quinone cofactor.[47]

Multifrequency EPR studies of the various forms of nitrous oxide reductase from the denitrifying *Pseudomonas stutzeri* reveal two distinct copper sites (A and C). The mononuclear centre (A) has a N_2S_2 coordination sphere similar to that of Cu_A proposed in cytochrome c oxidase, while centre (C) is binuclear and is represented by the mixed valence $S=1/2$ cluster. Centre (B) forms upon reduction yielding the blue form.[48] Magnetic susceptibility,[49] optical, low temperature MCD[50] and extended X-ray absorption fine structure[51] (EXAFS) measurements confirmed these assignments. EXAFS identified nitrogen and or oxygen atoms (Cu-N/O, 1.96-2.03 Å) and sulfur atoms (Cu-S, 2.20-2.25 Å) in the copper ion's coordination sphere in the native enzyme (as isolated), and when it was exposed to nitrous oxide and air.[51] Nitrous oxide reductase from *Wolinella succinogenes* was shown to contain a low spin iron(III) cytochrome c and an axial copper spectrum consistent with an exchange coupled Cu(I)-Cu(II), $S=1/2$ (half met) centre. Reduction of the enzyme followed by the addition of nitrous oxide caused these resonances to disappear and then subsequently reappear.[52]

Ceruloplasmin from the turtle *Caretta caretta* was purified by a single step procedure and found to contain 5.1 gatoms of EPR detectable copper *per* molecule. The enzyme was unusually stable to ageing and proteolysis when compared to other proteins.[53] EPR studies of the type II copper centres in human ceruloplasmin revealed that they occur in two stable forms and that both of these forms occur in the blood serum of control donors and patients.[54] A reinvestigation of the reaction between nitric oxide and ceruloplasmin indicated that the integrity of the trinuclear copper cluster, as monitored by the absence of the type II copper centres in the EPR spectrum, modulates the behaviour of the protein towards nitric oxide.[55]

Reaction of nitrite at pH 5.0 to 7.0 with deoxyhemocyanin of the mollusc *Helix pomatia* produced nitrosylhemocyanin ($Cu(I)_A$-NO^+-$Cu(II)_B$). This is in contrast to the formation of methemocyanin from deoxyhemocyanin in the crustacean *Astacus leptodactylus*.[56] The copper(II) substituted insulin hexamer dimer was found to undergo an analogous T to R conformational transition in solution that has been identified for the zinc- and cobalt-insulin hexamers. The copper sites were found to bind phenol and anions and in the case of phenylmethylthiolate (PMT) a type I copper centre was formed [$CuN_3(PMT)$].[57]

3 Iron Proteins

3.1 Non-Heme Iron Proteins.-

EPR and EXAFS spectra of nitrile hydratase from *Brevibacterium sp.* indicate that the low spin iron(III) ion is in a distorted octahedral environment of sulfur and nitrogen or oxygen donor atoms.[58]

The ligand field symmetry of the active site iron(III) ion in phenylalanine hydroxylase was found to be dependent upon the medium conditions (buffer ions) and the presence of ligands known to bind to the active site.[59] EPR and resonance raman spectra revealed dopamine as a ligand to the mononuclear high spin iron(III) ion in tyrosine hydroxylase. Addition of substrate resulted in a new rhombic EPR spectrum attributed to substrate binding to the iron site.[60] A tetrameric iron containing superoxide dismutase was purified from *Tetrahymena pyriformis* and shown to have similar optical and EPR properties to other iron containing superoxide dismutases.[61] Chlorocatechol dioxygenase from *P. putida* is able to oxygenate a wide range of substituted catechols bearing electron donating or multiple electron withdrawing groups in an intradiol manner. The similarity of the optical and EPR properties of this enzyme to those of catechol 1,2-dioxygenase and protocatechuate 3,4-dioxygenase suggests that the unique catalytic properties are a consequence of its substrate binding pocket.[62]

A combination of EPR and variable temperature variable field MCD reveals that the oxidation of the iron(II) ion in soybean lipoxygenase to iron(III) produces little change in the coordination sphere of the iron centre which includes at least two histidines in a cis-configuration.[63] Mössbauer and EPR spectra indicate that the catalytic iron(III) ion in lipoxygenase is reduced to iron(II) as linoleic acid is oxidised to a pentadienyl radical intermediate.[64,65] Peroxyl radicals have also been observed in this reaction.[65] 5- and 12-lipoxygenases isolated from porcine leukocytes were characterised by atomic absorption and EPR spectroscopy which showed the presence of a single high spin iron(III) centre per molecule.[66] Proteolytic cleavage of soybean lipoxygenase I by either trypsin or chymotrypsin produced a 30kDa C-terminal and a 60 kDa N-terminal fragment which behaved identically to the native enzyme in all aspects examined.[67]

ESEEM spectra of the copper(II) substituted isopenicillin N synthase revealed two nearly magnetically equivalent, equatorially coordinated imidazolate nitrogens, from histidine. Water was also shown to be a ligand which was displaced upon addition of substrate.[68] Substitution of this enzyme with cobalt(II) revealed resonances around g=4 consistent with a high spin octahedral cobalt(II) (S=3/2) metal binding site.[69] Site directed mutagenesis (2 Cys -> Ser) in conjunction with EPR, optical and Mössbauer spectroscopy was used to demonstrate that the source of thiolate coordination to the iron(II) ion during catalysis was from the substrate.[70]

Rapid freeze quench EPR and Mössbauer studies have identified the accumulation of an intermediate during the conversion of apo B2 subunit of *Escherichia coli* ribonucleotide reductase to native B2. The results indicated that this intermediate involved two high spin iron(III) centres and a radical possibly oxyl or hydroxyl from dioxygen and water.[71] The inactive metB2 form of ribonucleotide reductase contains the binuclear iron(III) centre but lacks the tyrosyl radical. Incubation of the inactive enzyme with hydrogen peroxide at pH 5.5 for 1.5 hours partially activates the enzyme by forming the tyrosyl radical.[72] A mixed valent iron(III)-iron(II) cluster form has been isolated and characterised by EPR spectroscopy. The spectra were shown to arise from an S=9/2 spin state of a ferromagnetically coupled iron(III)-iron(II) cluster.[73] Reduction of the tyrosyl radical and iron(III) ions in ribonucleotide reductase by mono-substituted hydrazines and hydroxylamines was found to occur by two mechanisms, dependent upon which centre was reduced first. The efficiency of these reducing agents suggests an alternative to the long range electron transfer hypothesis in that the conformational flexibility of the polypeptide chain in solution may allow small molecules to penetrate the protein and susbsequently react with the iron-radical centre.[74]

A comparison of the CD, NMR and EPR properties of recombinant mouse and herpes simplex mouse ribonucleotide reductase with those from *E. coli* shows that the antiferromagnetic coupling and ligand environment in the iron centre is nearly identical in all three species.[75] The intensity of the tyrosyl radical signal was shown to decrease after the treatment of promyelocytic leukemic HL60 cells with 2-formylpyridine- thiosemicarbazonatocopper(II). This was attributed to a decreased availability of iron for the activity of the B2 subunit of ribonucleotide reductase.[76] During its infectious cycle, vaccinia virus expresses a virus encoded ribonucleotide reductase which is distinct from the host cellular enzyme. The small subunit of this enzyme has been cloned into *E. coli* and the protein expressed into T7 RNA polymerase plasmid expression system. The protein produced exhibited similar properties to the native enzyme from the virus infected cell extract and the mouse enzyme.[77]

Mössbauer and EPR spectra of a model [Fe(III)Ni(II)BPMP(OPr)$_2$] (BPMP, the anion of 2,6-*bis*[(*bis*(2-pyridylmethyl)amino)methyl]-4-methylphenol; Pr, propyl) binuclear complex for the iron-oxo proteins indicate that the iron(III) ion is antiferromagnetically coupled (J=-24 cm^{-1}) to a high spin Ni(II) site yielding an S=3/2 ground state.[78] While the metal ions in the corresponding [Fe(II)-Fe(II)BPMP(O$_2$CR)$_2$], [Fe(II)-Zn(II)BPMP(O$_2$CR)$_2$] and [Fe(II)-Ga(III)BPMP(O$_2$CR)$_2$] (R, ethyl, phenyl) model complex are ferromagnetically coupled yielding g=16 resonances.[79]

EXAFS, Mössbauer and EPR studies of the binuclear iron centre in the hydroxylase component (protein A) of methane monooxygenase isolated from *Methylococcus capsulatus* reveal that the mixed valence cluster has J=-32 cm^{-1} indicating the presence of a hydroxo, alkoxo or monodentate carboxylate bridge.[80] The reduction potentials for the binuclear iron cluster have been determined by EPR spectroscopy; [Fe(III)-Fe(III)]/[Fe(III)-Fe(II)], 48 mV and [Fe(III)-Fe(II)]/[Fe(II)-Fe(II)], -135 mV versus NHE.[81] ENDOR resonances are observed along the principle axis directions g$_1$=1.94 and g$_3$=1.76 from at least nine different protons and two ineqiuvalent nitrogens. The ratio of the ligand hyperfine coupling constants is roughly 4:7, which equals the ratio of the expectation values of iron(II) and iron(III) in the ground state indicating that at least one histidine nitrogen is coordinated to each iron atom. At least three of the protons are exchangeable in deuterated water.[82] Variable temperature X-band EPR spectra of the [Fe(II)-Fe(II)] oxo bridged cluster indicate that the g=16 resonance arises from ferromagnetically coupled high spin iron(II) ions. In the strong field limit the resonance arises from $\Delta M_s=\pm 1$ transitions of an S=4 multiplet (J>>D, D=1.2 cm^{-1}) or in the weak field limit they arise from $\Delta M_s=0$ transitions (J=-0.75 cm^{-1}, $D_1=D_2$=-5cm^{-1}).[83] Kinetic, spectroscopic and chemical evidence for the formation of specific catalytically essential complexes between the three components of the soluble form of methane monooxygenase from *Methylosinus trichosporium* has been presented. In particular EPR data revealed changes in the electronic structure of the binuclear iron centre in the mixed valence and fully reduced forms when the molar ratio of component B to hydroxylase is two to one.[84]

Absorption, CD, MCD and EPR spectroscopy identified octahedral geometries for both the iron(II) and iron(III) centres in semi met,hemerythrin (reduction of methemerythrin), while in semi met,hemerythrin (oxidation of deoxyhemerythrin), iron(II) is five coordinate and iron(III) is six coordinate. J~-8 cm^{-1} for these derivatives is consistent with an hydroxide bridging the two iron centres.[85] Multifield saturation magnetisation, multifrequency EPR and optical spectra of the azide complex of deoxyhemerythrin yields J=-3.4 cm^{-1}, $|D_1|=|D_2|$=12 cm^{-1}, $E_1/D_1=E_2/D_2$=0.3 and average g-values, $g_1=g_2$=2.23 which is interpreted as two iron(II) ions antiferromagnetically coupled in the weak regime (J/D$_i$<1/3).[86] Two previously unknown isoforms of myohemerythrin (labelled I and II) have been isolated and characterised from the sipunculid worm, *Phascolopsis gouldii*.[87]

Reduced uteroferrin [Fe(III)-Fe(II)] yields an EPR spectrum which is consistent with weak antiferromagnetic coupling between the iron atoms. Substitution of the iron(II) ion with zinc produces EPR resonances consistent with a high spin iron(III) ion, S=5/2. The effect of anions, phosphate, arsenate and molybdate on the EPR spectra of the native and zinc substituted enzymes has provided new insights into the coordination chemistry of the binuclear iron site.[88]

Optical and EPR spectroscopy were used to characterise the binding of hydroxide, acetate and phosphate to the binuclear iron(III) centre in purple acid phosphatase isolated from bovine spleen. The results suggest a µ-carboxylato and two µ-aquo or -hydroxo bridges.[89] A new form of bovine spleen purple acid phosphatase has been isolated using a high salt medium. While the CD spectrum showed a more highly ordered structure, a range of other spectroscopic and biochemical techniques revealed properties which were identical to previously isolated enzymes.[90]

Two intermediates (a mononuclear iron(II) centre and a binuclear iron(II) - iron(III) centre) have been identified previously in the oxidative deposition of the iron(II) ion to form a hydrous iron(III) oxide mineral core in ferritin. EPR studies of these two intermediates revealed that the mononuclear site was not a precursor for the mixed valence binuclear complex. A value of the exchange coupling constant J=-4 cm^{-1} for the mixed valence binuclear iron complex was measured from the temperature dependence of the EPR spectrum.[91] ESEEM spectra of oxovanadium(IV) complexes with apoferritin from horse spleen indicate that a histidine and an aquo (pK$_a$~6.5) ligand are coordinated *cis* to the oxo group.[92] ENDOR measurements on this complex reveal couplings from two nitrogen nuclei, tentatively assigned to the N$_1$ and N$_3$ nitrogens of the imidazole group in histidine. A pair of exchangeable proton lines were assigned to the NH proton of the coordinated nitrogen.[93]

EPR and Mössbauer studies of *P. aeruginosa* show for the first time that this bacterium can accumulate iron in a bacterioferritin when grown under conditions of iron limitation and incubated with ferripyoverdine.[94]

Carbonate, oxalate and hybrid carbonate-oxalate complexes of iron(III) and copper(II) lactoferrin have been prepared and characterised with optical and EPR spectroscopy and X-ray crystallography. The results presented have led to a generalised model for synergistic anion binding by transferrrins.[95] Analysis of the ^{13}C ENDOR spectra from a frozen solution of a ^{13}C enriched cyanide adduct of transferrin revealed isotropic (A$_{iso}$ = -35.5 MHz) and dipolar (A$_{aniso,xx}$=4.4, A$_{aniso,yy}$=-2.35, A$_{aniso,zz}$=-2.12 MHz) coupling constants from one or two cyanide groups. A Fe-^{13}C distance of 2.09Å was determined from the latter coupling constants.[96] EPR spectroscopy was used to monitor the viability of chicken copper(II) ovotransferrin in XANES and EXAFS experiments which revealed monodentate and bidentate coordination of carbonate to the copper ions.[97]

Oxidation of linoleic acid in tetradecyltrimethylammonium bromide micelles was found to be induced by both iron(II) and iron(III) chelates in the presence of linoleic acid hydroperoxide and iron(II) chelates in the presence of hydrogen peroxide.[98] Mössbauer, EPR, EXAFS and magnetic susceptibility studies of the interaction of daunomycin with iron(III) reveals the formation of two species in equilibrium. Both species are magnetically ordered polynuclear aggregates but have different magnetic anisotropy.[99] EPR, EXAFS, XANES, CD and infrared spectra have identified a high spin iron(III) complex, most likely with graciliformin, in an acetone extract from the lichen *Cladonia deformis*.[100]

3.2 Heme Iron Proteins.

- A correlation between the porphyrin (π) to iron(III) ion (t_{2g}) charge transfer transitions (E_{ct}) and the low symmetry axial (Δ) and rhombic (V) crystal field parameters, determined from near infrared MCD and EPR spectroscopy respectively, has provided a protocol for the assignment of the axial ligands in low spin iron(III) chromophores.[101] The combination of these two methods has also enabled the identification of the heme macrocycle. For instance the porphyrin-containing nitrimyoglobin has been distinguished from the chlorin-containing hydroperoxidase II in *E. coli*.[102]

A short review of the importance of nitric oxide in biological systems has been presented by Traylor and Sharma.[103] In particular they suggest that the activation of guanylate cyclase, a heme protein important in relaxation of vascular smooth muscle, inhibition of platelet aggregation and neurotransmission, occurs by displacement of the proximal histidine by nitric oxide. *In vivo* production of nitric oxide (not NO_2^-) was demonstrated in the peritoneal cavity of rats using EPR spectroscopy and the trapping agent carbon monoxide-hemoglobin.[104] EPR resonances, similar to those found in macrophages, attributed to iron nitrosyl complexes were found in the blood and grafted tissue of rats during the rejection of allogenic (but not synergic) heart grafts.[105] Nitric oxide produced by, or acting on vascular cells was shown to be stabilised and stored as a dinitrosyl-iron(II) complex with protein thiols, and can be released from cells in the form of a low molecular weight dinitrosyl-iron(II)-dithiolate by intra- and extra-cellular thiols.[106] Low temperature photolysis of nitric oxide from the nitrosyl complexes of iron(III) myoglobin and the manganese porphyrin substituted protein was examined by EPR spectroscopy. Resonances ranging from 0 to 0.4T were attributed to a novel spin coupled pair with S=2 between the high spin metal centre and the photodissociated nitric oxide (S=1/2) trapped adjacent to the metal centre.[107] Chemical oxidation of iron(II) nitrosyl complexes of octaethylchlorin and octaethylbacteriochlorin yielded iron(II) nitrosyl π-cation radicals which exhibited valence isomerisation.[108]

Recombinant human myoglobin with the distal histidine residue replaced by Leu, Val or Gln were prepared by site directed mutagenesis and expression in *E. coli* in an attempt to understand the role of the distal histidine residue. EPR spectra of the oxy cobalt(II) and iron(III) nitric oxide mutant proteins showed that the orientation of oxygen or nitric oxide was dependent on the distal residue.[109] Replacement of histidine E_7, valine E_{11} and histidine F_8 with tyrosine, glutamic acid and tyrosine in sperm whale myoglobin was performed by site directed mutagenesis. Coordination of tyrosine and glutamate to the heme iron was determined by optical, EPR and resonance raman spectroscopy and a comparison with other native myoglobins.[110]

ESEEM spectroscopy was used to determine and compare the hyperfine and quadrupole coupling constants for N_ϵ of the proximal histidyl imidazole in oxy cobalt myoglobin and the monomeric oxy cobalt hemoglobin from *Glycera dibranchiata*. In adddition a deuterium coupling (0.6 MHz) was observed for the oxy cobalt myoglobin in deuterated water which was ascribed to the deuteron on the histidyl imidazole hydrogen bonded to dioxygen.[111] A combination of ESEEM spectroscopy and isotope substitution (^{15}N) has enabled the hyperfine and quadrupole interactions to be determined for the axially coordinated nitrogen of imidazole in the low spin myoglobin hydroxide complex (MbOH) and of pyridine and butylamine in two model compounds [Fe(III) TTP pyridine (OR⁻)] and [Fe(III) TTP butylamine (OR⁻)] (TPP, tetraphenylporphyrin; R, H or methyl) and the porphyrin nitrogens in MbOH.[112]

EPR and optical spectra have identified six coordinate low spin iron(III) *t*- or *n*-butylperoxide complexes with whale metmyoglobin.[113] EPR and CD spectroscopy were employed in the characterisation of the interaction of several heteropolytungstates ([SiW$_{11}$O$_{39}$]$^{8-}$, [NaSb$_9$W$_{21}$O$_{86}$]$^{18-}$ and [KAs$_4$W$_{40}$O$_{140}$]$^{27-}$) with metmyoglobin. Importantly, the formation of a hemichrome complex was demonstrated.[114]

A comparative study (EPR and optical absorption) of the nitric oxide derivative of iron(II) human, horse, buffalo, deer, mouflon, musk ox, ox and reindeer hemoglobin (Hb) in the absence of an allosteric factor at pH 6.5 show that human and horse Hb-nitric oxide are typical of the high affinity state, while the others are characteristic of the low affinity conformation.[115] Addition of inositol hexakisphosphate to the nitric oxide complexes of iron(II) horse and bovine hemoglobin was also shown to effect this conformational change.[116] EPR spectra of the nitrosyl hemoglobin from *Urechis caupo* provide evidence of a proximal strain similar to that encountered in T-state human hemoglobin.[117]

In addition to the major tetragonal high spin aqueous complex and the low spin hydroxide complex of methemoglobin, EPR spectroscopy has identified a rhombic high spin spectrum and two classes of low spin *bis*-histidine complexes.[118] Dehydration of human and bovine methemoglobin was monitored as a function of the high to low spin state equilibrium with EPR spectroscopy. The experiments revealed that dehydration induces displacement and torsion of the F-helix which drastically changes the coordination at the proximal site.[119]

The effect of 24 different amino acids in the α chain and 27 in the β chain of elephant hemoglobin, compared to human hemoglobin, on the electronic and magnetic properties and reactivity of the heme has been examined.[120] The three hemoglobins from the gill of the clam *Lucina pectinata* and their sulfide, cyanide, azide, nitrite and fluoride complexes have been characterised by optical and EPR spectroscopy.[121] Binding of methylthioglycolate to the monomeric methemoglobin from *G. dibranchiata* was monitored by EPR spectroscopy and showed the formation of two low spin iron(III) centres in a pH dependent equilibrium.[122] EPR and electronic absorption spectroscopy were employed in the characterisation of the spin states of *vitreoscilla (Vt)* hemoglobin, in its reduced and oxidised states and its complexes with cyanide, imidazole, dithionite and NADH plus NADH reductase.[123]

Site directed mutagenesis has been used to prepare all the possible tyrosine -> phenylalanine mutants in a recombinant myoglobin, including Tyr-103, -146 and -151, three double and the triple mutant. EPR and optical data show that all of the mutant proteins react with hydrogen peroxide to yield an oxo-iron(IV) intermediate and a protein radical. The presence of the protein radical centre in all mutants suggests that the radical is readily transferred from one amino acid to another.[124] Reaction of leghemoglobin with hydrogen peroxide was found to produce an oxo-iron(IV) intermediate and a tyrosine radical species which was attributed to tyrosine-132.[125] Mössbauer and EPR spectroscopy of the oxo-iron(IV) porphyrin cation radical complexes, were prepared by oxidising [Fe TMP] (TMP, tetramesitylporphyrin) and [TPP(2,6-Cl) Fe OSO$_2$CF$_3$] with *m*-chloro-peroxybenzoic acid, identified strong ferromagnetic coupling between the the oxo-iron(IV) iron (S=1) and the porphyrin radical (S=1/2) (J>80 cm^{-1}) in [Fe TMP] and weak coupling J~1 cm^{-1} in [TPP(2,6-Cl) Fe OSO$_2$CF$_3$].[126]

Reaction of iron(III) cytochrome *c* peroxidase from *Saccharomyces cerevisiae* with hydrogen peroxide produced compound I, an oxo-iron(IV) centre and a protein based free radical. A combination of site directed mutagenesis, EPR and ENDOR spectroscopy showed that the free radical signal was only affected when Trp-191 or residues in its vicinity (Met-230, Met-231 and Asp-235) were mutated.[127] Reaction of the iron(II) form of cytochrome *c* peroxidase with dioxygen results in the transient oxidation of the enzyme by one equivalent above the normal compound I oxidation state. Mutations which eliminated the radical signal in compound I also prevented the rapid oxidation of the iron(II) enzyme by dioxygen. The Trp-191 -> Phe, Gln, His; Asp-235 -> Asn mutants formed a iron(II)-dioxygen complex but could not be oxidized to the oxo-iron(IV) form.[128] The resting state of intestinal peroxidase only exhibits a high spin iron(III) EPR spectrum with pH dependent rhombicity. Addition of chloride shifts the equilibrium between acidic and neutral forms of the enzyme.[129] The effect of organic solvent on the structure and function of horseradish peroxidase has been examined with optical and EPR spectroscopy.[130]

Denaturation of myeloperoxidase with guanidine hydrochloride was monitored by optical absorption, EPR, resonance raman and its chlorinating activity. The results suggest that in the native enzyme amino acids reversibly perturb the π^* orbitals of the porphyrin causing an asymmetry in the electronic distribution of the macrocycle.[131] EPR spectra of the cyanide complex of myeloperoxidase in the presence and absence of halide substrates reveal that the electronic structure of the low spin iron(III) ion is modulated by halide binding to a protonated amino acid in the distal heme cavity.[132] EPR spectra have been reported for the chloride and nitrite complexes of the three myeloperoxidase isoenzymes purified from neutrophils.[133]

Reaction of (20R)-20-hydroperoxycholesterol with cytochrome P-450$_{scc}$ produced a transient radical species characterised by a single resonance, g=2.0035. The similarity of this resonance to that of the peroxidase compound I spectrum suggests that the EPR active species is an iron(IV) porphyrin II radical cation.[134] X-irradiation of the ternary complex of cytochrome P-450$_{cam}$ - substrate and dioxygen produces a new EPR active species with rhombic symmetry. Similar signals were found in X-irradiation experiments with oxy-myoglobin and oxy-hemoglobin suggesting that the added electron is resident on the FeO$_2$ complex.[135] Cytochrome P-460 purified to an electrophoretically homogeneous state was shown to contain three subunits and high spin heme centres and six cysteine residues.[136] Cytochrome P-450$_{11\beta}$ was purified as the 11-deoxycorticosterone-bound form from bovine adrenocortical mitochondria and was found to have an orthorhombically distorted high spin iron(III) EPR signal which is different from that observed for cytochrome P-450$_{succ}$. The iron(II)-carbon monoxide and -nitric oxide complexes of the two enzymes were similar.[137]

Cytochrome ba_3 from *Thermus thermophilus* reacts slowly with excess cyanide to produce a form in which Cu$_A$, cytochrome b and Cu$_B$ remain oxidised and cytochrome a_3 is reduced by one electron. One molecule of cyanide binds to the low spin iron(II) heme (S=0), while a second binds axially to Cu$_B$. ENDOR spectra revealed four nitrogen ligands in the equatorial plane of Cu$_B$.[138] NMR, EPR and optical studies of the monoheme c-550 isolated from cells of *Bacillus halodentrificans* identified three forms with pK$_a$ values of 6.0 and 11.0. Methionine coordination to the heme is implicated from the optical spectra.[139] Low temperature EPR and MCD spectra reveal a *bis*-histidine coordination of the heme iron in cytochrome b-558 from *B. subtilis*.[140] EPR spectra of resting neutrophils under aerobic conditions yielded resonances attributable to cytochrome b-558, myeloperoxidase, non heme iron, superoxide dismutase and a flavin radical. The anisotropic g values for the cytochrome b-558 indicate that the low spin heme is six coordinate with *bis*-imidazole coordination.[141] Low temperature EPR spectra of chromaffin granule membranes from bovine adrenal medulla revealed three different cytochrome b-561 signals, a high potential signal, g_z=3.11, a low potential signal, g_z=3.14 and a highly temperature sensitive heme with g_z=3.7. The presence of several forms of cytochrome b-561 may provide a structural basis for transmembrane electron transfer catalysed by this heme protein.[142]

The recombinant high molecular weight, multiheme cytochrome c from the sulfate reducing bacterium *Desulfovibrio vulgaris*, strain Hildenborough was found to contain one or two high spin hemes and fourteen or fifteen low spin hemes *per* molecule. Three different redox potentials were determined (0, -100 and -250 mV vs. NHE) and are consistent with the bis-histidyl coordinated hemes in cytochrome c_3.[143] Five cytochromes (c') isolated from the chemoheterotrophic bacteria *Achromobacter xylosoxidans* NCIB 11015 revealed EPR spectra from a mixed spin state, predominantly S=5/2 with a slight S=3/2 character.[144] Four heme components, attributable to cytochrome c-554 were identified by EPR spectroscopy in chlorosome depleted membranes of the green-gliding bacterium *Chloroflexus aurantiacus*.[145]

A comparison of the reductive titration of the hexaheme nitrite reductase (*W. succinogenes*) with methylviologen semiquinone and dithionite shows that a product of dithionite oxidation binds to the iron(II) heme causing a spin state change. A combination of absorption, CD, MCD and EPR spectroscopy suggests an active site consisting of a high spin five coordinate heme and a low spin six coordinate heme which are magnetically coupled, possibly bridged by histidine.[146] Mössbauer and EPR spectra have identified six distinct heme groups, one high spin and five low spin.[147] A novel triheme cytochrome from *D. vulgaris* which exhibits nitrite reductase activity was shown to contain a pentacoordinate high spin and two low spin hemes with *bis*-histidyl coordination.[148] Stopped flow and rapid freeze quenching were employed in the determination of the reaction mechanism for the reduction of nitrite to nitric oxide catalysed by nitrite reductase purified from *P. aeruginosa*.[149]

The nitrosyl derivative of *Arenicola marina* erythrocruorin exhibits a 'nine line' hyperfine EPR spectrum at pH 7.0 which is characteristic of nitric oxide *trans* to imidazole nitrogen. At low and high pH's the spectrum converts to a 'three line' spectrum attributed to the protonation of the distal histidine and cleavage of the iron(II) histidyl bond.[150]

Prostaglandin H synthase catalyses the formation of prostaglandin endoperoxide from arachidonic acid. This process involves distinct cyclooxygenase and peroxidase reactions. The current understanding of these processes has been reviewed by Smith *et al.*[151] N-acetylimidazole was found to inactivate apoprostaglandin H synthase but not the native enzyme indicating that the heme group prevents this inactivation and further, this modification prevents binding of the heme prosthetic group required for catalytic activity.[152] Addition of either ethylperoxide or 15-hydroperoxyeicosatetraenoic acid to the enzyme yielded a doublet in the EPR spectrum which was subsequently replaced by a singlet. These two species were assigned to two different tyrosyl radicals. Indomethacin, a selective inhibitor for this reaction, does affect the heme environment.[153] These results were also confirmed using arachidonic acid or 5-phenyl-4-pentyl-1-hydroperoxide. However addition of a peroxidase substrate, phenol, abolished the appearance of the doublet and suppressed the formation of the singlet. Substitution of the iron(III) protoporphyrin IX with manganese(III) protoporphyrin IX was also examined.[154]

E. coli produces two catalases, HPI and HPII, the latter containing either a dihydroporphyrin or chlorin heme prosthetic group. MCD and EPR spectra provided evidence for tyrosine being the proximal ligand.[155]

4 Iron Sulfur Proteins

Iron sulfur proteins have been classified by the International Union of Biochemistry into two classes, namely: Simple Iron Sulfur Proteins and Complex Iron Sulfur Proteins. The former class of proteins only contains iron sulfur clusters, while the second class of proteins contains additional prosthetic groups. Both clases of proteins may be involved in electron transfer or ezymatic catalysis.[5] Two recent reviews describe the methods by which the structure of an iron sulfur cluster can be determined and secondly the structure and function of iron sulfur cluster containing proteins.[156,157]

4.1 Simple Iron Sulfur Proteins.- EPR spectra of cobalt(II) substituted rubredoxin were typical of tetrahedrally coordinated cobalt(II) complexes and metalloproteins.[158] A new non-heme protein has been isolated from extracts of *D. desulfuricans* and *D. vulgaris* and found to have a molecular weight of 16 kDa and two iron atoms *per* molecule. Mössbauer and EPR data indicate the presence of two types of iron centre, a high spin Fe(III)S$_4$ centre similar to that found in desulforedoxin from *D. gigas* and an iron(II) site with nitrogen and / or oxygen ligands.[159]

The soluble [2Fe-2S] ferredoxin from *E. coli* was shown to have optical, CD, and EPR spectra similar to the animal ferredoxins and *P. putida* ferredoxin.[160] ^{2}H Mims pulsed ENDOR has been used to probe the hydrogen bonds in the [2Fe-2S]$^{1+}$ cluster in *Anabaena*

7120 ferredoxin and the binding of exogenous ligands (water and hydroxide) to aconitase.[161] The benefits of performing [14,15]N, ^{13}C, [57]Fe and [1,2]H ENDOR experiments at Q-band microwave frequencies has been demonstrated by characterising 10 proteins which contain [nFe-mS] (n,m=2,3,4) clusters.[162]

The facile transformation of iron sulfur clusters in metalloproteins as described by [3Fe-4S] + M^{2+} ⇌ [M3Fe-4S]$^{2+}$; (M=Fe,Zn,Cd) has been studied using cyclic voltammetry. EPR and MCD spectra of the oxidised and one electron reduced [Zn3Fe-4S]$^{2+;1+}$ and [Cd3Fe-4S]$^{2+;1+}$ clusters have identified S=2 and S=5/2 ground states. Zinc has an intrinsic preference for the fourth site over iron.[163] Thallium was shown to bind reversibly to both the oxidised and reduced states of the [3Fe-4S] cluster from *Desolfovibrio africanus* using EPR, MCD and voltammetry to monitor the oxidation state.[164] An alternative model for the antiferromagnetic coupling in the reduced [3Fe-4S] cluster has been proposed.[165] [57]Fe, [33]S and [14]N Q-band ENDOR studies on the [4Fe-4S]$^{1+}$ cluster in aconitase revealed four inequivalent iron sites of which one is sensitive to substrate binding (Fe$_A$), two types of sulfide with isotropic hyperfine coupling constants ~7.5 and ~9 MHz and a nitrogen signal which remained unchanged upon substrate binding.[166] [17]O, [1,2]H ENDOR spectra of the [4Fe-4S]$^{1+}$ cluster in the presence and absence of substrate suggest that an hydroxyl ion is coordinated to the Fe$_A$ site which becomes protonated upon substrate binding.[167] The [3Fe-4S] cluster of aconitase was identified as the major mitochondrial target for the inactivation of potato tuber mitochondria by hydrogen peroxide.[168]

Temperature dependent potentiometry, EPR and ENDOR spectroscopy were employed to probe the differences between the forms (S=1/2 and S=3/2 ground states) of the [4Fe-4S] cluster in the *Pyrococcus furiosus* ferredoxin.[169] This cluster has been further characterised with resonance raman and EPR spectroscopy in the oxidised and reduced states. Facile and quantitative interconversion of the [4Fe-4S] cluster to a [3Fe-4S] was also demonstrated.[170] EPR and MCD spectroscopy show that cyanide coordinates to the unique iron site in the reduced [4Fe-4S]$^{1+}$ cluster.[171]

Two mutant proteins of *Azotobacter vinelandii* ferredoxin I in which a ligand to the [4Fe-4S] cluster is removed (Cys-20A) and a free cysteine next to this cluster is deleted (Cys-24A) have been purified and shown to contain [3Fe-4S] and [4Fe-4S] clusters. The redox potential for the [4Fe-4S] cluster in the mutant proteins was quite different to that of the native protein. Reaction of ferricyanide with these proteins suggested that Cys-24 was involved in the formation of the cysteinyl disulfide radical.[172] Substitution of two iron atoms in the [4Fe-4S] cluster from *Clostridium pasteuranium* with cobalt(II) produced an EPR silent protein. Upon reduction with sodium dithionite an S=3/2 EPR spectrum consistent with high spin cobalt(II) was observed.[173]

The EPR spectra of the high potential iron sulfur (HPIP) protein from *Chromatium vinosum* consisted of a major component with g=2.02, 2.04, 2.12 and a minor component with g=2.04, 2.07, ~2.13. In the presence of 0.1 to 2 M NaCl, freezing was found to induce polymerisation of the protein, as evidenced by the intercluster spin-spin interactions in the EPR spectrum. Computer simulation of the X- and Q-band EPR spectra revealed that the g matrices in each dimer pair must be colinear and the internuclear distance between the clusters is ~13Å and ~16Å.[174]

S-, X-, P- and Q-band EPR spectroscopy, parallel-mode EPR spectroscopy and high resolution Mössbauer spectroscopy have been used to characterise the four oxidation states (3+, 4+, 5+ and 6+) of the putative [6Fe-6S] cluster in the six-iron-sulfur protein from *D. vulgaris*, strain Hildenborough. The fully oxidised protein (+6 state) appears to be diamagnetic. In the +5 state the cluster exists in two magnetic forms; 10% is low spin S=1/2, the remainder is in a high spin S=9/2 spin state. A one electron reduction yields a g~16 resonance, presumably from an S=4 spin state. Further reduction (+3 state) yields a mixture of two S=1/2 ground states.[175] In addition to a [2Fe-2S] cluster a 6 iron cluster has been characterised by Mössbauer and EPR spectroscopy in an iron sulfur protein purified to homogeneity from *D. desulfuricans*.[176] (ATCC 27774)

4.2 Complex Iron Sulfur Proteins.- Mason and Cammack have reviewed the ring-hydroxylating dioxygenases which are soluble multicomponent enzymatic systems, consisting of a reductase site, an electron transfer chain and an oxygenase site.[177]

Adenylate sulfate reductase, an iron sulfur flavoprotein, is a key enzyme in dissimilatory sulfate respiration. Optical and EPR studies have identified two distinct [4Fe-4S] clusters. Centre I has a high redox potential and is reduced by sulfite and AMP, while centre II has a very negative redox potential.[178,179] In addition, an isotropic signal with g=2.02 was observed at temperatures less than 20K and a spin-spin interaction was observed in the fully reduced enzyme.[179] Two distinct types of fumarase (A, C) were purified to homogeneity from aerobically grown *E. coli* W cells. The inactive form of fumarase A was shown to be EPR active with resonances at $g_{\parallel}$=2.00 and $g_{\perp}$=2.02, while the active form was EPR silent. These data suggest that this enzyme is a member of the [4Fe-4S] hydratases represented by aconitase.

Fumarase C did not contain any iron and was similar to mammalian enzymes.[180]

ESEEM studies of the reduced [2Fe-2S](I) and [4Fe-4S](II) clusters and the oxidised [3Fe-4S](III) cluster in fumarate reductase revealed hyperfine coupling from a peptide ^{14}N in clusters I and II.[181,182] Spin lattice relaxation times (T_1) for the oxidised and reduced protein confirmed the presence of a spin-spin interaction between the reduced clusters I and II.[181] Replacement of Cys-57, -62, -65 and -77 by serine only affected the properties of the [2Fe-2S] cluster and demonstrated that its unusually high redox potential was required for catalysis.[183] Replacement of Cys-65 by aspartic acid has virtually no effect on catalytic activity or the properties of the [2Fe-2S] cluster. Since aspartic acid occupies an eqiuvalent position in *E. coli* succinate dehydrogenase it is likely that this [2Fe-2S] cluster has an oxygen donor ligand.[184] The Cys-204 -> Ser, Cys-210 -> Ser and the Cys-214 -> Ser mutants of the fumarate reductase B protein produced enzymes with negligible activity, a consequence of the loss of the [3Fe-4S](III) and the [4Fe-4S](2) clusters. Although the Val-207 -> Cys mutant is a functional membrane bound enzyme, the [3Fe-4S](III) was converted to a [4Fe-4S](II) cluster.[185] Three proteins, (i) a cytochrome *b* dependent iron sulfur fumarate reductase, (ii) a [NiFe] hydrogenase and (iii) a protein linked to sulfur oxidoreductase activity were purified and characterised from *Spirillum* 5175 membranes in an attempt to identify the active constituents in the reduction of elemental sulfur to sulfide.[186]

Reductive optical and EPR titrations of trimethylamine dehydrogenase with sodium dithionite reveal that the equilibrium distribution of reducing equivalents between the flavin mononucleotide and [4Fe-4S] clusters in partially reduced enzyme is pH dependent.[187] EPR studies on the oxidised and reduced forms of glutamate synthase from *Azospirillum brasilense* have identified the presence of an [3Fe-4S]$^{1+}$ and two different [4Fe-4S]$^{2+,1+}$ clusters.[188]

Lysine-2,3-aminomutase from *C*. SB4 catalyses the interconversion of lysine and β-lysine and was found to contain 12 gatoms of iron and sulfide, 3.5 gatoms of cobalt and small amounts of copper and zinc. A high spin cobalt(II) EPR spectrum is observed at temperatures lower than 20K.[189] Further characterisation of this reaction has identified a radical signal with multiple hyperfine resonances attributed to lysine.[190]

A tetrameric dissimilatory bisulfite reductase was purified from a thermophilic sulfate reducing bacterium *D. thermophilus* (DSM 1276) and shown to contain a siroheme and two [4Fe-4S] clusters *per* subunit.[191] A new S=9/2 EPR signal, attributed to an iron sulfur cluster has been found in sulfite reductase. The observation of this signal argues against magnetic coupling between the siroheme and this unusual iron sulfur cluster.[192]

The structure and mechanistic details of iron containing hydrogenases have been presented by Adams.[193] Pulsed ENDOR and ESEEM spectroscopy have identified unusual nitrogen coordination to the iron sulfur cluster (H) in hydrogenase I from *C. pasteurianum*.[194]

The methylviologen reducing hydrogenase operon of *Methanobacterium thermoautotrophicum* was shown to produce a polyferredoxin which in the reduced state showed EPR spectra typical of [4Fe-4S] clusters.[195] Addition of tungsten to the growth medium of *Thermotoga maritima* was found to increase the cellular concentration of hydrogenase and its *in vitro* catalytic activity by up to 10-fold. Purification of the enzyme revealed that it contained twenty atoms of iron and eighteen atoms of acid-labile sulfide *per* molecule and trace amounts of nickel and selenium. It did not contain tungsten.[196]

After immunostimulation, murine macrophages oxidise L-arginine into nitric oxide which acts as an effector molecule. Coculture of these activated macrophages with murine L1210 leukemia cells and subsequent removal of the macrophages produced an EPR spectrum with g=2.041 and g=2.015 which is typical of a nitrosylated iron sulfur cluster.[197] Evidence from EPR studies has been presented that suggests that interleukin-1β induced inhibition of insulin secretion is dependent on the metabolism of L-arginine to nitric oxide.[198] A filamentous fungus *Phanerochaete chrysosporium* has been shown to convert glyceryl trinitrate into its di- and mononitrate derivatives, concurrently with the formation of nitric oxide.[199]

5 Nickel Containing Enzymes

A review on the catalysis by nickel in biological systems has been presented by Cammack.[200] In this review the composition of the protein, its catalytic properties and the structure of the active site nickel ion are described for urease, hydrogenase, methyl CoM reductase and carbon monoxide dehydrogenase (Acetyl-CoA synthase) and carbon monoxide oxidoreductase of photosynthetic bacteria.[200]

EXAFS spectra of the EPR active, oxidised forms (A and B) and the reduced (C) Ni centres in hydrogenase from *Thiocapsa roseopersicina* show 2±1 S(Cl) atoms at 2.2Å, 3±N(O) atoms at 2.0Å in all three forms. Additional long interactions in Form A (Ni-S(Cl),2.4Å) and B (Ni-S(Cl),2.50Å) were also observed.[201,202]

A ^{1}H and ^{2}H Q-band ENDOR study of the as isolated (Ni-A), hydrogen reduced (Ni-C) and reoxidised (Ni-A/Ni-B) states of *D. gigas* hydrogenase has identified two distinct types of exchangeable protons which may arise from the heterolytic cleavage of hydrogen.[203] Detailed redox titrations of the nickel and iron sulfur clusters in *D. vulgaris* hydrogenase show the enzyme is similar to that found in *D. gigas*.[204] Distinct redox behaviour of the nickel(III) nickel(II) transition is apparent in the ready and unready forms of *Chromatium vinosum* hydrogenase.[205] Reversible inactivation of nickel hydrogenases by oxygen was shown to involve the binding of dioxygen or a reduced adduct tightly in the vicinity, but not directly to the two forms of nickel(III) in hydrogenase from *Chromatium vinosum*. Reaction of carbon monoxide with hydrogen reduced enzyme showed that carbon monoxide was axially coordinated to the nickel ion and secondly that this is the site for the binding of hydrogen.[206]

Multifield saturation magnetisation, Mössbauer and EPR spectroscopy were used to characterise the nickel(II) and [4Fe-4S] cluster in the nickel-iron-selenium hydrogenase from *D. baculatus*.[207,208] In conjunction with previous EXAFS results, the nickel site was shown to be five coordinate and diamagnetic, low spin, S=0.[208]

Purification of carbon monoxide hydrogenase from *Methanothrix soehngenii* under anaerobic conditions was found to contain two EPR active centres, one of which partially disappeared upon the addition of substrate. In contrast, aerobic preparations produced different EPR active centres ([3Fe-4S]$^{1+}$ and two weakly coupled [4Fe-4S]$^{1+}$) which remained unchanged on the addition of substrates.[209]

Redox titration in conjunction with EPR spectroscopy was used to characterise the redox centres in carbon monoxide dehydrogenase from *C. thermoaceticum*.[210] In the absence of redox mediator dyes, redox titrations of the nickel ion in the hydrogenase from *C. thermoaceticum* and *Chromatium vinosum* revealed remarkably different behaviour.[211] ^{57}Fe, ^{61}Ni and ^{13}C Q-band ENDOR studies provided unambiguous evidence for the formation of a novel spin coupled cluster [Ni-3/4Fe-≥4S-C] during the reaction of carbon monoxide with carbon monoxide dehydrogenase. ^{13}C ENDOR spectra are compatible with carbon monoxide binding to either a nickel or an iron atom.[212] The Ni-Fe-C EPR active species has been shown to be catalytically competent by comparing its rate of formation with an isotope exchange reaction involving carbon monoxide and acetyl-CoA.[213] Further investigation of this protein has revealed resonances attributed to S=9/2 and S=5/2 spin states. ^{61}Ni hyperfine coupling was not resolved in enriched preparations of this enzyme.[214]

Preincubation of *M. thermoautotrophicum* cells with hydrogen enabled the purification of S-methyl-coenzyme M reductase with a high specific activity. The enzyme showed two nickel signals, designated MCR-red1 and MCR-red2 previously only observed in intact cells.[215] Purification and subsequent EPR studies of this enzyme from a medium enriched with [61]Ni revealed nickel hyperfine coupling and coupling to nitrogen nuclei from the porphyrin F-430 cofactor.[214] Spectroscopic characterisation (optical, MCD, EPR and resonance raman) of the two forms recently purified[216] show that they are similar but not identical.[217]

The oxidised form of nickel(II) substituted rubredoxin reveals an EPR spectrum similar to that observed in the nickel containing hydrogenases.[218]

6 Molybdenum and Tungsten Enzymes

6.1 Mononuclear Oxomolybdenum Enzymes.- Proton ENDOR spectra of the rapid, slow, inhibited and desulfo inhibited molybdenum(V) complexes in xanthine oxidase reveal coupling to at least nine protons or sets of protons with coupling constants in the range 0.08 to 4 MHz.[219] A comparison of the ^{95}Mo (I=5/2) and ^{33}S (I=3/2) hyperfine matrices for the rapid types I and II and slow molybdenum(V) centres in xanthine oxidase with those of the model complexes [Mo OXL]$^-$ and [Mo O(XH)L] (X=O,S; LH$_2$=N,N'-dimethyl-N,N'-*bis*(2-mercaptophenyl)-1,2-diaminoethane) confirm [Mo(V) O(SH)], [Mo(V) O(SH)(OH)] and [Mo(V) O(OH)] as the centres responsible for the rapid type 1 and 2 and slow molybdenum(V) signals respectively.[220]

The finding that bisulfate, an oxidation product of dithionite, competitively inhibits xanthine oxidase (K$_d$, 4.5 x 10^{-5}) necessitates a reevaluation of dithionite titrations of xanthine oxidase performed at neutral and low pH. EPR and optical spectra of this complex were reported.[221] Xanthine dehydrogenase was purified to homogeneity from the wild type and *rosy* mutant (Glu-89 -> Lys) strains of *Drosophila melanogaster* and was shown to be indistinguishable from xanthine oxidase, except the mutant lacked the [2Fe-2S] clusters. Although the mutant enzyme had xanthine:NAD$^+$ oxidoreductase activity comparable to the wild type enzyme, activity to certain oxidising substrates is lost, particularly phenazine methosulfate.[222] ESEEM spectra of the reduced [2Fe-2S]$^{1+}$ clusters in xanthine oxidase reveal hyperfine coupling to ^{14}N which was attributed to an amide nitrogen in the second coordination sphere.[223]

The pH dependence of the oxidation-reduction midpoint potentials for the

Mo(VI)/Mo(V), Mo(V)/Mo(IV) and FAD/FADH·⁻/FADH⁻ couples in *Chlorella vulgaris* nitrate reductase have been measured with EPR and CD spectroscopy. The results indicate that the uptake of four protons are coupled to the oxidation reduction reactions; FAD·⁻ ↔ FADH⁻, Mo(VI) ↔ Mo(V) ↔ Mo(IV) and the reduction of the heme group.[224,225] EPR spectra of partially reduced *Candida nitratophila* assimilatory NADPH nitrate reductase revealed a single molybdenum(V) site in which the unpaired electron is coupled to a single proton.[226] Two classes of iron sulfur clusters, based on their redox properties, have been found in nitrate reductase A purified from *E. coli*. The high potential class consists of magnetically isolated [3Fe-4S] and [4Fe-4S] clusters which are coupled by an anticooperative redox interaction. The second low potential class contains two [4Fe-4S] clusters. All of the [4Fe-4S]$^{1+}$ clusters are magnetically coupled in the fully reduced state.[227]

EPR studies of the electron transfer centres in dimethylsulfoxide reductase have identified four [4Fe-4S] clusters in subunit B with midpoint potentials between -50 and -330 mV. A molybdenum(V) EPR signal with similar characteristics to the high pH form of the molybdenum cofactor was also observed.[228] EPR spectra show that mutation of Cys-102 to Trp, Ser, Tyr and Phe in dimethylsulfoxide reductase causes a conversion of one of the [4Fe-4S] clusters into a [3Fe-4S] cluster.[229] The only cofactor in dimethylsulfoxide reductase from *R. sphaeroides* f. sp. *dentrificans* is molybdenum which allowed the measurement of the electronic absorption spectrum. The orthorhombic molybdenum(V) EPR spectrum showed coupling to an exchangeable proton, little sensitivity to pH, no detectable anion effects and was used to measure the Mo(VI)/Mo(V) and Mo(V)/Mo(IV) redox potentials.[230]

NAD⁺-coupled formate dehydrogenase has been purified to near-homogeneity from *Methylosinus trichosporium* OB3b and found to contain flavin (1.8±0.2), iron (46±6), inorganic sulfide (38±4) and molybdenum (1.5±0.1). Reaction of the enzyme with formate, NADH or sodium dithionite anaerobically produced 50% reduction of the iron and an EPR spectrum arising from five distinct redox active centres, one of which was molybdenum, was observed.[231]

6.2 Molybdenum Iron Enzymes.- A review of the metal containing clusters in the three genetically distinct classes of nitrogenase (molybdenum iron-, vanadium iron- and iron-component I) has been presented by Smith and Eady.[232] In particular the article focusses on the metal containing clusters and some of the unresolved aspects of the structure and function of the molybdenum containing nitrogenases. Both molybdenum and vanadium containing

2: Metalloproteins

nitrogenases have been identified in wild type and a *nifHDK*⁻ deletion mutant of *R. capsulatus*.[233]

ESEEM spectra of the purified molybdenum iron protein and whole cells revealed nitrogen coupling attributable to His-195 on the basis of site directed mutagenesis experiments. Moreover this residue was required for nitrogen fixation.[234] The effect of nitric oxide on the iron and the molybdenum iron protein of *A. vinelandii* nitrogenase has been examined. A combination of optical and EPR spectroscopy showed that the inactivation of the iron protein resulted from the oxidation of the [4Fe-4S] cluster with concomitant loss of the S=1/2 and S=3/2 EPR signals.[235] The approach to a steady state rate of hydrogen evolution from *Klebsiella pneumoniae* nitrogenase functioning under conditions of extremely low electron flux was accompanied by a 50% decrease in the amplitude of the molybdenum iron protein EPR signal.[236] Diazotrophic growth, molybdenum iron protein acetylene reduction and the intensity of the S=3/2 EPR signal were employed in the characterisation of the serine and alanine-70, -95 and -153 mutants of the β-subunit of *A. vinelandii* nitrogenase molybdenum iron protein.[237]

The linear electric field effect of the iron protein component of nitrogenase is consistent with a single [4Fe-4S] cluster. ESEEM spectra of this cluster revealed no interaction from ^{31}P of MgATP, though weak modulations from peptide nitrogens and protons were observed.[238] The dithionite reduced component I of the 'iron only' nitrogenase from *R. capsulatus* showed an EPR spectrum consisting of two components: a minor S=1/2 signal and a near stoichiometric S=3/2 signal with E/D=0.33.[239]

6.3 Tungsten Enzymes.- A red coloured tungsten containing protein has been purified and shown to contain approximately six iron, one tungsten and four acid labile sulfide atoms *per* molecule. The dithionite reduced protein revealed resonances ranging from 1.3 to 10.0. These were attributed to a magnetic interaction between an S=3/2 spin system and an unknown centre, whose redox potentials were -410 mV and -500 mV respectively.[240] This protein was shown to be an inactive form of an aldehyde ferredoxin oxidoreductase.[241]

When *M. wolfei* was grown on a medium deprived of molybdenum and supplemented with tungsten, a tungsten substituted molybdenum formylmethanofuran dehydrogenase was obtained. This enzyme was catalytically active and displayed an orthorhombic EPR signal attributable to tungsten.[242]

7 Vanadium Proteins

A review of the structure and function of vanadium in (i) the regulation of phosphate metabolizing enzymes, (ii) the halogenation of organic compounds by nonheme haloperoxidases, (iii) vanadium containing nitrogenases, (iv) amavadine and (v) the sequestration of vanadium by ascidians, has been presented by Rehder.[243]

EPR and ENDOR spectra of oxovanadium(IV) bound to the two metal binding sites in D-xylose isomerase from *S. rubiginosus* revealed that histidine was a ligand in the high affinity site while in the low affinity site only oxygen donor ligands were found.[244] Pyruvate kinase requires two divalent cations for activity. Substitution of magnesium for oxovanadium(IV) produced a variety of effects on various reactions catalysed by pyruvate kinase. EPR spectra showed that oxovanadium(IV) binds to the divalent cation site competitively with respect to magnesium and has a high affinity for bicarbonate. Direct coordination of pyruvate, oxalate and glycolate to the metal ion was obtained from the observation of ^{17}O ligand hyperfine coupling.[245]

Two isomomeric vanadium(IV) tunichrome complexes have been prepared from spermidine and trihydroxybenzoic acid and characterised by optical and EPR spectroscopy.[246] A comparison of the EPR spectra from humic and tannic acid oxovanadium(IV) complexes with model complexes over the pH range 2 to 6 indicates the presence of oxygen donor atoms from either carboxylate, salicylate or catecholate chelating groups.[247] Optical and EPR spectra reveal the formation of two (1:1) oxovanadium(IV) complexes with the trihydroxamic acid deferoxamine. The low pH form involves an octahedral vanadium(IV) ion coordinated by three hydroxamate ligands, one of which is displaced by oxygen at pH's greater than 2.8.[248]

Oxovanadium(IV) ions were retained in *Hansenula polymorpha* as mobile complexes associated with relatively low molecular weight ligands and immobilised complexes formed with high molecular weight compounds.[249] EPR studies of the binding of oxovanadium(IV) to native and modified soluble collagen from calf skin have identified an N_2O_2 coordination sphere for the metal ion.[250] A superoxide independent reduction of vanadate to oxovanadium(IV) by rat liver microsomes was found to involve a NAD(P)H dependent electron transport cytochrome P-450 system.[251]

2: *Metalloproteins*

8 Manganese Proteins

Manganese superoxide dismutase from *E. coli* has been prepared in homogeneous manganese(III) and manganese(II) redox forms and characterised using a combination of optical absorption, CD, MCD and EPR spectroscopy. The manganese(III) site has a distorted trigonal bipyramidal geometry and both redox forms are susceptible to binding of exogenous ligands.[252] The ground state magnetic properties of the oxidised (S=2, D=+2.44 cm^{-1}, E/D=0) and reduced (S=5/2, D=+0.50 cm^{-1}, E/D=0.027) forms of manganese superoxide dismutase from *T. thermophilus* were determined from saturation magnetisation and EPR spectroscopy.[253] Manganese(III) complexes with desferrioxamine and 1,4,8,11-tetraazacyclodecane displayed superoxide dismutase activity.[254]

EPR spectra of the binuclear manganese complex in the active site of catalase from *T. thermophilus* reveal three EPR active species [Mn(II)-Mn(II)], [Mn(II)-Mn(III)] and [Mn(III)-Mn(IV)]. The three species can be interconverted with the use of hydroxylamine, potassium iodide, air or potassium iodate.[255] Peroxide decomposition was found to involve the interconversion of the [Mn(II)-Mn(III)] and [Mn(III)-Mn(III)] clusters. Inhibition of catalase activity was a result of anions binding to the [Mn(II)-Mn(II)] cluster.[256,255]

ESEEM spectra of the manganese(II) binding sites in a series of lectins including concanavalin A, pea lectin (*Pisum sativum*), isolectin A (*Lens culinaris*), soybean agglutinin (*Glycine max*), *Erythrina indica* lectin and *Lotus tetragonolobus* isolectin A have identified a single nitrogen of a conserved histidine residue and two water molecules directly coordinated to the manganese(II) ion.[257] Q-band EPR spectra of a variant of Ha *ras* p21 - manganese(II) guanosine diphosphate complex have shown that four water molecules are coordinated to the manganese(II) ion. ^{17}O ligand hyperfine coupling was also observed when the β-phosphate of GDP was enriched with ^{17}O.[258]

Manganese accumulation in *Saccharomyces cerevisiae* was studied using room temperature EPR in media containing increasing amounts of MnSO$_4$.[259]

9 Chromium Substituted Proteins and Toxicity

Optical, EPR and resonance raman spectra were used to characterise the nitridochromium(V) protoporphyrin-IX complex and its complexes with apomyoglobin and apocytochrome *c* peroxidase.[260]

Although potassium tetraperoxochromium did not react with water or hydrogen peroxide to produce hydroxyl radicals, addition of the biological reductant NADH produced a chromium(V)-NADH complex which readily reacted with hydrogen peroxide to produce hydroxyl radicals.[261] EPR studies show that the diol moeity of NADPH coordinates to the chromium(V) ion when NADPH is used as a reductant.[262] EPR measurements provided evidence for the formation of chromium(V) intermediates and hydroxyl radicals in the enzymatic reduction of chromate by glutathione reductase.[263] Deferoxamine was found to inhibit the formation of chromium(V) intermediates in the reaction of chromium(VI) with biological reductants. This in turn inhibited the chromium(V)-mediated hydroxyl radical generation from hydrogen peroxide.[264]

10 Paramagnetic Metal Substituted Proteins

Substitution of the zinc(II) sites in horse liver alcohol dehydrogenase with copper(II) produces two different species. The predominant species corresponds to copper(II) bound at the active site and has type I copper(II) properties. The minor copper(II) species cannot be reduced nor does it interact with exogenous ligands. Its EPR spectrum is axially symmetric and reveals ligand hyperfine coupling to a single nitrogen atom.[265]

The metal selectivity of the metal thiolate adamantane type clusters in rabbit liver metallothionein has been studied by offering two ions, cobalt(II)/cadmium(II), zinc(II)/cadmium(II) or cobalt(II)/zinc(II). Homometallic cluster formation was observed for the cobalt(II)/cadmium(II) derivative while heterometallic clusters were observed in the other derivatives. The primary factor governing the cluster type is whether the protein tertiary structure can accommodate the various cluster sizes.[266] Titration of apometallothionein with cobalt(II) in water and dimethylsulfoxide showed that the first four titration steps were identical in both solvents indicating a thermodynamically controlled folding process. However, in the last three steps the completion of cluster formation between dimethylsulfoxide and water suggests a kinetically controlled process.[267]

Binding of kirromycin to the elongation factor Tu (Ef-Tu) caused broadening of the Q-band manganese resonances from the Ef-Tu-Mn, Ef-Tu-Mn-GDP and Ef-Tu-Mn-GDP.phosphate complexes indicating that it coordinates to the metal ion. ^{17}O ligand hyperfine coupling was used to identify the number of water molecules bound to the manganese ion in the Ef-Tu-Mn-GDP and Ef-Tu-Mn-GDP.phosphate complexes.[268] Using stereospecifically labelled (S_p)-(R_p)-[β-^{17}O GTP] it was found that only one diastereoisomer (S_p) produced detectable line broadening of the EPR spectrum of manganese at the active site of Ef-Tu. The complex giving rise to this signal was identified as the Ef-Tu-Mn-GDP.phosphate complex.[269]

Phosphonoacetohydroxamate (PhAH) a tight binding inhibitor of enolase is thought to mimic the *aci*-carboxylate form of the catalytic intermediate carbanion. EPR spectra of the enolase-PhAH complexes with two manganese ions bound at the active site revealed exchange coupling between the metal ions. Preparation of hybrid Mg(II)/Mn(II) complexes in conjunction with selective ^{17}O isotope enrichment of PhAH identified hydroxamate as a ligand of manganese(II) in the high affinity site and a phosphonate oxygen as a ligand of manganese(II) in the low affinity site with the carbonyl group forming a μ-oxo bridge.[270]

Both EPR and water proton longitudinal relaxation studies showed binding of one manganese(II) ion per 18 kDa subunit of *Trypanosoma cruzi* phosphofructokinase with a dissociation constant of 57±μM.[271] Three cobalt(II) ions were found to bind cooperatively in the active site of 3',5'-exonuclease. One of these sites appears to be five coordinate, while the other two have octahedral geometry.[272] The mechanism of the phosphodiesterase reaction catalysed by staphylococcal nuclease is thought to involve concerted general acid-base catalysis by Arg-87 and Glu-43. The mutual interactions between these two residues were investigated by comparing kinetic and thermodynamic properties of the single mutants Arg-87 -> Gly and Glu-43 -> Ser with those of the double mutant.[273]

11 Mitochondrial Enzymes

11.1 Complex I - NADH Ubiquinone Oxidoreductase.- Two related forms of complex I are synthesised in the mitochondria of *Neurospora crassa*. The large piericidin A sensitive form consisting of 23 different nuclear and 6 to 7 mitochondrially encoded subunits was found to contain iron sulfur clusters N-1, N-2, N-3 and N-4 in a concentration equivalent to that of the flavin concentration. The small piericidin A insensitive form has only 23 nuclear encoded subunits and was found to be missing iron sulfur cluster N-2. A model of the electron transfer pathway in the large form of complex I and an evolutionary pathway for the emergence of this form was presented.[274]

Gene disruption in the fungi *N. crassa* and *Aspergillus niger* was used to selectively prevent constitution of the membrane arm and the peripheral arm of complex I, thereby yielding a related enzyme which will potentially provide an insight into the structure and function of complex I.[275] A reevaluation of the EPR spectra from isolated adriamycin-perfused rat hearts has shown that the radical species is a peroxyl radical and not superoxide.[276]

The rotenone sensitive g=2 resonance, attributed to ubiquinone, is only observed in submitochondrial particles during coupled electron transfer from NADH to dioxygen and succinate to NAD⁺ when oligomycin is added. The signal was abolished upon the addition of uncouplers (carbonyl cyanide *m*-chlorophenyl hydrazone or gramicidin D).[277]

11.2 Complex II - Succinate Ubiquinone Oxidoreductase.- A review of electron transfer in succinate ubiquinone reductase and fumarate reductase has been presented by Salerno.[278] Succinate:menaquinone-7 oxidoreductase of the bacterium *B. subtilis* consists of a flavin containing polypeptide, an iron sulfur cluster containing peptide and a membrane spanning cytochrome *b*-558 polypeptide. A comparison of the optical, EPR and redox properties of the two protohemes IX in complex II with those in a purified preparation of cytochrome *b*-558 reveals minor differences, suggesting that the other two polypeptides affect the heme environment in complex II.[279]

Addition of HOCl and the myeloperoxidase-catalysed peroxidation of chloride to *E. coli* and *P. aeruginosa* were found to decrease the signal intensity of the iron sulfur clusters S-1 and S-2 in complex II. However, inhibition of respiration was not caused by the destruction of the iron sulfur clusters but at points in the respiratory chain between the iron sulfur clusters and the ubiquinone reductase site.[280]

11.3 Complex III - Ubiquinol Cytochrome *c* oxidoreductase.- EPR characteristics of cytochrome c_1, cytochromes *b*-565 and *b*-562, the Rieske iron sulfur cluster and an antimycin sensitive ubisemiquinone radical of cytochrome bc_1 complex from *R. sphaeroides* have been measured under a variety of conditions.[281]

The redox centres in a detergent solubilised, three subunit-containing cytochrome bc_1 complex from *R. rubrum* were found to be highly sensitive to the highly specific inhibitors, stigmatellin, myxothiazol, antimycin A and 5-undecyl-6-hydroxy-4,7-dioxobenzothiazole.[282] Zinc(II) was found to reversibly inhibit (by affecting quinol oxidation) the cytochrome *c* reductase activity of cholate solubilised or liposome-reconstituted bovine-heart cytochrome bc_1 complex. In addition to examining the effects of zinc on cytochrome bc_1 in the presence of inhibitors, zinc was found to shift the wavelength maxima of cytochrome *b*-566 and reduce the signal intensity of the Rieske iron sulfur cluster.[283]

Site directed mutagenesis of the *Saccharomyces cerevisiae* Rieske iron sulfur protein gene was used to convert the conserved cysteines-159, -164, -178 and -181 to serine and histidines-161 and -181 to arginine. The fully mutated protein was found to be inactive and lacking the iron sulfur cluster. In contrast when histidine-184 is replaced by arginine the iron sulfur cluster is assembled properly and the yeast can grow on nonfermentable carbon sources.[284] A similar study on this protein from *R. capsulatus* showed that Cys-133, His-135, Cys-153 and His-156 are potential ligands to the [2Fe-2S] cluster. A comparison of the amino acid sequence for this protein with that of dioxygenase suggests that Cys-138 and Cys-155 may form an internal disulfide bond which is important for the structure of the Rieske protein and the quinol oxidation site.[285]

Q-band ENDOR spectra of the ^{14}N and ^{15}N enriched Rieske protein in ubiquinol cytochrome *c* oxidoreductase from *R. capsulatus* identified two distinct histidine ligands, coordinated to the [2Fe-2S]$^{1+}$ cluster.[286] Midpoint potentials (+280 and +325 mV) for the Rieske [2Fe-2S] cluster in complex III were determined in the presence and absence of 5-n-undecyl-6-hydroxy-4,7-dioxobenzothiazole by EPR spectroscopy.[287] ESEEM spectra of the Rieske [2Fe-2S] clusters in the cytochrome bc_1 complex of *Rhodospirillum rubrum*, *R. sphaeroides R-26*, bovine heart mitochondria and the cytochrome $b_6 f$ complex in spinach chloroplasts reveal two classes of nitrogen hyperfine couplings attributed to the coordination of two histidines.[288] Singlet oxygen was shown to be the active agent in the hematoporphyrin promoted inactivation of ubiquinol cytochrome *c* reductase in beef heart mitochondria. The site of photoinactivation was shown to be the Rieske [2Fe-2S] cluster in which singlet oxygen destroyed the two histidine residues coordinated to the cluster.[289]

A proton ENDOR study of the bound Q_c^- ubisemiquinone in complex III has identified strongly coupled protons bound to the quinone oxygens, proton coupling from the quinone ring itself and weakly coupled protons from nearby amino acids.[290] Site directed mutagenesis of this complex was used to generate mutants which were resistant to Q_o site inhibitors. EPR was used to characterise the binding of ubiquinone/ubiquinol to these mutants.[291] The interaction of ubiquinone and the iron sulfur clusters in this complex was examined by monitoring the EPR lineshape of the [2Fe-2S] cluster as a function of the number of Q or QH$_2$ centres present, the different redox states of Q, the presence of inhibitors specific for the Q_o and Q_i sites and site specific mutations (Phe-144 -> Leu, Phe-144 -> Gly) which are known to be important for Q_o.[292] Spin relaxation enhancement by an external probe, holmium ethylenediaminetetracetic acid complexes, demonstrated that the semiquinone anion was 6 to 10 Å from the N side protein establishing that there are two separate quinone binding sites (Q_i and Q_o) within the cytochrome bc_1 complex of R. *capsulatus* on opposite sides of the membrane.[293]

11.4 Complex IV - Cytochrome c oxidase.- B. *subtilis* was found to express two different aa_3 cytochrome oxidases, denoted caa_3-605 and aa_3-600. The former resembles cytochrome c oxidases from PS3 having cytochrome c fused to subunit II and a normal Cu_A site, while the latter exhibits little cytochrome c oxidase activity, catalyses the oxidation of quinols and has no Cu_A EPR signal.[294] The terminal oxidase in the aerobic respiratory system of the hydrocarbonoclastic marine bacterium P. *nautica* 617 was found to contain a high spin and a low spin type b heme. These hemes remained unresolved in optical spectra and throughout potentiometric titrations.[295] Only a- and b-type cytochrome oxidases are found in the electron transport chains in the thermoacidophilic bacterium *Sulfolobus acidocaldarius*. Although the a-type oxidase has an absorption band at 604-605 nm, which is typical of aa_3 type cytochromes, it has a number of distinct differences suggesting the presence of a novel type of cytochrome aa_3 oxidase.[296]

Direct current plasma atomic emission spectrometry of seven preparations of cytochrome c oxidase confirms a stoichiometry of five copper, four iron, two zinc and two magnesium atoms. The three additional copper atoms (Cu_x) could be removed by either monomerisation of the enzyme or depletion of subunit III indicating that Cu_x is bound to subunit III.[297] Cytochrome c oxidase exists in two forms, one of which exhibits slow intramolecular electron transfer, designated the resting form. In the resting form a g'=12 resonance is observed attributed to a binuclear heme a_3 Cu_B centre. A novel g'=2.95 EPR signal was attributed to the EPR active species giving rise to the g'=12 resonance on the basis of their rates of decay, temperature dependence and microwave power saturation.[298,299]

Addition of either sodium formate, formamide, formaldehyde or sodium nitrite to the highly homogeneous preparation of cytochome c oxidase causes the Soret band to shift 4 to 6 nm. However, only formate produces measurable intensity in the g'=12 region, which by analogy with cyanide binding suggests that formate binds to Cu_B.[300] Although the optical absorption spectrum of wheat germ cytochrome c oxidase is blue shifted compared to the eukaryotic enzymes, the EPR spectrum of the Cu_A site was simlar to the maize and beef heart enzymes indicating that the copper environment is not altered.[301] A reductive titration of the cyanide inhibited cytochrome c oxidase from *Paracoccuus denitrificans* shows that the change in the manganese EPR spectrum is associated with the reduction of the Cu_A site. The change reflects a rearrangement of the manganese(II) ions octahedral environment and is indicative of a redox linked conformational transition in the enzyme.[302] The method of time resolved spin probe oximetry has been applied to monitor the oxygen consumption by cytochrome c oxidase[303] and photosynthetic systems.[304]

The cytochrome o complex, a ubiquinol-bo oxidase in an aerobic respiratory pathway of E. coli, was purified and shown to have four subunits, two equivalents of iron and one equivalent of copper. The absence of copper(II) EPR signals suggests that Cu_A is absent in this enzyme; however subunit II, which does not contain Cu_A, was found to be essential for catalytic activity.[305] EPR spectroscopy was used to show that the low and high spin hemes were oriented with their planes perpendicular to the membrane plane.[306] Proton translocation coupled to oxidation of ubiquinol by dioxygen was studied in spheroplasts of two mutant strains of E. coli, one of which expresses cytochrome d but not cytochrome bo and the other expressing only the latter. Dioxygen pulse experiments revealed that cytochrome d catalyses the separation of protons and electrons of ubiquinol oxidation and cytochrome bo acts as a proton pump.[307] A formate complex of this enzyme shows resonances at g'=12 and g'=2.9 which have only previously been observed for the resting and formate complex of cytochrome c oxidase.[308]

12 Photosynthetic Proteins

The photosynthetic electron transfer chain, shown diagrammatically in a previous review article in this series,[309] contains a large number of redox and EPR active centres. A guide describing the spectral features of each signal, the structure of the EPR active species and its involvement in the photosynthetic electron transfer chain has been published.[310]

The reaction centre light harvesting B-875 complex from the purple bacterium *Rhodocyclus gelatinosus* was shown to contain four hemes in approximately equal stoichiometry. Their redox potentials and relative orientation in isolated chromatophore membranes have also been determined.[311] EPR studies of the reaction centre photochemistry in *Heliobacterium chlorum* identified two iron sulfur clusters and two triplet states[312], while the cation and triplet states of *R. capsulatus* reaction centres were examined by a combination of site directed mutagenesis and EPR spectroscopy.[313]

12.1 Photosystem II Reaction Centre (PSII).- A comparison of the EPR spectra of Tyr_D in PSII and model tyrosine radicals which had been deuterated at selected positions on the phenol ring and at the β-methylene carbon showed that Tyr_D was indeed a tyrosine radical. The unpaired electron spin density in both species is essentially identical with high ρ values at the carbons *ortho* and *para* to the phenolic oxygen. The spectral differences betwen Tyr$_D$ and the model radicals were attributed to a variation in the orientation of the β-methylene group.[314] EPR and optical data indicate that the order of susceptibility of PSII components to photodamage when dioxygen evolution is impaired is Chl/Car > Tyr$_Z$ > Tyr$_D$ >> P-680, Pheo, Q$_A$.[315]

A comparison of the EPR properties of the PSII complex consisting of a 47 kDa chlorophyll binding protein (CP47), the reaction centre proteins D1 and D2 and cytochrome *b*-559 with the D1 - D2 - cytochrome *b*-559 complex indicates that retention of the Cp47 protein produces a more native, but quinone depleted PSII reaction centre.[316] Preparations of PSII core particles without bicarbonate have been obtained using the detergents *n*-octyl β-D-glucopyranoside or *n*-heptyl β-D-thioglucoside.[317] Reduction of Tyr$_D$ occurs with concomitant oxidation of cytochrome *b*-559 in PSII of thylakoid membranes stored in the dark at low temperatures.[318]

Saturation recovery EPR spectroscopy was used to determine a distance of greater than or equal to 38 Å between the Tyr$_D$ in PSII and the nonheme iron(II) site.[319] Distances between Tyr$_D$ and the manganese cluster (24-27 Å), P-680 and the manganese cluster (21-26 Å) and Tyr$_D$ and Q$_A$Fe(II) (26-33 Å) were obtained from measurements of the spin lattice relaxation times of the various centres under different conditions.[320] Although the g=1.8 Q$_A$ iron-semiquinone signal formed in formate treated plant PSII membranes is similar to the signal in purple bacteria, the g=1.9 Q$_A$ iron-semiquinone signal formed in plant PSII when bicarbonate remains bound exhibits a unique temperature behaviour.[321] EPR studies of Q$_A$ and Q$_B$ in PSII from *Phormidium laminosum* and their analogues in higher plant PSII reveal an interaction between the Q$_A$ and Q$_B$ iron-semiquinones (Q$_A^-$ - Fe(II) - Q$_B^-$).[322,323] Continuous illumination of dithionite treated *Ectothiorhodospira* sp. reaction centres resulted in the photobleaching of Φ_A producing Φ_A^-. Further illumination produced Φ_A and Φ_B with the concomitant reduction of Φ_A^-.[324] Thermoluminescence studies have provided evidence for light induced oxidation of tyrosine and histidine residues in manganese depleted PSII particles.[325]

The electron conduit, Z, from the manganese cluster to the primary chlorophyll donor has been identified as Tyr-161 within the D1 polypeptide by site directed mutagenesis. In another preparation the Tyr-161 -> Phe revealed a radical whose lineshape was different from that of the Z$^+$ radical, suggesting that this mutant contains a redox active amino acid.[326]

Fluorescence and EPR spectroscopy have been used to characterise the formation of chlorophyll triplet states (Chl⁺ and P-680⁺) in the reaction centre of PSII.[327] The tetrapyrrolic plane of this chlorophyll was tilted at 30° to the plane of the membrane. A similar orientation was found in D1 - D2 - cytochrome b-559 particles indicating that the triplet is confined to the reaction centre. In conjunction with previous optical data it was proposed that P-680 was a single chlorophyll molecule oriented at 30° to the membrane.[328]

Multiple electron flow through PSII and apparently the S states was shown to occur without detectable dioxygen evolution when the samples were treated with lauroylcholine chloride. The source of electrons in this case was shown to be the lipid analog, in which a peroxide type radical was observed.[329]

12.2 Oxygen Evolving Complex (OEC).- The photosynthetic oxidation of water to dioxygen under proton release takes place *via* a sequence of four univalent redox steps in a manganese active site. The structure, redox properties, entry of the substrate water and problems of the stoichiometry of proton release coupled with individual redox steps, have been reviewed.[330] A highly active OEC has been prepared from the transformable cyanobacterium *Synechocystis* 6803 and characterised by EPR spectroscopy.[331,332,333]

Parallel polarisation EPR sensitive to $\Delta M_S=0$ transitions was used to detect an S=1 EPR signal from an exchange coupled manganese centre in the resting state (S_1) of the photosynthetic manganese OEC.[334] EPR spectroscopy showed that inhibition of photosynthetic water oxidation was analogous to the binding of the inhibitor hydrazine to the water oxidising complex followed by photoreversible reduction of manganese (loss of the S_1 -> S_2 reaction).[335] The irreversible inhibition of dioxygen evolution in the S_1 state of PSII by *tris*(hydroxymethyl)aminomethane occurs by the reduction of the manganese complex to a highly reduced form, containing labile manganese(II) ions. This reaction occurs at the chloride sensitive site previously identified from the reversible inhibition of dioxygen evolution by amines.[336]

The g=4 and g=2 multiline EPR signals arising from the manganese OEC in the S_2 state have been studied in oriented PSII membranes[337] and PSII membranes treated with ammonium chloride[338] and ammonia[337]. The observed hyperfine coupling is highly anisotropic, such that when the membrane normal is perpendicular to the applied field no hyperfine structure is observed. Satisfactory simulation of the EPR spectrum was obtained assuming a single exchange coupled tetranuclear manganese cluster with a spin state of S=3/2 or S=5/2.[337] EPR spectra of ¹⁴N and ¹⁵N labelled preparations of the OEC from PSII showed that substructure on the multiline manganese(II) EPR spectrum does not arise from ligand hyperfine to nitrogen. However, ESEEM spectra revealed weak hyperfine coupling from a nitrogen atom.[339] Multifrequency (S-, X- and P-band) EPR spectra of the OEC reveal g-anisotropy in the g=4 resonance which is only consistent with a transition in the middle Kramers doublet of an S=5/2 spin system. Computer simulation yields a zero field splitting, D=0.43 cm^{-1} and E/D = 0.25. Increased values of D and E/D are observed upon the addition of fluoride or ammonia.[340] Microwave power saturation and temperature dependence (4-20K) of the g=2 and 4 resonances are consistent with them arising from S=1/2 and S=3/2 spin states of a total spin S=7/2 spin system.[341] Biochemical and EPR spectroscopic characterisation of the Asp-170 -> Asn mutant D1 polypeptide indicate that this residue is important in the stabilisation or assembly of the manganese cluster.[342]

Previous experiments on manganese depleted PSII have identified a histidine residue which is photooxidised and charge recombination between the oxidised histidine and Q_A^- emits the thermoluminescence A_T-band.[343] Using this band as a marker for oxidised histidine and the EPR signals II$_f$ and II$_s$ for tyrosine oxidation, the redox active histidine was found to be essential for the photooxidation of coordinated manganese. A scheme for electron transfer on the donor side of PSII involving histidine oxidation via Z$^+$ was proposed.[344] Trypsin digestion experiments showed that this histidine was localised in a domain which provided ligands to the manganese cluster.[345]

The effect of extracting weakly bound calcium(II) by low pH treatment of the OEC in PSII membranes was examined by EPR and XANES spectroscopy. The results indicated that removal of calcium induces both structural and functional modifications.[346] Washing PSII with sodium chloride and ethylenediaminetetraacetic acid also produced these changes which could be reversed by the addition of calcium.[347] In addition to a modified multiline manganese S$_2$ state EPR signal which has at least twenty six resonances with a smaller hyperfine coupling[348], a new radical intermediate (S$_3$) was observed that was conformationally coupled to the manganese cluster.[349] A description of the factors responsible for the development of this stable and modified S$_2$ signal and an S$_3$ signal has been discussed by Boussac et al.[350] Originally the S$_3$ EPR active species was assigned to an oxidised histidine magnetically coupled to the manganese cluster on the basis of its optical properties.[351] However, there is some debate in the literature as to whether this is an oxidised histidine or Tyr$_D^+$.[352,353] Inhibition of Tyr$_Z$ photooxidation by the depletion of calcium and chloride was found to occur after the formation of S$_3$ and S$_2$-Tyr-Z$^+$.[354] Treatment of calcium depleted PSII membranes with ammonia and continuous light produces an asymmetric ($\Delta B_{1/2}$=10 mT)

EPR signal whose origin was thought to arise from a perturbed S_3 state.[355]

Lanthanides were shown to compete with calcium for the binding sites on the oxidising side of PSII. Low temperature EPR spectra of both diamagnetic and paramagnetic lanthanides revealed a close interaction between the lanthanides and Tyr_D.[356] The ability of calcium(II) to retard the rate of NH_2OH inhibition (reduction of the manganese cluster) arises from its ability to stabilise the ligation environment of the manganese cluster.[357] New evidence from EPR spectroscopy on the chloride requirement for photosynthetic dioxygen evolution has indicated that chloride facilitates oxidation of the manganese cluster by the PSII Tyr-Z^+ radical.[358]

ENDOR and pulsed field swept EPR of the bis(μ-oxo-)-Mn(III)-Mn(IV) complexes of 1,4,8,11-tetraazacyclotetradecane and $tris$(2-methyl(pyridyl))amine reveal nitrogen hyperfine couplings attributable to a Fermi contact interaction with the nitrogen 2s orbital from the nitrogens axially coordinated to the manganese(III) ions.[359]

12.3 The Cytochrome b_6 - f Complex.- The Rieske [2Fe-2S] cluster in B. PS3 a gram positive thermophilic eubacterium, was found to have a midpoint potential of +165±10 mV and an orthorhombic EPR spectrum which was similar to other centres involved in electron transport using a menaquinone pool but different to other Rieske centres using ubiquinone or plastoquinone.[360] Under nonsaturating conditions the Rieske centre was found to have an identical orientation to its cytochrome bc_1 counterpart. Saturating the EPR signal caused drastic changes to the orientation of the cluster.[361]

12.4 Photosystem I Reaction Centre (PSI).- Electron spin polarised (ESP) EPR studies of spinach PSI particles depleted of vitamin K_1 and following reconstitution by quinone support the hypothesis that the ESP EPR signals arise from a radical pair consisting of the oxidised PSI primary donor, P-700$^+$ and reduced vitamin K_1.[362] Native PSI samples and preparations depleted in the A_1-acceptor site, reconstituted with quinones, vitamin K_1, duroquinone and perdeuterated duroquinone and napthoquinone showed that only the vitamin K_1 derivative produced sequential electron transfer. Consequently, the native A_1-acceptor is vitamin K_1.[363] The ESP EPR signal was absent in PSI samples depleted of the A_1-acceptor site. Restoration of the ESP EPR signal required an acceptor with a potential greater than -750 mV (vitamin K_1, -710 mV), two or more aromatic rings (ie. a napthoquinone) or a benzoquinone with a substituted alkyl tail.[364] Under highly reducing conditions flash absorption spectroscopy identified two kinetic phases corresponding to a recombination reaction between P-700$^+$ and A_1^- forming ^{3}P-700 and the decay of this triplet.[365]

Optical, EPR and electrophoretic data indicate that although PsaC protein can bind loosely to the PSI core protein of *Synechococcus* sp. PCC 6301, the presence of PsaD leads to a stable isolatable PSI complex which is indistinguishable from the native complex.[366] Site directed mutagenesis was used to produce a complementary pair of mutant PsaC proteins in which specific cysteine ligands of the [4Fe-4S] clusters (F_A and F_B) were changed to aspartic acid. In the mutant in which F_B is intact F_A is converted to a [3Fe-4S] cluster while in the other mutant protein F_A is intact and F_B is converted into a cluster whose identity is unknown.[367] A low temperature single crystal EPR study of PSI has determined the anisotropic g matrices and their orientation with respect to the crystallographic axes.[368] EPR spectra of the iron sulfur clusters F_A and F_B in spinach and *Synechococcus* sp. 6301 PSI complexes are slightly different. Reconstitution of the holoproteins, isolated from thylakoid membranes by solvent extraction, was performed by incubating the holoprotein with iron(III) chloride and sodium sulfide in the presence of β-mercaptoethanol for twelve hours under anaerobic conditions. The reconstituted proteins displayed similar EPR spectra to the native PSI complexes, while a cross species spinach iron sulfur protein - *Synechococcus* core protein complex yielded an EPR spectrum similar to the spinach complex.[369]

A photosynthetic reaction centre complex prepared from the green sulfur bacterium *Chlorobium* was shown to have similar charge separation properties to intact membranes and to contain the F_A and F_B iron sulfur clusters.[370] The [4Fe-4S] clusters F_A and F_B have been shown to be essential components in the reduction of $NADP^+$ by the PSI complex.[371] Reconstitution of the iron sulfur clusters in this reaction centre restored both of the recombination reactions which allowed a definitive assignment of the iron sulfur clusters. The two kinetic phases were associated with back reactions between $P-840^+$ and the iron sulfur centre F_X and $F_A F_B$.[372,373]

Acknowledgement.- I would like to thank Sue Sumpter for her assistance in typing this review and my wife Lyn whose encouragement, patience and proofreading skills were invaluable during the writing of this review.

References

1   H. Beinert, *J. Inorg. Biochem.*, 1991, **44**, 173.
2   *J. Inorg. Biochem.*, 1992, 47.
3   *Pure and Appl. Chem.*, 1992, **64** (6), 781.
4   *J. Inorg. Biochem.*, 1991, **43**, (2-3), 75.
5   Nomenclature Committe of the International Union of Biochemistry, *FEBS Lett.*, 1991, **200**, 599.

6   G. Malmstrom, *Biol. Metals*, 1990, 3 64.
7   G.W. Canters, A. Lommen, M. van de Kamp and C.W.G., Hoitink, *Biol. Metals*, 1990, **3**, 67.
8   M.M. Werst, C.E. Davoust and B.M. Hoffman, *J. Am. Chem. Soc.*, 1991, **113**, 1533.
9   D.J.R. Brook, R.C. Haltiwanger and T.H. Koch, *J. Amer. Chem. Soc.*, 1991, **113**, 5910.
10 R. Miyamoto, Y. Ohba and M. Iwaizumi, *Inorg. Chem.*, 1990, **29**, 3234.
11 C.W.G. Hoitink and G.W. Canters, *J. Biol. Chem.*, 1992, **267**, 13836.
12 R. Guzzi and L. Sportelli, *J. Inorg. Biochem.*, 1992, **45**, 39.
13 A.M. Nerissian, V.Z. Melkonyan and R.M. Nalbandyan, *Biochim. Biophys. Acta*, 1991, **1076**, 337.
14 J.D. McManus, D.C. Brune, J. Han, J. Sanders-Loehr, T.E. Meyer, M.A. Cusanovich, G. Tollin and R.E. Blankenship, *J. Biol. Chem.*, 1992, **267**, 6531.
15 M.L. Brader, D. Borchardt and M.F. Dunn, *J. Am. Chem. Soc.*, 1992, **114**, 4480.
16 G.T. Babcock, M.K. El-Deeb, P.O. Sandusky, M.M. Whittaker and J.W. Whittaker, *J. Am. Chem. Soc.*, 1992, **114**, 3727.
17 K. Clark, J.E. Penner-Hahn, M.M. Whittaker and J.W. Whittaker, *J. Am. Chem. Soc.*, 1990, **112**, 6433.
18 M.M. Whittaker and J.W. Whittaker, *J. Biol. Chem.*, 1990, **265**, 9610.
19 L. Banci, I. Bertini, D. Cabelli, R.A. Hallewell, C. Luchinat and M.S. Viezzoli, *Inorg. Chem.*, 1990, **29**, 2398.
20 Y. Lu, E.B. Gralla, J.A. Roe and J.S. Valentine, *J. Am. Chem. Soc.*, 1992, **114**, 3560.
21 M. Paci, A. Desideri, M. Sette and G. Rotilio, *Arch. Biochem. Biophys.*, 1991, **286**, 222.
22 M.R. Ciriolo, A. Desideri, M. Paci and G. Rotilio, *J. Biol. Chem.*, 1990, **265**, 11030.
23 T. Schechinger, W. Hiller, C. Maichle, J. Strähle and U. Weser, *Biol. Metals*, 1988, **1**, 112.
24 F. Jiang, J. McCracken and J. Peisach, *J. Am. Chem. Soc.*, 1990, **112**, 9035.
25 M.J. Colaneri, J.A. Potenza, H.J. Schugar and J. Peisach, *J. Am. Chem. Soc.*, 1990, **112**, 9451.
26 B. Kalyanaraman, W.E. Antholine and S. Parthasarathy, *Biochim. Biophys. Acta*, 1990, **1035**, 286.
27 A. Pezeshk and V. Pezeshk, *J. Inorg. Biochem.*, 1991, **42**, 267.
28 W. Korytowski and T. Sarna, *J. Biol. Chem.*, 1990, **265**, 12410.
29 H. Sakurai, A. Fukudome, R. Tawa, M. Kito, S. Takeshima, M. Kimura, N. Otaki, K. Nakajima, T. Hagino, K. Kawano, S. Hirai and S. Suzuki, *Biochem. Biophys. Res. Comm.*, 1992, **184**, 1393.
30 M. Chikira, T. Sato, W.E. Antholine and D.H. Petering, *J. Biol. Chem.*, 1991, **266**, 2859.
31 B. Mondovi and P. Riccio, *Biol. Metals*, 1990, **3**, 110.
32 J.Z. Pedersen, S. El-Sherbini, A. Finazzi-Agrò and G. Rotilio, *Biochemistry*, 1992, **31**, 8.
33 J. McCracken, J. Peisach, C.E. Cote, M.A. McGuirl and D.M. Dooley, *J. Am. Chem. Soc.*, 1992, **114**, 3715.
34 S.N. Gacheru, P.C. Trackman, M.A. Shah, C.Y. O'Gara, P. Spacciapoli, F.T. Greenway and H.M. Kagen, *J. Biol. Chem.*, 1990, **265**, 19022.
35 L. Casella, M. Gullotti, G. Pallanza, A. Pintar and A. Marchesini, *Biol. Metals*, 1991, **4**, 81.
36 J.L. Cole, L. Avigliano, L. Morpurgo and E.I. Solomon, *J. Amer. Chem. Soc.*, 1991, **113**, 9080.
37 M.T. Graziani, P. Loreti, L. Morpurgo, I. Savini, L. Avigliano, *Inorg. Chim. Acta*, 1990, **173**, 261.
38 T. Sakurai, *Inorg. Chim. Acta*, 1992, **195**, 255.
39 T. Sakurai, *Biochem. J.*, 1992, **284**, 681.
40 J.-B. Li, D.R. McMillin and W.E. Antholine, *J. Am. Chem. Soc.*, 1992, **114**, 725.
41 J.L. Cole, P.A. Clark and E.I. Solomon, *J. Am. Chem. Soc.*, 1990, **112**, 9534.
42 K.A. Meadows, M.M. Morie-Bebel and D.R. McMillin, *J. Inorg. Biochem.*, 1991, **41**, 253.
43 T. Sakurai and S. Suzuki, *J. Inorg. Biochem.*, 1990, **40**, 197.
44 J. Lu, C.J. Bender, J. McCracken, J. Peisach, J.C. Severns and D.R. McMillin, *Biochemistry*, 1992, **31**, 6265.
45 J.C. Severns and D.R. McMillin, *Biochemistry*, 1990, **29**, 8592.
46 T. Sakurai, *J. Inorg. Biochem.*; 1991, **41**, 277.
47 E. Karhunen, M.-L. Niku-Paavola, L. Viikari, T. Haltia, R.A. van der Meer and J.A. Duine, *FEBS Lett.*, 1990, **267**, 6.
48 P.M.H. Kroneck, J. Riester, W.G. Zumft and W.E. Antholine, *Biol. Metals*, 1990, **3**, 103.
49 D.M. Dooley, J.A. Landin, A.C. Rosenzweig, W.G. Zumft and E.P. Day, *J. Am. Chem. Soc.*, 1991, **113**, 8978.
50 J.A. Farrar, A.J. Thomson, M.R. Cheesman, D.M. Dooley and W.G. Zumft, *FEBS Lett.*, 1991, **294**, 11.
51 C.K. SooHoo, T.C. Hollocher, A.F. Kolodziej, W.H. Orme-Johnson and G. Bunker, *J. Biol. Chem.*, 1991, **266**, 2210.
52 C.-S. Zhang, T.C. Hollocher, A.F. Kolodziej and W.H. Orme-Johnson, *J. Biol. Chem.*, 1991, **266**, 2199.
53 G. Musci, M. Carbonaro, A. Adriani, A. Lania, A. Galtieri and L. Calabrese, *Arch. Biochem. Biophys.*, 1990, **279**, 8.
54 V.V. Rylkov, M.Y. Tarasiev and K.A. Moshkov, *Eur. J. Biochem.*, 1991, **197**, 185.
55 G. Musci, S.D. Marco, M.C. Bonaccorsi di Patti and L. Calabrese, *Biochemistry*, 1991, **30**, 9866.

56 J.-P. Tahon, G. Maes, C. Vinckier, R. Witters, T. Zeegers-Huyskens, M. De Ley and R. Lontie, *Biochem. J.*, 1990, **271**, 779.
57 M.L. Brader, D. Borchardt and M.F. Dunn, *Biochemistry*, 1992, **31**, 4691.
58 M.J. Nelson, H. Jin, I.M. Turner, Jr., G. Grove, R.C. Scarrow, B.A. Brennan and L. Que, Jr., *J. Am. Chem. Soc.*, 1991, **113**, 7072.
59 A. Martínez, K.K. Andersson, J. Haavik and T. Flatmark, *Eur. J. Biochem.*, 1991, **198**, 675.
60 K.K. Andersson, C. Vassort, B.A. Brennan, L. Que, Jr., J. Haavik, T. Flatmark, F. Gros and J. Thibault, *Biochem. J.*, 1992, **284**, 687.
61 D. Barra, M.E. Schinina, F. Bossa, K. Puget, P. Durosay, A. Guissani and A.M. Michelson, *J. Biol. Chem.*, 1990, **265**, 17680.
62 J.B. Broderick and T.V. O'Halloran, *Biochemistry*, 1991, **30**, 7349.
63 Y. Zhang, M.S. Gebhard, and E.I. Solomon, *J. Am. Chem. Soc.*, 1991, **113**, 5162.
64 M.O. Funk, Jr., R.T. Carroll, J.F. Thompson, R.H. Sands and W.R. Dunham, *J. Am. Chem. Soc.*, 1990, **112**, 5375.
65 M.J. Nelson, S.P. Seitz and R.A. Cowling, *Biochemistry*, 1990, **29**, 6897.
66 P.M.H. Kroneck, C. Cucurou, V. Ullrich, N. Ueda, H. Suzuki, T. Yoshimoto, S. Matsuda and S. Yamamoto, *FEBS Lett.*, 1991, **287**, 105.
67 S. Ramachandran, R.T. Carroll, W.R. Dunham and M.O. Funk, Jr., *Biochemistry*, 1992, **31**, 7700.
68 F. Jiang, J. Peisach, L.-J. Ming, L. Que Jr. and V.J. Chen, *Biochemistry*, 1991, **30**, 11437.
69 L.-J. Ming, L. Que Jr., A. Kriauciunas, C.A. Frolik and V.J. Chen, *Biochemistry*, 1991, **30**, 11653.
70 A.M. Orville, V.J. Chen, A. Kriauciunas, M.R. Harpel, B.G. Fox, E. Münck and J.D. Lipscomb, *Biochemistry*, 1992, **31**, 4602.
71 J.M. Bollinger, Jr., J. Stubbe, B.H. Huynh and D.E. Edmondson, *J. Am. Chem. Soc.*, 1991, **113**, 6289.
72 M. Sahlin, B.-M. Sjöberg, G. Backes, T. Loehr and J. Sanders-Loehr, *Biochem. Biophys. Res. Comm.*, 1990, **167**, 813.
73 M.P. Hendrich, T.E. Elgren, L. Que, Jr., *Biochem. Biophys. Res. Comm.*, 1991, **176**, 705.
74 C. Gerez and M. Fontecave, *Biochemistry*, 1992, **31**, 780.
75 G.J. Mann, A. Gräslund, E.-I. Ochiai, R. Ingemarson and L. Thelander, *Biochemistry*, 1991, **30**, 1939.
76 J. Narasimhan, W.E. Antholine, C.R. Chitambar and D.H. Petering, *Arch. Biochem. Biophys.*, 1991, **289**, 393.
77 M.L. Howell, J. Sanders-Loehr, T.M. Loehr, N.A. Roseman, C.K. Mathews and M.B. Slabaugh, *J. Biol. Chem.*, 1992, **267**, 1705.
78 T.R. Holman, C. Juarez-Garcia, M.P. Hendrich, L. Que, Jr., and E. Münck, *J. Am. Chem. Soc.*, 1990, **112**, 7611.
79 A.S Borovik, M.P. Hendrich, T.R. Holman, E. Münck, V. Papaefthymiou, and L. Que, Jr., *J. Am. Chem. Soc.*, 1990, **112**, 6031.
80 J.G. DeWitt, J.G. Bentsen, A.C. Rosenzweig, B. Hedman, J. Green, S. Pilkington, G.C. Papaefthymiou, H. Dalton, K.O. Hodgson and S.J. Lippard, *J. Am. Chem. Soc.*, 1991, **113**, 9219.
81 K.E. Liu and S.J. Lippard, *J. Biol. Chem.*, 1991, **266**, 12836.
82 M.P. Hendrich, B.G. Fox, K.K. Andersson, P.G. Debrunner and J.D. Lipscomb, *J. Biol. Chem.*, 1992, **267**, 261.
83 M.P. Hendrich, E. Münck, B.G. Fox and J.D. Lipscomb, *J. Am. Chem. Soc.*, 1990, **112** 5861.
84 B.G. Fox, Y. Liu, J.E. Dege and J.D. Lipscomb, *J. Biol. Chem.*, 1991, **266**, 540.
85 J.M. McCormick, R.C. Reem and E.I. Solomon, *J. Am. Chem. Soc.*, 1991, **113**, 9066.
86 M.P. Hendrich, L.L. Pearce, L. Que, Jr., N. D. Chasteen and E.P. Day, *J. Am. Chem. Soc.*, 1991, **113**, 3039.
87 R.C. Long, J.-H. Zhang, D.M. Kurtz, Jr., A. Negri, G. Tedeschi and F. Bonomi, *Biochim. Biophys. Acta*, 1992, **1122**, 136.
88 S.S. David and L. Que, Jr., *J. Am. Chem. Soc.*, 1990, **112**, 6455.
89 M. Dietrich, D. Münstermann, H. Suerbaum and H. Witzel, *Eur. J. Biochem.*, 1991, **199**, 105.
90 J.B. Vincent, M.W. Crowder and B.A. Averill, *Biochemistry*, 1991, **30**, 3025.
91 P.M. Hanna, Y. Chen and N.D. Chasteen, *J. Biol. Chem.*, 1991, **266**, 886.
92 G.J. Gerfen, P.M. Hanna, N. D. Chasteen, and D.J. Singel, *J. Am. Chem. Soc.*, 1991, **113**, 9513.
93 P.M. Hanna, N.D. Chasteen, G.A. Rottman and P. Aisen, *Biochemistry*, 1991, **30**, 9210.
94 E.V. Mielczarek, S.C. Andrews and R. Bauminger, *Biol. Metals*, 1992, **5**, 87.
95 M.S. Shongwe, C.A. Smith, E.W. Ainscough, H.M. Baker, A.M. Brodie and E.N. Baker, *Biochemistry*, 1992, **31**, 4451.
96 P.A. Snetsinger, N. D. Chasteen, and H. van Willigen, *J. Am. Chem. Soc.*, 1990, **112** 8155.
97 R.C. Garrett, R.W. Evans, S.S. Hasnain, P.F. Lindley, R. Sarra, *Biochem. J.*, 1991, **280**, 151.
98 K. Fukuzawa, T. Fujii and K. Mukai, *Arch. Biochem. Biophys.*, 1991, **290**, 489.
99 B.F. Matzanke, E. Bill, C. Butzlaff, A.X. Trautwein, H. Winkler, C. Hermes, H.-F. Nolting, R. Barbieri and U. Russo, *Eur. J. Biochem.*, 1992, **207**, 747.

100	L.A. Alagna, T. Prosperi, A.G. Tomlinson and H. Kjøsen, *Biochim. Biophys. Acta*, 1990, **1036**, 71.
101	P.M. Gadsby and A.J. Thomson, *J. Am. Chem. Soc.*, 1990, **112**, 5003.
102	Q. Peng, R. Timkovich, P.C. Loewen and J. Peterson, *FEBS Lett.*, 1992, **309**, 157.
103	T.G. Traylor and V.S. Sharma, *Biochemistry*, 1992, **31**, 2847.
104	H. Kosaka, M. Watanabe, H. Yoshihara, N. Harada and T. Shiga, *Biochem. Biophys. Res. Comm.*, 1992, **184**, 1119.
105	J.R. Lancaster, Jr., J.M. Langrehr, H.A. Bergonia, N. Murase, R.L. Simmons and R.A. Hoffman, *J. Biol. Chem.*, 1992, **267**, 10994.
106	A. Mülsch, P. Mordvintcev, A.F. Vanin and R. Busse, *FEBS Lett.*, 1991, **294**, 252.
107	H. Hori, M. Ikeda-Saito, G. Lang and T. Yonetani, *J. Biol. Chem.*, 1990, **265**, 15028.
108	S. Ozawa, H. Fuji and I. Morishima, *J. Am. Chem. Soc.*, 1992, **114**, 1548.
109	M. Ikeda-Saito, R.S. Lutz, D.A. Shelley, E.J. McKelvey, R. Mattera and H. Hori, *J. Biol. Chem.*, 1991, **266**, 23641.
110	K.D. Egeberg, B.A. Springer, S.A. Martinis, S.G. Sligar, D. Morikis and P.M. Champion, *Biochemistry*, 1990, **29**, 9783.
111	H.C. Lee, M. Ikeda-Saito, T. Yonetani, R.S. Magliozzo and J. Peisach, *Biochemistry*, 1992, **31**, 7274.
112	R.S. Magliozzo and J. Peisach, *Biochemistry*, 1992, **31**, 189.
113	J. Jinno, M. Shigematsu, K. Tajima, H. Sakurai, H. Ohya-Nishiguchi and K. Ishizu, *Biochem. Biophys. Res. Comm.*, 1991, **176**, 675.
114	G. Chottard and N. El Ajouz and G. Herve, *Biochim. Biophys. Acta*, 1992, **1122**, 113.
115	P. Ascenzi, M. Coletta, A. Desideri, R. Petruzzelli, F. Polizio, M. Bolognesi, S.G. Condò and B. Giardina, *J. Inorg. Biochem.*, 1992, **45**, 31.
116	P. Ascenzi, M. Coletta, A. Desideri, F. Polizio, S.G. Condò and B. Giardina, *J. Inorg. Biochem.*, 1990, **40**, 157.
117	T.J. DiFeo, A.W. Addison and J.J. Stephanos, *Biochem. J.*, 1990, **269**., 739.
118	A. Levy, P. Kuppusamy and J.M. Rifkind, *Biochemistry*, 1990, **29**, 9311.
119	L.M. Neto, M. Tabak and O.R. Nascimento, *J. Inorg. Biochem.*, 1990, **40**, 309.
120	J.J. Stephanos and A.W. Addison, *Eur. J. Biochem.*, 1990, **189**, 185.
121	D.W. Kraus, J.B. Wittenberg, L. Jing-Fen and J. Peisach, *J. Biol. Chem.*, 1990, **265**, 16054.
122	J.J. Stephanos and A.W. Addison, *J. Inorg. Biochem.*, 1990, **39**, 351.
123	P.M.H. Kroneck, W. Jakob, D.A. Webster and R. DeMaio, *Biol. Metals*, 1991, **4**, 119.
124	A. Wilks and P.R. Ortiz de Montellano, *J. Biol. Chem.*, 1992, **267**, 8827.
125	M.J. Davies and A. Puppo, *Biochem. J.*, 1992, **281**, 197.
126	E. Bill, X.-Q. Ding, E.L. Bominaar, A.X. Trautwein, H. Winkler, D. Mandon, R. Weiss, A. Gold, K. Jayaraj, W.E. Hatfield and M.L. Kirk, *Eur. J. Biochem.*, 1990, **188**, 665.
127	L.A. Fishel, M.F. Farnum, J.M. Mauro, M.A. Miller, J. Kraut, Y. Liu, X.-L. Tan and C.P. Scholes, *Biochemistry*, 1991, **30**, 1986.
128	M.A. Miller, D. Bandyopadhyay, J.M. Mauro, T.G. Traylor and J. Kraut, *Biochemistry*, 1992, **31**, 2789.
129	M. Ikeda-Saito and S. Kimura, *Arch. Biochem. Biophys.*, 1990, **283**, 351.
130	K. Ryu and J.S. Dordick, *Biochemistry*, 1992, **31**, 2588.
131	R. Wever, W.A. Oertling, H. Hoogland, B.G.J.M. Bolscher, Y. Kim and G.T. Babcock, *J. Biol. Chem.*, 1991, **266**, 24308.
132	H.C. Lee, K.S. Booth, W.S. Caughey and M. Ikeda-Saito, *Biochim. Biophys. Acta*, 1991, **1076**, 317.
133	C.E. Cooper and E. Odell, *Biochem. Soc. Trans.*, 1992, **20**, 108S.
134	C. Larroque, R. Lange, L. Maurin, A. Bienvenue and J.E. van Lier, *Arch. Biochem. Biophys.*, 1990, **282**, 198.
135	R. Davydov, R. Kappl, J. Hüttermann and J.A. Peterson, *FEBS Lett.*, 1991, **295**, 113.
136	M. Numata, T. Saito, T. Yamazaki, Y. Fukumori and T. Yamanaka, *J. Biochem.*, 1990, **108**, 1016.
137	M. Tsubaki, Y. Ichikawa, Y. Fujimoto, N.-T. Yu and H. Hori, *Biochemistry*, 1990, **29**, 8805.
138	K.K. Surerus, W.A. Oertling, C. Fan, R.J. Gurbiel, O. Einarsdòttir, W.E. Antholine, R.B. Dyer, B.M. Hoffman, W.H. Woodruff and J.A. Fee, *Proc. Natl. Acad. Sci. USA*, 1992, **89**, 3195.
139	L.M. Saraiva, G. Denariaz, M.-Y. Liu, W.J. Payne, J. LeGall and I. Moura, *Eur. J. Biochem.*, 1992, **204**, 1131.
140	H. Fridén, M.R. Cheesman, L. Hederstedt, K.K. Andersson and A.J. Thomson, *Biochim. Biophys. Acta*, 1990, **1041**, 207.
141	I. Ueno, S. Fujii, H. Ohya-Nishiguchi, T. Iizuka and S. Kanegasaki, *FEBS Lett.*, 1991, **281**, 130.
142	D.Sh. Burbaev, I.A. Moroz, Yu.A. Kamenskiy and A.A. Konstantinov, *FEBS Lett.*, 1991, **283**, 97.
143	M. Bruschi, P. Bertrand, C. More, G. Leroy, J. Bonicel, J. Haladjian, G. Chottard, W.B.R. Pollock and G. Voordouw, *Biochemistry*, 1992, **31**, 3281.
144	T. Yoshimura, S. Suzuki, T. Kohzuma, H. Iwasaki and S. Shidara, *Biochem. Biophys. Res. Comm.*, 1990, **169**, 1235.
145	P. van Vliet, D. Zannoni, W. Nitschke and A.W. Rutherford, *Eur. J. Biochem.*, 1991, **199**, 317.

146 R.S. Blackmore, P.M.A. Gadsby, C. Greenwood and A.J. Thomson, *FEBS Lett.*, 1990, **264**, 257.
147 C.C. Costa, J.J.G. Moura, I. Moura, M.Y. Liu, H.D. Peck, Jr., J. LeGall, Y. Wang and B.H. Huynh, *J. Biol. Chem.*, 1990, **265**, 14382.
148 J. Tan and J.A. Cowan, *Biochemistry*, 1990, **29**, 4886.
149 M.C. Silvestrini, M.G. Tordi, G. Musci and M. Brunori, *J. Biol. Chem.*, 1990, **265**, 11783.
150 F.H.A. Kadir, M.T. Wilson, J. Al-Basseet and A.J. Thomson, *J. Inorg. Biochem.*, 1991, **43**, 9.
151 W.L. Smith, T.E. Eling, R.J. Kulmacz, L.J. Marnett and A.-L. Tsai, *Biochemistry*, 1992, **31**, 3.
152 H.-J. Scherer, R. Karthein, S. Strieder and H.H. Ruf, *Eur. J. Biochem.*, 1992, **205**, 751.
153 R.J. Kulmacz, Y. Ren, A.-L. Tsai and G. Palmer, *Biochemistry*, 1990, **29**, 8760.
154 G. Lassmann, R. Odenwaller, J.F. Curtis, J.A. DeGray, R.P. Mason, L.J. Marnett and T.E. Eling, *J. Biol. Chem.*, 1991, **266**, 20045.
155 J.H. Dawson, A.M. Bracete, A.M. Huff, S. Kadkhodayan, C.M. Zeitler, M. Sono, C.K. Chang and P.C. Loewen, *FEBS Lett.*, 1991, **295**, 123.
156 R. Cammack, *Adv. Inorg. Chem.*, 1992, **38**, 281.
157 A.J. Thomson, J. Breton, S.J. George, J.N. Butt, F.A. Armstrong, E.C. Hatchikian, *Biochem. Soc. Trans.*, 1991, **19**, 594.
158 I. Moura, M. Teixeira, J. LeGall and J.J.G. Moura, *J. Inorg. Biochem*, 1991, **44**, 127.
159 I. Moura, P. Tavares, J.J.G. Moura, N. Ravi, B.H. Huynh, M.-Y. Liu and J. LeGall, *J. Biol. Chem.*, 1990, **265**, 21596.
160 D.T. Ta and L.E. Vickery, *J. Biol. Chem.*, 1992, **267**, 11120.
161 C. Fan, M.C. Kennedy, H. Beinert and B.M. Hoffman, *J. Am. Chem. Soc.*, 1992, **114**, 374.
162 A.L.P. Houseman, B.-H. Oh, M.C. Kennedy, C. Fan, M.M. Werst, H. Beinert, J.L. Markley and B.M. Hoffman, *Biochemistry*, 1992, **31**, 2073.
163 J.N. Butt, F.A. Armstrong, J. Breton, S.J. George, A.J. Thomson and E.C. Hatchikian, *J. Am. Chem. Soc.*, 1991, **113**, 6663.
164 J.N. Butt, A. Sucheta, F.A. Armstrong, J. Breton, A.J. Thomson and E.C. Hatchikian, *J. Am. Chem. Soc.*, 1991, **113**, 8948.
165 I. Bertini, F. Briganti and C. Luchinat, *Inorg. Chim. Acta*, 1990, **175**, 9.
166 M.M. Werst, M.C. Kennedy, A.L.P. Houseman, H. Beinert and B.M. Hoffman, *Biochemistry*, 1990, **29**, 10533.
167 M.M. Werst, M.C. Kennedy, H. Beinert and B.M. Hoffman, *Biochemistry*, 1990, **29**, 10526.
168 F. Verniquet, J. Gaillard, M. Neuburger and R. Douce, *Biochem. J.*, 1991, **276**, 643.
169 J.-B. Park, C. Fan, B.M. Hoffman and M.W.W. Adams, *J. Biol. Chem.*, 1991, **266**, 19351.
170 R.C. Conover, A.T. Kowal, W. Fu, J.-B. Park, S. Aono, M.W.W. Adams and M.K. Johnson, *J. Biol. Chem.*, 1990, **265**, 8533.
171 R.C. Conover, J.-B. Park, M.W.W. Adams and M.K. Johnson, *J. Am. Chem. Soc.*, 1991, **113**, 2799.
172 S.E. Iismaa, A.E. Vàzquez, G.M. Jensen, P.J. Stephens, J.N. Butt, F.A. Armstrong and B.K. Burgess, *J. Biol. Chem.*, 1991, **266**, 21563.
173 L. Skjeldal, K.K. Andersson, D. Grace and T. Ljones, *Biol. Metals*, 1989, **2**, 135.
174 W.R. Dunham, W.R. Hagen, J.A. Fee, R.H. Sands, J.B. Dunbar and C. Humblet, *Biochim. Biophys. Acta*, 1991, **1079**, 253.
175 A.J. Pierik, W.R. Hagen, W.R. Dunham and R.H. Sands, *Eur. J. Biochem.*, 1992, **206**, 705.
176 I. Moura, P. Tavares, J.J.G. Moura, N. Ravi, B.H. Huynh, M.-Y. Liu and J. LeGall, *J. Biol. Chem.*, 1992, **267**, 4489.
177 J.R. Mason and R. Cammack, *Ann. Rev. Microbiol.*, 1992, **46**, 277.
178 J. Lampreia, G. Fauque, N. Speich, C. Dahl, I. Moura, H.G. Trüper and J.J.G. Moura, *Biochem. Biophys. Res. Comm.*, 1991, **181**, 342.
179 J. Lampreia, I. Moura, M. Teixeira, H.D. Peck, Jr., J. LeGall, B.H. Huynh and J.J.G. Moura, *Eur. J. Biochem.*, 1990, **188**, 653.
180 Y. Ueda, N. Yumoto, M. Tokushige, K. Fukui and H. Ohya-Nishiguchi, *J. Biochem.*, 1991, **109**, 728.
181 J.K. Shergill, R. Cammack and J.H. Weiner, *J. Chem. Soc. Faraday Trans.*, 1991, **87**, 3199.
182 J.K. Shergill, J.H. Weiner and R. Cammack, *Biochem. Soc. Trans.*, 1991, **19**, 256S.
183 M.T. Werth, G. Cecchini, A. Manodori, B.A.C. Ackrell, I. Schröder, R.P. Gunsalus and M.K. Johnson, *Proc. Natl. Acad. Sci. USA*, 1990, **87**, 8965.
184 M.T. Werth, H. Sices, G. Cecchini, I. Schröder, S. Lasage, R.P. Gunsalus and M.K. Johnson, *FEBS Lett.*, 1992, **299**, 1.
185 A. Manodori, G. Cecchini, I. Schröder, R.P. Gunsalus, M.T. Werth and M.K. Johnson, *Biochemistry*, 1992, **31**, 2703.
186 A. Zöphel, M.C. Kennedy, H. Beinert and P.M.H. Kroneck, *Eur. J. Biochem.*, 1991, **195**, 849.
187 R.J. Rohlfs and R. Hille, *J. Biol. Chem.*, 1991, **266**, 15244.
188 M.A. Vanoni, D.E. Edmondson, G. Zanetti and B. Curti, *Biochemistry*, 1992, **31**, 4613.
189 R.M. Petrovich, F.J. Ruzicka, G.H. Reed and P.A. Frey, *J. Biol. Chem.*, 1991, **266**, 7656.

190 M.D. Ballinger, G.H. Reed and P.A. Frey, *Biochemistry*, **31**, 949.
191 G. Fauque, A.R. Lino, M. Czechowski, L. Kang, D.V. DerVartanian, J.J.G. Moura, J. LeGall and I. Moura, *Biochim. Biophys. Acta*, 1990, **1040**, 112.
192 A.J. Pierik and W.R. Hagen, *Eur. J. Biochem.*, 1991, **195**, 505.
193 M.W.W. Adams, *Biochim. Biophys. Acta*, 1990, **1020**, 115.
194 H. Thomann, M. Bernardo and M.W.W. Adams, *J. Am. Chem. Soc.*, 1991, **113**, 7044.
195 R. Hedderich, S.P.J. Albracht, D. Linder, J. Koch and R.K. Thauer, *FEBS Lett.*, 1992, **298**, 65.
196 A. Juszczak, S. Aono, M.W.W. Adams, *J. Biol. Chem.*, 1991, **266**, 13834.
197 J.-C. Drapier, C. Pellat and Y. Henry, *J. Biol. Chem.*, 1991, **266**, 10162.
198 J.A. Corbett, J.R. Lancaster, Jr., M.A. Sweetland and M.L. McDaniel, *J. Biol. Chem.*, 1991, **266**, 21351.
199 D. Servent, C. Ducrocq, Y. Henry, A. Guissani and M. Lenfant, *Biochim. Biophys. Acta*, 1991, **1074**, 320.
200 R. Cammack, *Catalysis by Nickel in Biological Systems*: Bioinorganic Catalysis, (J. Reeijk, ed.) Marcel Dekker, New York, 1992, p 189.
201 J.P. Whitehead, G.J. Colpas, C.Bagyinka and M.J. Maroney, *J. Am. Chem. Soc.*, 1991, **113**, 6288.
202 M.J. Maroney, G.J. Colpas, C. Bagyinka, N. Baidya and P.K. Mascharak, *J. Am. Chem. Soc.*, 1991, **113**, 3962.
203 C. Fan, M. Teixeira, J. Moura, I. Moura, B.-H. Huynh, J. LeGall, H.D. Peck Jr. and B.M. Hoffman, *J. Am. Chem. Soc.*, 1991, **113**, 20.
204 M. Asso, B. Guigliarelli, T. Yagi and P. Bertrand, *Biochim. Biophys. Acta*, 1992, **1122**, 50.
205 J.M.C.C. Coremans, J.W. van der Zwaan and S.P.J. Albracht, *Biochim. Biophys Acta*, 1992, **1119**, 157.
206 J.W. van der Zwaan, J.M.C.C. Coremans, E.C.M. Bouwens and S.P.J. Albracht, *Biochim. Biophys. Acta*, 1990, **1041**, 101.
207 M.Teixeira, I. Moura, G. Fauque, D.V. DerVartanian, J. LeGall, H.D. Peck, Jr., J.J.G. Moura and B.H. Huynh, *Eur. J. Biochem.*, 1989, **189**, 381.
208 C.-P. Wang, R. Franco, J.J.G. Moura and E.P. Day, *J. Biol. Chem.*, 1992, **267**, 7378.
209 M.S.M. Jetten, W.R. Hagen, A.J. Pierik, A.J.M. Stams and A.J.B. Zehnder, *Eur. J. Biochem.*, 1991, **195**, 385.
210 W. Shin, P.R. Stafford and P.A. Lindahl, *Biochemistry*, 1992, **31**, 6003.
211 J.M.C.C. Coremans, C.J. van Garderen and S.P.J. Albracht, *Biochim. Biophys. Acta*, 1992, **1119**, 148.
212 C. Fan, C.M. Gorst, S.W. Ragsdale and B.M. Hoffman, *Biochemistry*, 1991, **30**, 431.
213 C.M. Gorst and S.W. Ragsdale, *J. Biol. Chem.*, 1991, **266**, 20687.
214 M.S.M. Jetten, A.J. Pierik and W.R. Hagen, *Eur. J. Biochem.*, 1991, **202**, 1291.
215 S. Rospert, R. Böcher, S.P.J. Albracht and R.K. Thauer, *FEBS Lett.*, 1991, **291**, 371.
216 S. Rospert, D. Linder, J. Ellerman and R.K. Thauer, *Eur. J. Biochem.*, 1990, **194**, 871.
217 M.C. Brenner, L. Ma, M.K. Johnson and R.A. Scott, *Biochim. Biophys. Acta*, 1992, **1120**, 160.
218 I. Mus-Veteau, D. Diaz, J. Gracia-Mora, B. Guigliarelli, G. Chottard and M. Bruschi, *Biochim. Biophys. Acta*, 1991, **1060**, 159.
219 B.D. Howes, N.M. Pinhal, N.A. Turner, R.C. Bray, G. Anger, A. Ehrenberg, J.B. Raynor and D.J. Lowe, *Biochemistry*, 1990, **29**, 6120.
220 G.L. Wilson, R.J. Greenwood, J.R. Pilbrow, J.T. Spence and A.G. Wedd, *J. Am. Chem. Soc.*, 1991, **113**, 6803.
221 K.M. Fish, V. Massey, R.H. Sands and W.R. Dunham, *J. Biol. Chem.*, 1990, **265**, 19665.
222 R.K. Hughes, B. Bennett and R.C. Bray, *Biochemistry*, 1992, **31**, 3073.
223 R. Cammack, A. Chapman, J. McCracken and J. Peisach, *J. Chem. Soc. Faraday Trans.*, 1991, **87**, 3203.
224 C.J. Kay, L.P. Solomonson and M.J. Barber, *Biochemistry*, 1990, **29**, 10823.
225 C.J. Kay, L.P. Solomonson and M.J. Barber, *J. Biol. Chem.*, 1986, **261**, 5799.
226 C.J. Kay, M.J. Barber, L.P. Solomonson, D. Kau, A.C. Cannons and C.R. Hipkin, *Biochem. J.*, 1990, **272**, 545.
227 B. Guigliarelli, M. Asso, C. More, V. Augier, F. Blasco, J. Pommier, G. Giordano and P. Bertrand, *Eur. J. Biochem.*, 1992, **207**, 61.
228 R. Cammack and J.H. Weiner, *Biochemistry*, 1990, **29**, 8410.
229 R.A. Rothery and J.H. Weiner, *Biochemistry*, 1991, **30**, 8296.
230 N.R. Bastian, C.J. Kay, M.J. Barber and K.V. Rajagopalan, *J. Biol. Chem.*, 1991, **266**, 45.
231 D.R. Jollie and J.D. Lipscomb, *J. Biol. Chem.*, 1991, **266**, 21853.
232 B.E. Smith and R.R. Eady, *Eur. J. Biochem.*, 1992, **205**, 1.
233 K. Schneider, A. Müller, U. Schramm and W. Klipp, *Eur. J. Biochem.*, 1991, **195**, 653.
234 H. Thomann, M. Bernardo, W.E. Newton and D.R. Dean, *Proc. Natl. Acad. Sci. USA*, 1991, **88**, 6620.
235 M.R. Hyman, L.C. Seefeldt, T.V. Morgan, D.J. Arp and L.E. Mortenson, *Biochemistry*, 1992, **31**, 2947.
236 K. Fisher, D.J. Lowe and R.N.F. Thornely, *Biochem. J.*, 1991, **279**, 81.
237 H.D. May, D.R. Dean and W.E. Newton, *Biochem. J.*, 1992, **277**, 457.
238 T.V. Morgan, J. McCracken, W.H. Orme-Johnson, W.B. Mims, L.E. Mortenson and J. Peisach, *Biochemistry*, 1990, **29**, 3077.

239	A. Müller, K. Schneider, K. Knüttel and W.R. Hagen, *FEBS Lett.*, 1992, **303**, 36.
240	S. Mukund and M.W.W. Adams, *J. Biol. Chem.*, 1990, **265**, 11508.
241	S. Mukund and M.W.W. Adams, *J. Biol. Chem.*, 1991, **266**, 14208.
242	R.A. Schmitz, S.P.J. Albracht and R.K. Thauer, *FEBS Lett.*, 1992, **309**, 78.
243	D. Rehder, *BioMetals*, 1992, **5**, 3.
244	R. Bogumil, J. Hüttermann, R. Kappl, R. Stabler, C. Sudfeldt and H. Witzel, *Eur. J. Biochem.*, 1991, **196**, 305.
245	K.A. Lord and G.H. Reed, *Arch. Biochem. Biophys.*, 1990, **281**, 124.
246	E. Kime-Hunt, K. Spartalian, S. Holmes, M. Mohan and C.J. Carrano, *J. Inorg. Biochem.*, 1991, **41**, 125.
247	M. Branca, G. Micera, A. Dessí and D. Sanna, *J. Inorg. Biochem.*, 1990, **39**, 109.
248	R.J. Keller, J.D. Rush and T.A. Grover, *J. Inorg. Biochem.*, 1991, **41**, 269.
249	M.A. Zoroddu, R.P. Bonomo, A.J. Di Bilio, E. Berardi and M.G.L. Meloni, *J. Inorg. Biochem.*, 1991, **43**, 731.
250	R.P. Ferrari, *Inorg. Chim. Acta*, 1990, **176**, 83.
251	X. Shi and N.S. Dalal, *Arch. Biochem. Biophys.*, 1992, **295**, 70.
252	J.W. Whittaker and M.M. Whittaker, *J. Am. Chem. Soc.*, 1991, **113**, 55288.
253	J. Peterson, J.A. Fee and E.P. Day, *Biochim. Biophys. Acta*, 1991, **1079**, 161.
254	J.D. Rush, Z. Maskos and W.H. Koppenol, *Arch. Biochem. Biophys.*, 1991, **289**, 97.
255	S.V. Khangulov, V.V. Barynin, N.V. Voevodskaya and A.I. Grebenko, *Biochim. Biophys. Acta*, 1990, **1020**, 305.
256	S.V. Khangulov, V.V. Barynin and S.V. Antonyuk-Barynina, *Biochim. Biophys. Acta*, 1990, **1020**, 25.
257	J. McCraken, J. Peisach, L. Bhattacharyya and F. Brewer, *Biochemistry*, 1991, **30**, 4486.
258	D.G. Latwesen, M. Poe, J.S. Leigh and G.H. Reed, *Biochemistry*, 1992, **31**, 4946.
259	F. Galiazzo, J.Z. Pedersen, P. Civitareale, A. Schiesser and G. Rotilio, *Biol. Metals*, 1989, **2**, 6.
260	H. Hori, M. Tsubaki, N.-T. Yu and T. Yonetani, *Biochim. Biophys. Acta*, 1991, **1077**, 392.
261	X. Shi and N.S. Dalal, *Arch. Biochem. Biophys.*, 1990, **281**, 90.
262	X. Shi, N.S. Dalal and V. Vallyathan, *Arch. Biochem. Biophys.*, 1991, **290**, 381.
263	X. Shi and N.S. Dalal, *J. Inorg. Biochem.*, 1990, **40**, 1.
264	X. Shi, X. Sun, P.M. Gannett and N.S. Dalal, *Arch. Biochem. Biophys.*, 1992, **293**, 281.
265	G. Formicka, M. Zeppezauer, F. Fey and J. Hüttermann, *FEBS Lett.*, 1992, **309**, 92.
266	M. Good, R. Hollenstein and M. Vašák, *Eur. J. Biochem.*, 1991, **197**, 655.
267	D. Rakshit and M. Vašák, *J. Biol. Chem.*, 1992, **267**, 235.
268	H.R. Kalbitzer and A. Wittinghofer, *Biochim. Biophys. Acta*, 1991, **1078**, 133.
269	H.R. Kalbitzer, J. Feuerstein, R.S. Goody and A. Wittinghofer, *Eur. J. Biochem.*, 1990, **188**, 355.
270	R.R. Poyner and G.H. Reed, *Biochemistry*, 1992, **31**, 7166.
271	J.A. Urbina, X. Ysern and A.S. Mildvan, *Arch. Biochem. Biophys.*, 1990, **278**, 187.
272	H. Han, J.M. Rifkind and A.S. Mildvan, *Biochemistry*, 1991, **30**, 11104.
273	D.J. Weber, A.K. Meeker and A.S. Mildvan, *Biochemistry*, 1991, **30**, 6103.
274	D.-C. Wang, S.W. Meinhardt, U. Sackmann, H. Weiss and T. Ohnishi, *Eur. J. Biochem.*, 1991, **197**, 257.
275	U. Weidner, N. Nehls, R. Schneider, W. Fecke, H. Leif, A. Schmiede, T. Friedrich, R. Zensen, U. Schulte, T. Ohnishi and H. Weiss, *Biochim. Biophys. Acta*, 1992, **1101**, 177.
276	B. Kalyanaraman and J.E. Baker, *Biochem. Biophys. Res. Comm.*, 1990, **169**, 30.
277	A.B. Kotlyar, V.D. Sled, D.Sh. Burbaev, I.A. Moroz and A.D. Vinogradov, *FEBS Lett.*, 1990, **264**, 17.
278	J.C. Salerno, *Biochem. Soc. Trans.*, 1991, **19**, 599.
279	C. Hägerhäll, R. Aasa, C. von Wachenfeldt and L. Hederstedt, *Biochemistry*, 1992, **31**, 7411.
280	J.K. Hurst, W.C. Barrette, Jr., B.R. Michel and H. Rosen, *Eur. J. Biochem.*, 1991, **202**, 1275.
281	J.P. McCurley, T. Miki, L. Yu and C.-A. Yu, *Biochim. Biophys. Acta*, 1990, **1020**, 176.
282	S. Güner, D.E. Robertson, L. Yu, Z.-H. Qiu, C.-A. Yu and D.B. Knaff, *Biochim. Biophys. Acta*, 1991, **1058**, 269.
283	M. Lorusso, T. Cocco, A.M. Sardanelli, M. Minuto, F. Bonomi and S. Papa, *Eur. J. Biochem.*, 1991, **197**, 555.
284	L.A. Graham and B.L. Trumpower, *J. Biol. Chem.*, 1991, **266**, 22485.
285	E. Davidson, T. Ohnishi, E.Atta-Asafo-Adjei and F. Daldal, *Biochemistry*, 1992, **31**, 3342.
286	R.J. Gurbiel, T. Ohnishi, D.E. Robertson, F. Daldal and B.M. Hoffman, *Biochemistry*, 1991, **30**, 11579.
287	K.M. Andrews, A.R. Crofts and B.R. Gennis, *Biochemistry*, 1990, **29**, 2645.
288	R.D. Britt, K. Sauer, M.P. Klein, D.B. Knaff, A. Kriauciunas, C.-A. Yu, L. Yu and R. Malkin, *Biochemistry*, 1991, **30**, 1892.
289	T. Miki, L. Yu and C.-A. Yu, *Biochemistry*, 1991, **30**, 230.
290	J.C. Salerno, M. Osgood, Y. Liu, H. Taylor and C.P. Scholes, *Biochemistry*, 1990, **29**, 6987.
291	D.E. Robertson, F. Daldal and P.L. Dutton, *Biochemistry*, 1990, **29**, 11249.
292	H. Ding, D.E. Robertson, F. Daldal and P.L. Dutton, *Biochemistry*, 1992, **31**, 3144.
293	S.W. Meinhardt and T. Ohnishi, *Biochim. Biophys. Acta*, 1992, **1100**, 67.

294 M. Lauraeus, T. Haltia, M. Saraste and M. Wilkström, *Eur J. Biochem.*, 1991, **197**, 699.
295 S. Arnaud, F. Malatesta, B. Guigliarelli, J.-P. Gayda, P. Bertrand, R. Miraglio and M. Denis, *Eur. J. Biochem.*, 1991, **198**, 349.
296 S. Anemüller and G. Schäfer, *Eur. J. Biochem.*, 1990, **191**, 297.
297 L.P. Pan, Z. Li, R. Larsen and S.I. Chan, *J. Biol. Chem.*, 1991, **266**, 1367.
298 C.E. Cooper and J.C. Salerno, *J. Biol. Chem.*, 1992, **267**, 280.
299 C.E. Cooper, A.J. Moody, P.R. Rich, J.M. Wrigglesworth and N. Ioannidis, *Biochem. Soc. Trans.*, 1991, **19**, 259S.
300 J.R. Schoonover and G. Palmer, *Biochemistry*, 1991, **30**, 7541.
301 W.E. Peiffer, R.T. Ingle and S. Ferguson-Miller, *Biochemistry*, 1990, **29**, 8696.
302 T. Haltia, *Biochim. Biophys. Acta*, 1992, **1098**, 343.
303 J. Jiang, J.F. Bank, W. Zhao and C.P. Scholes, *Biochemistry*, 1992, **31**, 1331.
304 K. Strzalka, T. Walczak, T. Sarna and H.M. Swartz, *Arch. Biochem. Biophys.*, 1990, **281**, 312.
305 K.C. Minghetti, V.C. Goswitz, N.E. Gabriel, J.J. Hill, C.A. Barassi, C.D. Georgiou, S.I. Chan and R.B. Gennis, *Biochemistry*, 1992, **31**, 6917.
306 J.C. Salerno and W.J. Ingledew, *Eur. J. Biochem.*, 1991, **198**, 789.
307 A. Puustinen, M. Finel, T. Haltia, R.B. Gennis and M. Wikström, *Biochemistry*, 1991, **30**, 3936.
308 M.W. Calhoun, R.B. Gennis and J.C. Salerno, *FEBS Lett.*, 1992, **309**, 127.
309 G.R. Hanson in "Specialist Periodical Reports - Electron Spin Resonance", M.C.R. Symons, Ed., Royal Society of Chemistry, 1991, **12B**, p. 129.
310 A.-F. Miller and G.W. Brudvig, *Biochim. Biophys. Acta*, 1991, **1056**, 1.
311 W. Nitschke, I. Agalidis and A.W. Rutherford, *Biochim. Biophys. Acta*, 19992, **1100**, 49.
312 W. Nitschke, P. Sétif, U. Liebl, U. Feiler and A.W. Rutherford, *Biochemistry*, 1990, **29**, 11079.
313 E.J. Bylina, S.V. Kolaczkowski, J.R. Norris and D.C. Youvan, *Biochemistry*, 1990, **29**, 6203.
314 B.A. Barry, M.K. El-Deeb, P.O. Sandusky and G.T. Babcock, *J. Biol. Chem.*, 1990, **265**, 20139.
315 D.J. Blubaugh, M. Atamian, G.T. Babcock, J.H. Golbeck and G.M. Cheniae, *Biochemistry*, 1991, **30**, 7586.
316 J. Petersen, J.P. Dekker, N.R. Bowlby, D.F. Ghanotakis, C.F. Yocum and G.T. Babcock, *Biochemistry*, 1990, **29**, 3226.
317 S.J. Bowden, B.J. Hallahan, S.V. Ruffle, M.C.W. Evans and J.H.A. Nugent, *Biochim. Biophys. Acta*, 1991, **1060**, 89.
318 I. Vass, Z. Deák, C. Jegerschöld and S. Styring, *Biochim. Biophys. Acta*, 1990, **1018**, 41.
319 D.J. Hirsh, W.F. Beck, J.B. Innes and G.W. Brudvig, *Biochemistry*, 1992, **31**, 532.
320 Y. Kodera, K. Takura and A. Kawamori, *Biochemistry*, 1992, **1101**, 23.
321 J.H.A. Nugent, D.C. Doetschman and D.J. Maclachlan, *Biochemistry*, 1992, **31**, 2935.
322 A.R. Corrie, J.H.A. Nugent and M.C.W. Evans, *Biochim. Biophys. Acta*, 1991, **1057**, 384.
323 B.J. Hallahan, S.V. Ruffle, S.J. Bowden and J.H.A. Nugent, *Biochim. Biophys. Acta*, 1991, **1059**, 181.
324 T. Mar and G. Gingras, *Biochim. Biophys. Acta*, 1991, **1056**, 190.
325 S.I. Allakhverdiev, V.V. Klimov and S. Demeter, *FEBS Lett.*, 1992, **297**, 51.
326 G.H. Noren and B.A. Barry, *Biochemistry*, 1992, **31**, 3335.
327 I. Vass and S. Styring, *Biochemistry*, 1992, **31**, 5957.
328 F.J.E. van Mieghem, K. Satoh and A.W. Rutherford, *Biochim. Biophys. Acta*, 1991, **1058**, 379.
329 G. Ananyev, T. Wydrzynski, G. Renger and V. Klimov, *Biochim. Biophys. Acta*, 1992, **1100**, 303.
330 G. Renger and T. Wydrzynski, *Biol. Metals*, 1991, **4**, 73.
331 G.H. Noren, R.J. Boerner and B.A. Barry, *Biochemistry*, 1991, **30**, 3943.
332 D.L. Kirilovsky, A.G.P. Boussac, F.J.E. van Mieghem, J.-M.R.C. Ducruet, P.R. Sétif, J. Yu, W.F.J. Vermaas and A.W. Rutherford, *Biochemistry*, 1992, **31**, 2099.
333 F. Nilsson, K. Gounaris, S. Styring and B. Andersson, *Biochim. Biophys. Acta*, 1992, **1100**, 251.
334 S.L. Dexheimer and M.P. Klein, *J. Am. Chem. Soc.*, 1992, **114**, 2821.
335 J. Tso, V. Petrouleas and G.C. Dismukes, *Biochemistry*, 1990, **29**, 7759.
336 K.W. Rickert, J. Sears, W.F. Beck and G.W. Brudvig, *Biochemistry*, 1991, **30**, 7888.
337 D.H. Kim, R.D. Britt, M.P. Klein and K. Sauer, *Biochemistry*, 1992, **31**, 541.
338 D.H. Kim, R.D. Britt, M.P. Klein and K. Sauer, *J. Am. Chem. Soc.*, 1990, **112**, 9389.
339 V.J. DeRose, V.K. Yachandra, A.E. McDermott, R.D. Britt, K. Sauer and M.P. Klein, *Biochemistry*, 1991, **30**, 1335.
340 A. Haddy, W.R. Dunham, R.H. Sands and R. Aasa, *Biochim. Biophys. Acta*, 1992, **1099**, 25.
341 R.J. Pace, P. Smith, R. Bramley and D. Stehlik, *Biochim. Biophys. Acta*, 1991, **1058**, 161.
342 R.J. Boerner, A.P. Nguyen, B.A. Barry and R.J. Debus, *Biochemistry*, 1992, **31**, 6660.
343 T.-A. Ono and Y. Inoue, *FEBS Lett.*, 1991, **278**, 183.
344 T.-A. Ono and Y. Inoue, *Biochemistry*, 1991, **30**, 6183.
345 T.-A. Ono and Y. Inoue, *Biochim. Biophys. Acta*, 1992, **1099**, 185.
346 T.-A. Ono, M. Kusunoki, T. Matsushita, H. Oyanagi and Y. Inoue, *Biochemistry*, 1991, **30**, 6836.

347 T.-A. Ono and Y. Inoue, *Biochim. Biophys. Acta*, 1990, **1020**, 269.
348 J. Tso, M. Sivaraja and G.C. Dismukes, *Biochemistry*, 1991, **30**, 4734.
349 J. Tso, M. Sivaraja, J.S. Philo and G.C. Dismukes, *Biochemistry*, 1991, **30**, 4740.
350 A. Boussac, J.-L. Zimmermann and A.W. Rutherford, *FEBS Lett.*, 1990, **277**, 69.
351 A. Boussac, J.L. Zimmermann, A.W. Rutherford and J. Lavergne, *Nature*, 1990, **347**, 303.
352 B.J. Hallahan, J.H.A. Nugent, J.T. Warden and M.C.W. Evans, *Biochemistry*, 1992, 31, 4562.
353 A. Boussac and A.W. Rutherford, *Biochemistry*, 1992, 31, 7441.
354 A. Boussac, P. Sétif and A.W. Rutherford, *Biochemistry*, 1992, **31**, 1224.
355 L.-E. Andréasson and K. Lindberg, *Biochim. Biophys. Acta*, 1992, **1100**, 177.
356 A. Bakou, C. Buser, G. Dandulakis, G. Brudvig and D.F. Ghanotakis, *Biochim. Biophys. Acta*, 1992, **1099**, 131.
357 R. Mei and C.F. Yocum, *Biochemistry*, 1991, **30**, 7836.
358 M. Baumgarten, J.S. Philo and G.C. Dismukes, *Biochemistry*, 1990, 29, 10814.
359 X.L. Tan, Y. Gultneh, J.E. Sarneski and C.P. Scholes, *J. Am. Chem. Soc.*, 1991, 113, 7853.
360 U. Liebl, S. Pezennec, A. Riedel, E. Kellner and W. Nitschke, *J. Biol. Chem.*, 1992, **267**, 14068.
361 A. Riedel, A.W. Rutherford, G. Hauska, A. Müller and W. Nitschke, *J. Biol. Chem.*, 1991, **266**, 17838.
362 R. Rustandi, S.,W. Snyder, L.L. Feezel, T.J. Michalski, J.R. Norris, M.C. Thurnauer and J. Biggins, *Biochemistry*, 1990, **29**, 8030.
363 I. Sieckman, A. van der Est, H. Bottin, P. Sétif and D. Stehlik, *FEBS Lett.*, 1991, **284**, 98.
364 R.R. Rustandi, S.W. Synder, J. Biggins, J.R. Norris and M.C. Thurnauer, *Biochim. Biophys. Acta*, 1992, **1101**, 311.
365 P. Sétif and K. Brettel, *Biochim. Biophys. Acta*, 1990, **1020**, 232.
366 N. Li, J. Zhao, P.V. Warren, J.T. Warden, D.A. Bryant and J.H. Golbeck, *Biochemistry*, 1991, **30**, 7863.
367 J. Zhao, N. Li, P.V. Warren, J.H. Golbeck and D.A. Bryant, *Biochemistry*, 1992, 31, 5093.
368 K. Brettel, I. Sieckmann, P. Fromme, A. van der Est and D. Stehlik, *Biochim. Biophys. Acta*, 1992, **1098**, 266.
369 T. Mehari, K.G. Parrett, P.V. Warren and J.H. Golbeck, *Biochim. Biophys. Acta*, 1991, **1056**, 139.
370 U. Feiler, W. Nitschke and H. Michel, *Biochemistry*, 1992, 31, 2608.
371 J.A. Hanley, J. Kear, G. Bredenkamp, G. Li, P. Heathcote and M.C.W. Evans, *Biochim. Biophys. Acta*, 1992, **1099**, 152.
372 M. Miller, X. Liu, S.W. Snyder, M.C. Thurnauer and J. Biggins, *Biochemistry*, 1992, 31, 4354.
373 W. Nitschke, U. Feiler and A.W. Rutherford, *Biochemistry*, 1990, 29, 3834.

3
EPR and ENDOR in the Lanthanides

BY J. M. BAKER

1 Introduction

The feature which differentiates electron magnetic resonance in compounds of the lanthanide ions (and actinide ions) from those of most d group ions and free radicals is that there is a large contribution from the orbital angular momentum. This gives rise to g-values very different from the free spin g-value, and it gives rise to rapid spin-lattice relaxation which means that electron magnetic resonance is observable only at low temperatures, below about 20K. For this reason, even in a journal on ESR, I have labelled the electron magnetic resonance EPR (electron paramagnetic resonance), as much more than electron spin is involved. Also, because readers of the journal are probably much more familiar with electron magnetic resonance in radicals and 3d ions, it is appropriate to give a background on the magnetic properties of the lanthanide ions, which has been done in sections 2 and 3.

No comprehensive catalogue of EPR data on lanthanides has been compiled since 1959,[1] and that was part of a general review of all transition metal data. Selected data for lanthanides, and a general discussion of magnetic and resonance properties of lanthanides, have been given by Abragam and Bleaney[2] and in the Handbook on the Physics and Chemistry of the Rare Earths.[3] This author's intention to complete an up to date comprehensive catalogue was overwhelmed by the mass of material to be catalogued and shortage of time. It would be a valuable service to those interested in resonance in lanthanides if someone were to make such a catalogue. What follows are general considerations about EPR and ENDOR of lanthanides in sections 4 and 5 respectively, illustrated by selected data in the later sections: lanthanide systems measured by magnetic resonance in section 6, including materials with structural phase transitions and with low symmetry sites; and interactions between magnetic ions in section 7.

2 Properties of the lanthanide ions

In their compounds, the lanthanides form primarily ionic bonds with their ligands. In solids, the trivalent lanthanide ions Ln^{3+}, and also the divalent ions where they have been found, have the configuration $1s^2 2s^2 2p^6 3s^2 3p^6 3d^{10} 4s^2 4p^6 4d^{10} 4f^n 5s^2 5p^6$. The physical feature responsible for many of the distinctive properties of the lanthanides is that the unfilled 4f subshell is deep within the atom, inside $5s^2 5p^6$. This largely shields the subshell from chemical bonding, and from crystal fields due to neighbouring ions. It also ensures that the 4f electrons see a large effective nuclear charge and have compact wavefunctions. This leads to relatively large spin-orbit coupling and hyperfine interactions. So the 4f sub-shell satisfies the conditions for LS coupling: the spins and orbital angular momenta of the $4f^n$ configuration couple to give a range of terms characterised by L and S, each of which is split by spin-orbit coupling into a series of values of J between L+S and |L-S|. Table 1 lists the configurations, together with the values of L,S and J for the ground state, and the Lande g-factor for these quantum numbers. Many lanthanide salts are strongly coloured because of optical transitions among the terms of the $4f^n$ configuration. Although, in principle, electric dipole transitions are not allowed between these levels, such transitions are made strong by either static or dynamic low symmetry electrostatic crystal fields, which admix configurations with the opposite parity.

Table 1 Parameters for the ground state of ions with n unpaired 4f electrons

n	ion	S	L	J	g_J	ionic radius (nm)	$<r^2>_{4f}^{1/2}$ (nm)
0	La^{3+}					0.118	
1	Ce^{3+}, Pr^{4+}	1/2	3	5/2	6/7	0.114	0.065
2	Pr^{3+}	1	5	4	4/5	0.114	0.062
3	Nd^{3+}	3/2	6	9/2	8/11	0.112	0.059
4	Pm^{3+}	2	6	4	3/5		
5	Sm^{3+}	3	5	5/2	2/7	0.109	0.056
6	Sm^{2+}, Eu^{3+}	3	3	0	0	0.107	
7	$Eu^{2+}, Gd^{3+}, Tb^{4+}$	7/2	0	7/2	2	0.106	0.052
8	Tb^{3+}	3	3	6	3/2	0.104	
9	Dy^{3+}	5/2	5	15/2	4/3	0.103	0.050
10	Dy^{2+}, Ho^{3+}	2	6	8	5/4	0.102	
11	Ho^{2+}, Er^{3+}	3/2	6	15/2	6/5	0.100	0.048
12	Tm^{3+}	1	5	6	7/6	0.099	
13	Tm^{2+}, Yb^{3+}	1/2	3	7/2	8/7	0.098	0.046
14	Lu^{3+}					0.097	
	Y^{3+}					0.102	

Table 1 also lists ionic radii for the trivalent ions (including that for Y^{3+}), for which different authors give somewhat different values, because it is an ill defined quantity: the data given is for 8-fold coordination with O^{2-}.[4] The values of the rms radius of the 4f electrons, $<r^2>_{4f}^{1/2}$, is given for comparison.[5] The radius of the lanthanide ions decreases gradually by 22% from lanthanum to lutecium. This change is sufficient to alter the size of the coordination sphere, so that lanthanum has room for nine oxygen ligands whereas lutecium has room for only eight. This causes a change of crystal structure for some salts between light and heavy lanthanides. Although the bonding to ligands is primarily ionic, there is some evidence of covalent character with the non-spherically symmetrical $4f^n$ sub-shell, which may influence ligand positions: some solids have very irregular ligand positions [section 6.4]. It is the positions of the ligands which primarily influences the symmetry and magnitude of the crystal field [section 3]. The radial distribution of the $4f^n$ sub-shell decreases more rapidly than the ionic radius with increasing Z (see table 1), because $4f^n$ lies closer to the nucleus than $5s^26p^6$. Hence $4f^n$ sees a steadily increasing Z_{eff} as Z increases, whereas for $5s^25p^6$ the increase in Z is partially offset by the increase in the number of 4f electrons. This affects covalency and overlap of the 4f electrons with ligands [sections 3 and 4.5], and super-exchange between neighbouring paramagnetic ions [section 7].

In solids and liquids, the crystalline electric field set up by electrostatic interaction with the ligands and other ions in the environment is usually at least an order of magnitude smaller than the spin-orbit coupling, which ensures that the quantum number J derived using Hund's rules is a good quantum number. The orbital angular momentum is not quenched, as it is in compounds of the 3d group. The crystal field produces only a small admixture of excited J states; its main effect is to split up the (2J+1) components of the lowest J manifold by about 10^{-2} eV.

These properties of lanthanide ions were clear from the optical spectrum and the magnetic susceptibility, which were measured early in the century: the magnetic measurements and their interpretation were described in Van Vleck's classic book [6] in 1932. The magnetic susceptibility at room temperature conforms well with that expected for a Boltzmann distribution over the (2J+1) levels of a system with Lande g-factor g_J. This agreement is because at room temperature (kT ~ 2.5 x 10^{-2} cm^{-1}) all (2J+1) energy levels are populated. As the temperature is lowered and the upper levels become depopulated, the magnetic properties show dramatic variation with temperature, which depends upon the details of the energies and eigenstates.

The energy level pattern, and degeneracies which remain, depends upon the symmetry of the crystal field, the lower the symmetry the less the degeneracy. For ions $4f^n$ with odd n, the levels have even degeneracy in the absence of an applied magnetic field; hence, even for sites of low symmetry there are (J+1/2) doublets: the so called Kramers' degeneracy is imposed by time reversal symmetry. There is no such symmetry requirement for even n, and in low symmetry there may be (2J+1) singlet states. Even in cubic symmetry, there are not generally degeneracies higher than 4 for odd n or 3 for even n. The crystal field energies may be measured through splittings of the optical transitions between terms. Application of a magnetic field produces further Zeeman splittings, which give information about the degeneracies and eigenstates.

A detailed understanding of the crystal field as revealed by optical spectroscopy has been built up over the last 60 years (for a review of this and magnetic properties, see volume 11 of the Handbook of the Physics and Chemistry of the Rare Earths).[7]

In addition to producing these various effects on the energy levels and eigenstates, the orbital content of the wavefuction also gives rise to strong coupling between the 4f electrons and the lattice through the crystal field. Modulation of the crystal field by lattice vibrations gives rise to matrix elements between the crystal field states; this causes all of them to have short lifetimes, except the lowest [section 3.1].

Information about the energies and eigenfunctions of the excited states of the lowest J manifold is limited because, although there are several ways in which information may be found in principle, they are difficult to use in practice because of very rapid relaxation rates to the ground state caused by the large orbital content of the wavefunction.

All states at various temperatures contribute to the magnetic susceptibility and specific heats of materials containing lanthanide ions, but only in special cases, such as one excited state of much lower energy that the others, has unambiguous information been obtained from such measurements.

At temperatures high enough for there to be appreciable optical absorption from the excited state, the line broadening due to rapid relaxation rates makes useful measurement difficult. Even fluorescent transitions from a metastable state excited by optical absorption at low temperatures are often too wide for useful measurement.

Inelastic neutron scattering and Brillouin scattering have been used to measure energies of the excited crystal field levels. The former technique requires large samples which do not need to be single crystals.

3 The crystal field

The crystal field is the electrostatic interaction of the 4f electrons in the lanthanide ions with the electrons and nuclei of their surroundings, which to a first approximation may be regarded as adding an electrostatic potential to the hamiltonian for the 4f electrons. This model is described, and its limitations discussed, in reference [2]. Although the crystal field acts upon the orbital motion of the 4f electrons, it may to a good approximation be projected upon J for the lowest J manifold and, using the Wigner-Ekart theorem, may be expressed in terms of vector operators in

3: EPR and ENDOR in the Lanthanides

total angular momentum $\mathbf{J}$. This type of theory was first developed by Stevens.[8] The hamiltonian for the crystal field has the form:

$$H_{cf} = \Sigma_{n,m} B_n{}^m O_n{}^m(J), \tag{1}$$

where $O_n{}^m(J)$ are angular momentum operators of order n, with matrix elements $<J,M_J|O_n{}^m(J)|J,M_J+m>$, and $B_n{}^m$ are coefficients describing the magnitude and symmetry of the crystal field. Only values of n = 2,4 and 6 contribute in first order to the energy of the 4f electrons; m may have all integer values between -n and n. The values of the parameters $B_n{}^m$ depend upon the choice of axes with respect to which the vector operators $O_n{}^m$ are specified. For sites which have axes of symmetry, the selection of the symmetry axes for the specification of the vector operators renders many of the $B_n{}^m$ equal to zero, and so greatly simplifies the hamiltonian. The higher the symmetry, the fewer the parameters required. Just from the symmetry of the site, which parameters $B_n{}^m$ are not zero is known, but the magnitudes are more difficult to calculate. The factors which contribute to $B_n{}^m$ are indicated by the original Stevens approach [8] where

$$B_n{}^m = A_n{}^m \theta_n <r^n>, \tag{2}$$

where: $A_n{}^m$ depends upon the potential set up by the surrounding ions; θ_n is a multiplicative "Stevens" factor, relating to a reduced matrix element depending upon L,S and J, taking account of the projection of an operator involving orbital angular momentum upon J; and $<r^n>$ is the mean value of r^n for the distribution of 4f electrons in the ion. The operators $O_n{}^m(J)$ and their matrix elements have been listed by several authors. Unfortunately a number of variants of this theory have been used by different authors, not merely differences of notation, which makes it difficult to compare the work of these authors without considerable care. Additional potential confusion is caused by the similar looking spin hamiltonian used to describe the fine structure in S-state ions [equation (10) in section 4.1]

The simplest crystal field corresponds to cubic symmetry, set up either by four, six or eight ligands at the corners of a tetrahedron, octahedron or cube respectively. Unlike for the 3d group, complete specification of the crystal field requires two parameters B_4 and B_6, and the energy and eigenvalues depend upon B_4/B_6.[9]

Theoretical calculations of $B_n{}^m$ is not trivial. Many publications make estimates assuming that the ligands are point charges at known lattice positions, which set up an electrostatic potential to be superimposed upon that within the lanthanide ion from its own nucleus and electrons. Such estimates are of limited value, as the overlap of wavefunctions invalidates a purely point charge model. Several attempts have been made to calculate the effects of overlap and covalency. Also recently, Newman has introduced a new type of phenomenological model which involves only a few parameters, known as the Superposition Model.[10,11,12] It is based upon the supposition that whatever the details of the mechanism, the crystal field set up by a number of ligands is the sum of their effects individually. This is convenient to use because the crystal field set up by a ligand distant $\underline{R}_i$ is axially symmetric about $\underline{R}_i$, and so may be expressed;

$$H_{cf} = \overline{A}_2 O_2^0(J) + \overline{A}_4 O_4^0(J) + \overline{A}_6 O_6^0(J), \tag{3}$$

where the $O_n^0(J)$ are expressed relative to the direction $\underline{R}_i$. This may be re-expressed in relation to any cartesian axis system in the crystal, for which $\underline{R}_i$ is specified by (R_i, θ_i, ϕ_i), in terms of known rotation parameters $K_n^m(\theta_i, \phi_i)$.[12] This allows the crystal field set up by any number of ligands at known positions $\underline{R}_i$ to be written in terms of a common set of axes in the form of equation (1) with:

$$B_n^m = \Sigma_i K_n^m(\theta_i, \phi_i) \overline{A}_n(R_i). \tag{4}$$

The crystal field set up by ligands at the same distance R involves only the three parameters $\overline{A}_2$, $\overline{A}_4$ and $\overline{A}_6$, but allowance for different ligand distances may be made by assuming that:

$$\overline{A}_n(R) = \overline{A}_n(R_o)(R_o/R)^{t_n}, \tag{5}$$

introducing three new empirical parameters t_n. The values of the parameters $\overline{A}_n$ and t_n for a particular type of ligand must be evaluated empirically from crystal fields measured in a material of known crystal structure; but once that is done, this theory may be used to calculate a crystal field at any site with the same type of ligand in a known crystal structure. It may also be used to express changes in the crystal field when the crystal is strained [section 3.1]. Although more sophisticated theories of the crystal field have been developed by theoreticians,[13] the superposition model has overtaken the point charge model as that used by experimentalists to interpret their data.

One difficulty with application of the theory to the systems usually met in EPR is that the crystal field is required for an ion which is essentially an impurity in a diamagnetic host lattice, for which the ligand distances $\underline{R}_i$ are slightly different by an unknown amount from the values appropriate to the host, which are the parameters which may be known from X-ray crystallography.

3.1 Orbit-lattice interaction.- The coupling between a lanthanide ion and its ligands through the crystal field depends upon their relative coordinates (R_i, θ_i, ϕ_i). Hence, any change in these coordinates, due to static strain, thermal vibrations or sound waves, causes a change in the matrix elements of the crystal field hamiltonian. The hamiltonian H_{cf} may be expanded as a Taylor series in terms of local strain $e(\Gamma, p)$ [or of local symmetry coordinates $\sigma(\Gamma, p, \lambda)$],[14] and the term which is linear in strain is called the orbit-lattice interaction:

$$H_{OL} = \Sigma_{\Gamma, n, p} V(\Gamma, n) O(\Gamma, n, p) e(\Gamma, p). \tag{6}$$

Γ are the irreducible representations of the group of symmetry operations around the site of the paramagnetic ion, and $e(\Gamma, p)$ are the strains which transform as the pth component of that irreducible representation. $O(\Gamma, n, p)$ are linear combinations of angular momentum operators of order n, which using the same procedure as for $O_n^m(J)$ in H_{cf} may be expressed in terms of vector operators in $\underline{J}$. $V(\Gamma, n)$ are coefficients with the units of energy, which in general, it is as difficult to calculate as the analogous crystal field parameters B_n^m. However, they may be deduced using the superposition model, as the explicit dependence of the contribution to H_{cf} of any ligand upon (R_i, θ_i, ϕ_i) is known.

The uniform strain produced by an applied stress, hydrostatic or uniaxial, must be factorised into components of $e(\Gamma,p)$ in order to calculate its effect on a lanthanide ion through the matrix elements of H_{OL}. Such stress may produce both displacements of crystal field levels and changes in the eigenstates. The former are generally rather difficult to measure, as the displacements are very small relative to the unstressed energy separations [but see section 4.2]. The latter may be measured using EPR [see section 4.2].

H_{OL} may also be used to express the coupling between the lanthanide ion and the broad range of phonon modes described by k-vectors, in terms of which the local symmetry coordinates $\sigma(\Gamma,\lambda,p)$ may be expressed. The effect of this is to give matrix elements of H_{OL} coupling the various crystal field states involving creation and annihilation operators for the various phonon modes. Exact calculation of their effect would require detailed knowledge of the phonon modes, but the Debye approximation can be used to make an estimate. They give rise to transitions between the crystal field states with either the emission, absorption or scattering of a phonon. This limits the lifetime of a paramagnetic ion in an excited state and gives a mechanism for the establishment of thermal equilibrium over the crystal field states.

The first order effect of the orbit-lattice interaction H_{OL} on a Kramers doublet is an equal shift of both levels, which does not affect the EPR. In second order, cross terms between H_{OL} and the Zeeman operator lead to transitions between the Zeeman levels of the ground doublet.

The modulation of the crystal field due to lattice vibrations, discussed above, leads in second order to transitions between the Zeeman levels of the ground doublet, with emission or absorption of a phonon, or inelastic scattering of a phonon. This gives a mechanism for exchange of energy with the lattice which corresponds to spin-lattice relaxation. When the Zeeman levels are separated by hf, emission or absorption of a phonon is more likely in the region $kT \sim hf$: the direct process, when $T_1^{-1} \propto T$. Inelastic scattering is more likely for $kT >> hf$: the Raman process, for which $T_1^{-1} \propto T^n$, where n depends upon the ion, but may be as large as 9. Resonance between the phonon energy and that of an excited crystal field state at Δ gives rise to the Orbach process for which $T_1^{-1} \propto \exp(-\Delta/kT)$. These processes are discussed, with many examples, by Jeffries.[15] For systems with an Orbach relaxation process, the energy Δ of the first excited state can often be determined. However, even this determination is complicated by the difficulty of separating an exponential dependence upon T for the Orbach process from a T^9 dependence for Raman process.

4 Electron Paramagnetic Resonance (EPR)

EPR was observed in lanthanide salts soon after its application to electronic systems in 1946, in the ground state of neodymium ethyl sulphate at 20K.[16]

In principle, one could measure EPR transitions between the Zeeman components of any of the populated states, ie. of the lowest J manifold. There are a few examples of EPR from thermally populated excited states, whose energy may be deduced from the temperature variation of intensity, proportional to the Boltzmann population. However, because the relaxation for excited states is so rapid, EPR is usually measurable only for the lowest states. For some systems, EPR in highly excited states may be measured either by populating them using optical excitation from the ground state, or by ODMR.

The relaxation rate for most lanthanides is usually very high at room temperature; above 20K the rapid spin-lattice relaxation dominates the EPR line width $\delta f \sim T_1^{-1}$. Only for ions with a half filled sub-shell, $4f^7$ (Eu^{2+}, Gd^{3+}, Tb^{4+}), L = 0, S = 7/2, the so called S-state ions, is the spin-lattice relaxation time relatively long, as there is very little orbital content of the wavefunction for the lattice to interact with, and EPR may be observed at room temperature and above. At low temperatures the resonance is heavily saturated, but may be observed in dispersion. Hence, only EPR from the S-sate ions is observable at room temperature. However, as T_1 is a rapid function of temperature, by cooling the other lanthanides below about 20K the line width becomes small enough for EPR to be observable.

With the one or two rare exceptions, where it is possible to observe the first excited state, EPR is observable only between the Zeeman components of the ground state. Because of this restriction to transitions between the lowest of the levels left by the crystal field, the information contributed by EPR to the overall picture is rather limited. However, the information about this limited aspect is very detailed and precise, and it is for this reason that EPR has been able to make a significant contribution to the understanding of the magnetic properties.

Information about the magnetic properties of the ground state is also provided by measurements of the magnetic susceptibility at temperatures low enough that it is the only state populated, but even then it may be an aggregation of contributions from differently oriented sites whose individual contributions cannot be separated. The great advantage of EPR is that differently oriented sites exhibit different resolved spectra. A further complication of using magnetic susceptibility measurements is that for magnetically concentrated salts the magnetic interactions may lead to an ordered state, and the magnetic susceptibility of a system diluted sufficiently not to order may be too small to measure.

3: EPR and ENDOR in the Lanthanides

4.1 The spin hamiltonian.- The eigenfunctions of a lanthanide ion in a crystal field, which is assumed to be small compared with the energy separation of different spin-orbit states characterised by J, have the general form:

$$\phi = \Sigma_M \alpha(M_J)|J,M_J>. \tag{7}$$

In an applied magnetic field B such that $\mu_B g_J JB$ is small compared with the separation between the crystal field states, one may describe the energy levels of one of the sets of m states, degenerate in the absence of B, by a spin hamiltonian with fictitious spin S' such that m = (2S'+1). For most systems containing an odd number of electrons, m = 2, S' = 1/2. An applied magnetic field causes the levels to diverge as a linear function of B, so that at some value of B the separation is equal to a given microwave quantum hf; so if the transition is allowed it could be observed by EPR at frequency f. For such Kramers' doublets, the spin hamiltonian has the form:

$$H = \mu_B \underline{S}' \cdot \underline{g} \cdot \underline{B}, \tag{8}$$

where the principal components g_1, g_2, g_3 of the **g**-matrix may be very different from the free spin value g_e = 2.0023. This may lead to a very anisotropic EPR spectrum, both in the field required for resonance and in the intensity of the lines. As the maximum possible g-value for a doublet (Jg_J for pure $M_J = \pm J$) is 20 (for Dy^{3+}: J = 15/2, g_J = 4/3), and the minimum possible value is 0, EPR lines for the lanthanides may occur over a wide range of values of B. In the early days of EPR, when it was difficult to obtain highly purified samples of lanthanide salts, one could find EPR lines from different lanthanides in the same crystal all over the range of field. A recording of the spectrum "looked like the Alps" in the view of one prominent theorist in the field.

If the parameters of the crystal field are known, the coefficients $\alpha(M_J)$ in equation (7) may be calculated. The **g**-matrix may be calculated by projecting the Zeeman operator $g_J \mu_B \underline{B} \cdot \underline{J}$ on the eigenstates (7). Unfortunately, there is not enough information in the **g**-matrix of the ground doublet alone to make this calculation backwards to deduce the crystal field parameters. For this one requires information about energies and **g**-matrices of some of the excited states as well.

If the symmetry of the site is high, it may be found from the symmetry of the **g**-matrix of the ground state. Cubic symmetry gives isotropic g. 3-fold, 4-fold or 6-fold symmetry leads to an axially symmetric **g**-matrix. Orthorhombic symmetry leads to a **g**-matrix with three different principal g-values, with principal axes along the symmetry axes. However, lower symmetry would lead to a similar **g**-matrix, but with no special relation between principal directions and the crystal field.

Even for a low symmetry site, information may be obtained if the relative orientation of the principal axes and the crystallographic axes is known. This may be found by EPR on a crystal oriented by X-ray diffraction, or from its morphology, and carefully transferred to the EPR spectrometer. Or, if the crystal contains symmetry related sites, the coincidence of spectra from these sites can be used to identify crystallographic symmetry planes and directions.

For cubic symmetry, some Kramers' ions have four-fold degenerate ground states, so m = 4 and S' = 3/2.. These Γ_8 quartets are described by the following spin hamiltonian:

$$H = g\mu_B \underline{S}'.\underline{B} + f\mu_B[S'^3_x B_x + S'^3_y B_y + S'^3_z B_z]. \tag{9}$$

The principle of applying the spin hamiltonian formalism to sets of m states which are not degenerate in the absence of B may be extended to higher values of m, provided that the overall spread of energies is not much larger than the microwave quantum, and also that the set of m energy levels is well separated from other levels of the J manifold.

For S-state ions, with S = 7/2, one might expect an eight-fold degeneracy for B = 0. In practice the crystal field, through high order processes, does produce small separation of the eight levels (typically 10^{-4}eV ~ 1 cm^{-1} ~ 30 MHz); so the spin hamiltonian contains fine structure terms describing this splitting:

$$H = \mu_B \underline{S}.\underline{g}.\underline{B} + \Sigma_{n,m} b_n^m O_n^m(S), \tag{10}$$

where $O_n^m(S)$ are spin angular momentum operators of order n, with matrix elements $<S,M_S|O_n^m(S)|S,M_S+m>$, and b_n^m are coefficients determined by the magnitude and symmetry of the crystal field. Only values of n = 2,4 and 6 occur, but m may have all integer values between -n and n. The remarks made about the crystal field parameters B_n^m in section 3 also apply to the fine structure parameters. For cubic symmetry n = 4 terms greatly dominate n = 6 terms, and n = 2 terms are zero. However, for lower symmetry, the n = 2 terms usually dominate, and the fine structure terms $\Sigma_m b_2^m O_2^m$ may be written in the form $\underline{S}.\underline{D}.\underline{S}$ to a first approximation. The remarks made above about the relationship between the symmetry of the g-matrix and the crystal field are also applicable to the D-matrix. The transitions between the levels described by equation (10) are highly anisotropic, though to a first approximation they are isotropic in a plane perpendicular to a 3-, 4- or 6-fold symmetry axis. Even though they are small, the n = 4 terms do allow an identification of 3-fold and 4-fold symmetry, and n = 6 terms can identify 6-fold symmetry by producing a small n-fold angular variation in a plane normal to the n-fold axis. In equation (10) the Zeeman operator has been written formally in terms of a g-matrix, because high order effects of the crystal field also produce very small departures from the free spin g-value, which can be anisotropic. However for many S-state systems the g-value may be regarded as isotropic and close to g_e.

The use of S in equation (10) rather than S' is deliberate, because the S-state ion has L = 0, so the magnetic properties are due to real spin and not fictitious spin. Having made the point that the spin hamiltonian for other lanthanides involves fictitious spin, we will hereafter follow the practice of most of the literature by dropping the prime on S'.

The detail of the origin of the fine structure, and the relationship between the fine structure parameters b_n^m in equation (10) and the crystal field parameters B_n^m in equation (1), has been the subject of many theoretical calculations. That the parameters are related suggests that the superposition model could be used for relating b_n^m for different arrangements of a particular ligand.[17] The analogues of equations (4) and (5) are:

$$b_n^m = \Sigma_i K_n^m(\theta_i,\phi_i)\overline{b}_n(R_i); \quad \overline{b}_n(R) = \overline{b}_n(R_o)(R_o/R)^{t_n}. \tag{11}$$

Whether a site has a centre of symmetry can be established by applying a uniform electric field E. For a site with a centre of symmetry, the EPR lines shift as a quadratic function of E, but for a site without a centre of symmetry, the lines split linearly in E. Mims has reviewed the information to be derived from electric field effects in EPR, including several examples of their application to lanthanides.[18] However, that there can be complications in interpreting such effects is shown by the fact that the electric field effect for Gd^{3+} in a K^+ site in $KTaO_3$ is found to be different from that for Fe^{3+} at the same site: some possible reasons for this difference are discussed, but somewhat inconclusively.[19]

4.2 The spin-lattice hamiltonian and the effect of external stress.-In section 3.1 it was stated that there are cross terms between H_{OL} and the Zeeman operator which give rise in second order to coupling between the components of a Kramers doublet. This may be expressed in terms of a spin hamiltonian operating within the ground state spanned by S, called the spin-lattice hamiltonian:

$$H_{s\text{-}l} = \Sigma_{\Gamma,p} A(\Gamma) O\{\Gamma,p(B,S)\} e(\Gamma,p), \tag{12}$$

where $O\{\Gamma,p(B,S)\}$ are operators linear in B and S.[20]

Static strain produced by external stress F, leads to a shift in g-value which is linear in F, from which the coefficients $A(\Gamma)$ may be found experimentally.[20]

The effect of strain produced by phonons may be calculated in the same way as was described for H_{OL} in section (3.1) leading to expressions for the various spin-lattice relaxation processes in terms of the coefficients $A(\Gamma)$.[21] Hence, except for the approximation of using the Debye model for the phonon modes, one may calculate spin lattice relaxation rates using coefficients measured by applying external static stress.[20]

Although the effects of H_{OL} for static stress on the crystal field levels are usually too small to measure in comparison with the level separations at zero stress, it is possible to use EPR to measure such effects if the levels are sufficiently close together for zero stress. An example of this is provided by the Γ_8 quartet in a cubic site, whose component doublets separate linearly in applied stress, which could also be written in spin-hamiltonian form within $S = 3/2$. These effects have been measured for Dy^{3+} in CaF_2 and Er^{3+} in MgO.[22]

External stress also changes the coefficients b_n^m describing the fine structure of S-state ions. This has been written in terms of parameters $G^{(n)}$ in the following spin hamiltonian:[23]

$$H_{s\text{-}1} = \Sigma_{\Gamma,n,p} G_\Gamma^{(n)} O_{\Gamma,p}^{(n)} e_{\Gamma,p},\qquad(13)$$

where the notation has been slightly modified to be more consistent with that used in equations (6) and (12). In view of the many coefficients which would be required for a site of low symmetry, most work on S-state ions has been performed on cubic systems. Here, for uniaxial stress, the cubic symmetry is lowered, and the n = 2 terms in $H_{s\text{-}1}$ are dominant, although n = 4 terms have been shown to be significant.[23] By studying the effect of uniaxial stress as a function of temperature, it has been shown that there is a contribution to $G^{(2)}(T)$ from zero point vibrations. Fitting the temperature dependence to an Einstein model with characteristic frequency f_E produces a particularly simple form:[24]

$$G_\Gamma^{(2)}(T) = G_\Gamma^{(2)}(RL) + K_\Gamma^{(2)}\coth(hf_E/2kT).\qquad(14)$$

Use of the Debye model produces a similar T dependence. Using this sort of relationship, the experimental data may be decomposed to find the rigid lattice contribution $G_\Gamma^{(2)}(RL)$ and the vibrational parameter $K_\Gamma^{(2)}$. Theoretical models[25] have shown both the importance of including relativistic effects in the wavefunction for the S-state ion, and of considering two-phonon first-order mechanisms of the orbit-lattice interaction, whose contribution is shown to be more than an order of magnitude larger than the one-phonon second-order mechanism. The correct order of magnitude and sign was obtained for the coupling parameters, but as a point charge model of the orbit-lattice interaction was used, a more realistic model for H_{OL} could modify this conclusion.

For hydrostatic stress, cubic symmetry is retained, so there are no n = 2 terms, and the fine structure is dominated by b_4. Using EPR, under hydrostatic pressure P, one can measure $(\partial b_4/\partial P)_T$. b_4 also varies as a function of temperature, but there are two contributions to this, one from the change of mean ligand distance R_i due to thermal contraction, and another due to vibronic effects through the spin-lattice interaction. These two contributions become apparent if one attempts to deduce the power k in the relation $b_4 \propto R^{-k}$ from $(\partial b_4/\partial P)_T$ using the measured compressibility to find how R depends upon P, and from $(\partial b_4/\partial T)_P$ using the measured thermal expansivity to find how R depends upon T. These lead to different values of k, say k_P and k_T. Although one might expect the local compressibility and expansivity around paramagnetic impurities doped in low dilution to be different from these parameters for the bulk host crystal, it has been suggested that their ratio is the same as for the bulk sample.[26] That k_T is much larger than k_P suggests that there is an additional mechanism contributing to $(\partial b_4/\partial T)_P$ which is attributed to vibronic effect. Measurements of this sort have been made for Eu^{2+} and Gd^{3+} both in several alkaline earth fluorides (MF_2)[27] and several fluoroperovskites (MNF_3).[28] For MF_2 the values of k were typically 10, whereas for MNF_3 they were much larger, typically 20 - 30. Also for MNF_3 the vibronic effects are much more pronounced, amounting to about 70% of the total, compared with less than 50% for MF_2. High values of k have also been found for Gd^{3+} in

$NH_4Ln(SO_4)_2 \cdot 4H_2O$, for various lanthanides Ln.[29] The theory discussed above for n = 2 terms cannot be applied directly to discuss the data, but a semi-empirical model gives a satisfactory account provided an appropriate value for the unknown Debye θ is assumed.[30]

4.3 Line width.- The resolution of EPR of different lanthanides depends upon the line width, which is limited by spin-spin interaction with neighbouring electronic and nuclear magnetic moments.

We have already stressed that the compactness of the 4f wavefunctions causes a relatively small crystal field interaction with the ligands. The compactness of the wavefunction has two other effects upon the interactions of the lanthanides with their surroundings: (a) it means that for nuclei beyond the ligand shell, the magnetic moment of the 4f electrons may be regarded as located at the nucleus, giving approximately dipolar interactions [see section 4.5]; (b) it means that exchange interactions between neighbouring lanthanide ions is much smaller than such interactions between 3d ions, so that in many cases magnetic dipole-dipole or other interactions make similar or even larger contributions [see section 7]. Also it means that the temperatures at which undiluted lanthanide salts go into an ordered magnetic state are much lower than for 3d salts, often in the helium temperature range.

Hyperfine structure [section 4.4], which arises from an interaction between any nucleus with a nuclear spin and magnetic moment and the unpaired electron, was first observed in EPR by Penrose[31] in 1949 following a suggestion by Gorter that the line width in EPR could be greatly reduced by separation of the magnetic ions by making a dilute mixed crystal with an isomorphous diamagnetic material. The chemical similarity of the lanthanides makes for long homologous series of salts, so this technique of dilution with a diamagnetic salt (of lanthanum, lutecium or yttrium) is particularly appropriate to lanthanides. All of the lanthanides, except cerium, have some isotopes with non-zero nuclear spin, and so exhibit hyperfine structure (Table 2). The relative abundance and nuclear spins and moment ratios makes the hyperfine pattern a characteristic fingerprint which can be used to identify an unknown spectrum. This can be quite useful for recognising the spectrum of impurities, which readily occur as the chemical properties of the lanthanides are so similar. In the 1950s the nuclear spin of several lanthanide isotopes was determined for the first time from their hyperfine structure in EPR.

The line width in samples where the electronic magnetic moments are well separated is likely to be limited by interaction with neighbouring nuclei. Only rarely has hyperfine interaction with neighbouring nuclei (transferred, or super-, hyperfine structure) been well resolved in lanthanide EPR: another difference from the 3d group, caused by the compactness of the 4f wavefunctions. Information about the transferred hyperfine interactions has come mainly from ENDOR [section 5.2].

Some lattices have been used as hosts for lanthanide ions in which there are very few nuclei with magnetic moments, eg. CaO, SrO, CeO_2, ThO_2. Such material rarely gives the very narrow lines one might expect, because the anisotropic nature of lanthanide spectra makes them extremely susceptible to broadening due to local distortion of the lattice due to strains, and dislocations. Although most magnetic resonance since the early work of Bagguley et al[32] has been performed on dilute systems to reduce the line width, it has been fruitful to do some work on undiluted lanthanide salts. One early example is neodymium ethyl sulphate[33], and more recently lanthanide nicotinate dihydrates[34] where the structure gives two or one, respectively, near neighbours for any lanthanide site which allows the interactions between the near neighbours to give a resolved structure, unobscured by the line broadening effects of interaction with more distant neighbours. Another example is in ordered antiferromagnetic systems where the internal field is the same at each site on any sublattice, so that the spin-spin broadening is reduced. Magnetic resonance has been measured using probe ions in such systems [section 7]. A further example is of systems with a large g-value in a high magnetic field at a temperature low enough for nearly all of the atoms to be in the same ground state: this resembles a ferromagnetic state. Here the internal field at all sites is the same, provided that the sample has an ellipsoidal shape so that the macroscopic magnetisation does not lead to a macroscopically inhomogeneous field. Another way to reduce line width in an undiluted paramagnetic salt with modest g-value is to use very high magnetic field and very high microwave frequency. Janssen has used fields up to 6T with resonance frequencies up to 1200GHz generated by a far infra-red laser.[35]

4.4 Hyperfine structure.- All of the lanthanides except Ce contain isotopes with non-zero nuclear spin. The compactness of the $4f^n$ wavefunctions, and the relatively large Z_{eff}, leads to large hyperfine interactions. Table 2 lists the values of these hyperfine interactions. Hence, the hyperfine structure is an important aspect of lanthanide EPR spectra.

This large hyperfine structure leads to specific heat anomalies in the 1mK to 1K range, and has been exploited for magnetic cooling and nuclear orientation.[36] At X-band and higher frequencies, the electronic Zeeman interaction is larger than the hyperfine interaction. Here, transitions between nuclear states may be observed as ENDOR [see section 5]. At lower frequencies, nuclear magnetic resonance transitions between the hyperfine levels at low fields have been measured in the 100 MHz to 3GHz range.[37]

There are three terms introduced to the spin hamiltonian by nuclei with non-zero nuclear spin: (a) the magnetic hyperfine interaction, (b) the electric quadrupole interaction, (c) the nuclear Zeeman interaction. In the free ion these are described by the last three terms respectively in the following hamiltonian:

$$H = g_J\mu_B\underline{B}.\underline{J} + a\underline{J}.\underline{I} + H_{EQ} - g_N\mu_N\underline{B}.\underline{I}, \qquad (15)$$

3: *EPR and ENDOR in the Lanthanides*

where $H_{EQ} = [e^2q_JQ/2I(2I-1)J(2J-1)]*[3(\underline{I}.\underline{J})^2 + (3/2)\underline{I}.\underline{J} - I(I+1)J(J+1)]$.

This may be projected on the eigenstates of the crystal field to give the following terms in the spin hamiltonian:

$$H_s = \mu_B\underline{B}.\underline{g}.\underline{S} + \underline{S}.\underline{A}.\underline{I} + \underline{I}.\underline{P}.\underline{I} - g'_N\mu_N\underline{B}.\underline{I}. \qquad (16)$$

These terms arise respectively from the first order effect of the terms in (15), but there are also second order contributions to the latter two terms[38] [also see section 4.6 equation (26)]. It follows that $\underline{g}$ and $\underline{A}$ must have the same principal directions, and that for the ith principal direction $g_i/A_i = g_J/a$, provided that the ground state may be written:

$$\phi = \Sigma_M \alpha(M_J)|J,M_J>, \qquad (17)$$

with no admixture of excited state with different J. Conversely, departure from the equality of A_i/g_i for the principal directions may be used as an indication of the admixture of excited J states.

Table 2 Nuclear spins and hyperfine parameters for lanthanide ions

ion	isotope	% abundance	nuclear spin	a/h(MHz)	a/$g_J\mu_B$(mT)
Pr^{3+}	141	100	5/2	+ 1093(10)	91.3
Nd^{3+}	143	12.3	7/2	- 220.3(2)	21.6
	145	8.3	7/2	- 136.9(1)	13.4
Pm^{3+}	147	radioactive	7/2	559(6)	199.8
Sm^{3+}	147	15.0	7/2	- 240(3)	60.0
	149	13.9	7/2	- 194(3)	48.5
Eu^{3+}	151	47.8	5/2	S-state	3.50
	153	52.2	5/2	S-state	1.48
Gd^{3+}	155	14.7	3/2	S-state	0.36
	157	15.7	3/2	S-state	0.48
Tb^{3+}	159	100	3/2	+ 530(5)	25.2
Dy^{3+}	161	19.0	5/2	- 109.5(22)	5.78
	163	24.9	5/2	+ 152.4(30)	8.17
Ho3+	165	100	7/2	+ 812.1(10)	46.4
Er^{3+}	167	22.9	7/2	- 125.3(12)	7.46
Tm^{3+}	169	100	1/2	- 393.5	24.1
Yb^{3+}	171	14.4	1/2	+ 887.2(15)	55.5
	173	16.2	5/2	- 243.3(4)	15.2

The separation in values of magnetic field for the hyperfine lines in first order is $\delta B = A/g\mu_B$, so that the same value of A_i/g_i corresponds to the same separation of hyperfine lines in field units (right hand column in table 2): note that this is independent of the microwave frequency of measurement. This constancy is preserved in different materials with different crystal fields, so that the hyperfine structure from a given lanthanide is easily recognisable.

4.5 Transferred (Super-) hyperfine structure.- There is also a hyperfine interaction between the unpaired electron and those nuclei of neighbouring diamagnetic ions which have non-zero nuclear spin, which has been called transferred hyperfine structure and superhyperfine structure by various authors. This adds the following terms to the spin hamiltonian (16):

$$H = \Sigma_i [\underline{S}.\underline{A}_i.\underline{I}_i + \underline{I}_i.\underline{P}_i.\underline{I}_i + g_{Ni}\mu_N\underline{I}_i\underline{B}], \qquad (18)$$

the summation being over all nuclei.

The magnitude of $\underline{A}_i$ decreases rapidly with the distance $\underline{R}_i$ between the electron and the nucleus, so only near nuclei are of interest: more distant nuclei simply contribute to the line width [section 4.3]. $\underline{A}_i$ may be factorised into an isotropic and a traceless part:

$$\underline{S}.\underline{A}.\underline{I} = A_s \underline{S}.\underline{I} + \underline{S}.\underline{A}_t.\underline{I}, \qquad (19)$$

where we have dropped the subscript i on the understanding that we are discussing a specific nucleus.

For lanthanides, as only a small fraction of unpaired electron is transferred from the ligands, and the 4f wavefunction is compact, one may calculate the dipolar contribution to the hyperfine interaction by taking the magnetic moment of the 4f wavefunction as centred at the lanthanide nucleus. For a nucleus characterised by I, g_N at distance $\underline{R}$ relative to an ion with an anisotropic g-matrix, the dipolar contribution to the hyperfine interaction may be written:

$$(\mu_o/4\pi)(g_N\mu_N\mu_B/R^3)*\{g_1(1-3l^2)S_xI_x + g_2(1-3m^2)S_yI_y + g_3(1-3n^2)S_zI_z$$
$$- 3g_1(lmS_xI_y + lnS_xI_z) - 3g_2(mlS_yI_x + mnS_yI_z) - 3g_3(nlS_zI_x + nmS_zI_y)\} \qquad (20)$$

In this equation (l,m,n) are the direction cosines of $\underline{R}$ relative to the principal axes of $\underline{g}$. For an isotropic g-matrix, this reduces to the simpler form:

$$A_d(S_zI_z - \underline{S}.\underline{I}), \text{ for z along } \underline{R} \text{ and } A_d = (\mu_o/4\pi)gg_N\mu_B\mu_N/R^3. \qquad (21)$$

When the bond with the ligand has a three-fold or higher rotational symmetry, the whole traceless part of the interaction $\underline{S}.\underline{A}_t.\underline{I}$ may be written $(A_p + A_d)(3S_zI_z - \underline{S}.\underline{I})$, by choosing z along $\underline{R}$.

The non-dipolar components A_s and A_p for ligands are typically quite large, as shown for example by measurements in alkaline earth fluorides.[39] These have been commonly analyzed theoretically for a cluster comprising the lanthanide and its ligands by (a) a configuration interaction method, or (b) a molecular orbital method.[40] A detailed example of the latter is reviewed by Anderson[39] where the covalent bonding and overlap is calculated between the 4f electrons and linear combinations of atomic orbitals on the ligands which have the symmetry of

3: *EPR and ENDOR in the Lanthanides* 147

the site. This is particularly straightforward for sites of high symmetry, when all of the ligands are the same and at the same distance; eg. in CaF_2, when the Ln^{n+} site is at the centre of a cube of eight F⁻ ligands.[41] These calculations lead to positive values for A_s and A_p. For S-state ions, A_s arises from unpaired s-electrons on the ligand and A_p from unpaired p-electrons, except for core polarisation contributions, but for anisotropic systems the breakdown between s- and p-electron contributions is less straightforward.

An alternative method of calculation proposed by McGarvey[42,43] considers the bonding with a single ligand using orbitals quantised with respect to the bond axis, producing a simple system with axial symmetry. This resembles the simplification used in the superposition model for the crystal field [section 3].

Freeman and Watson[44] accounted for the negative measured signs of A_s and A_p for S-state ions by invoking a contribution from exchange polarisation of the $5s^2 5p^6$ closed sub-shells on Ln^{3+} by interaction with the unfilled 4f sub-shell. Even though the polarisation is small, it has a potent effect on the transferred hyperfine structure because of the large overlap between $5s^2 5p^6$ with the ligand orbitals. Furthermore, the exchange sucks in the $5s^2 5p^6$ orbitals which are aligned parallel to the 4f spin, leaving orbitals with oppositely directed spin to interact more strongly with the ligand, hence giving a contribution of opposite sign to the direct overlap with 4f electrons. This calculation did not include relativistic effects, which have been shown to be important in more recent calculations.[45]

A difficulty of interpretation is that the ligand distance R is not precisely known, as the incorporation of Ln^{n+} into a host lattice inevitably produces some local distortion. Baberschke[46] proposed a method of estimating this for a series of isomorphous hosts like the alkaline earth fluorides, by postulating that $A_p \propto A_s$. Taken together with the postulation that there would be zero distortion when the Ln^{n+} ion has the same ionic radius as the ion it replaces, eg Eu^{2+} in SrF_2, there is enough information to calculate A_p, and hence A_d and **R** for other hosts.

Baker[47] proposed two modifications of this model: (a) assuming that A_s and A_p had a power law dependence upon R with different exponents n_s and n_p, and (b) allowing for an internal electric field, owing both to distortion and extra charge on Ln^{n+}, polarising the ligands by admixture of excited 3p and 3s states (for F⁻ or O^{2-} ligands). This gives a consistent interpretation of A_s and A_p for Eu^{2+} and Gd^{3+} at cubic sites in alkaline earth fluorides, which was later extended to include Gd^{3+} at cubic sites in $CsCdF_3$, $CsCaF_3$[48] and $RbCdF_3$.[49]

In recent theoretical calculations, several mechanisms not previously considered have been shown to give significant contributions to A_s and A_p,[50] including: (a) transfer of ligand electrons into unfilled lanthanide shells such as 5d and 6s, both of which leave a net negative spin density on the ligand, and (b) excitation of an electron from 5p to 4f with simultaneous transfer of a ligand electron into the 5p hole.

Several calculations for these various mechanisms have been made for various Ln^{n+} ions, for comparison with measured parameters in various host lattices. The agreement between theory and experiment is only modest, and the tables of contributions of different signs from the different mechanisms leave one with a feeling that the theory is very complex and could be overlooking other significant mechanisms.

4.6 non-Kramers' ions.-In general, one would not expect to observe EPR in non-Kramers' ions, as the crystal field might be expected to leave $(2J+1)$ well separated singlet states. However, if the symmetry of the crystal field is high enough, one may find doublets. For example, pure axial symmetry [like equation (3), which would obtain for a square anti-prism, ie. ligands at the corners of a cube with one face rotated by 45° about its normal] would produce J doublets $|\pm M_J>$ and one singlet $|0>$. Lower symmetry terms $B_n^m O_n^m(J)$ admix $|M_J>$ with $|M_J \pm m>$. This can lead to two different types of doublet.

All of the possibilities for non-Kramers' ions are illustrated by considering Pr^{3+} (J = 4) at a site of four-fold symmetry, where the only crystal field terms with m not equal to zero are B_n^4, which leads to states:

(a) $\cos\theta|\pm 3> + \sin\theta|\mp 1>$
(b) $\sin\theta|\pm 3> - \cos\theta|\mp 1>$
(c) $|2s>$
(d) $|2a>$
(e) $|4a>$
(f) $\cos\phi|4s> + \sin\phi|0>$
(g) $\cos\phi|0> - \sin\phi|4s>$,

where $|ns> = (2)^{-1/2}\{|n> + |-n\}$ and $|na> = (2)^{-1/2}\{|n> - |-n>\}$.

States (a) and (b) are doublets, degenerate for B = 0: these are type I doublets. States (c), (d) and (e) are well separated singlets. States (f) and (g) are also singlets, but sometimes ϕ is small, when the two states are nearly degenerate for B = 0. Then these two states may be described in S = 1/2 formalism by the spin hamiltonian:

$$H = g_z\mu_B S_z B_z + A_z S_z I_z + \delta S_x. \tag{22}$$

These are type II doublets. The magnetic hyperfine term has been included in (22), but quadrupole and nuclear Zeeman terms have been omitted for clarity. EPR transitions may be excited between these levels by a component of microwave magnetic field parallel to the z-axis, provided that hf > δ. The transition probability is largest for hf = δ, at B = 0. As hf increases, the field B for resonance varies as:

$$(hf)^2 = (g_z\mu_B B_z + A_z M_I)^2 + \delta^2, \tag{23}$$

and the transition probability decreases as:

$$I \propto \delta^2/(hf)^2. \tag{24}$$

For small values of B the energy levels vary quadratically with B, but at high values of B the variation becomes linear. An example of a type II doublet as the ground state is provided by Tb^{3+} in lanthanum ethyl sulphate.[51]

One might expect type I doublets to be described by equation (22) with $\delta = 0$, when the energy levels vary linearly with B, but the transition probability between them is zero. In practice EPR has been observed in such systems, when a type I doublet is the ground state, because the symmetry of the site is lowered by static or dynamic distortions, which lead to matrix elements of the crystal field which admix and lift the degeneracy of the doublets. This may be described by equation (22), where δ has a range of values. For a given hf, the intensity $I(\delta)$ varies as equation (24), so that the line shape is unusual, and the shape and width depends upon the details of the distribution of values of δ. An example of his behaviour in a type I doublet as ground state is provided by Pr^{3+} in lanthanum ethyl sulphate.[51]

It is possible to excite paraelectric transitions between the components of both type I and type II doublets, even for $\delta = 0$, using the electric vector of the microwave field, through terms which should be added to the spin hamiltonian (22): $\mu_B(G_x E_x S_x + G_y E_y S_y)$, where E_x and E_y are components of the applied electric field.[52,53]

As type I or type II doublets are quite common occurrences, their EPR has been observed for non-Kramers' ions in many materials. It is characteristic of a non-Kramers ion that a doublet, even for a crystal field of low symmetry, has only one principal component of the **g**-matrix and the **A**-matrix [equation (22)], though the z-axis for S may have no obvious connection with crystal field axes, and the z-axis for I may be different from that for S. Putting the latter point another way, if the same axis is chosen for I and S, the hyperfine terms may become $A_{zz} S_z I_z + A_{zx} S_z I_x + A_{zy} S_z I_y$.[54]

If the ground state is a singlet, no EPR will be observable, unless by accident there is another level within a distance hf of the ground state. Type II doublets correspond to such a situation, except that the "accident" is not unlikely, as such doublets arise in situations where the states correspond to $|\pm M_J\rangle$, such that crystal field terms B_n^m cannot produce first order admixture, because $m < 2M_J$: although admixtures can be produced in second and higher order, which leads to small values of δ.

However, in many materials, particularly if the site symmetry is low, the ground state is a singlet separated by more than a few cm^{-1} from the next highest state. Such a system exhibits Van Vleck temperature independent paramagnetism and a quadratic Zeeman effect in the optical spectrum. These arise from matrix elements of the Zeeman hamiltonian between the ground and excited states, $|g>$ and $|e>$ respectively. The energy shift ΔW_g and the magnetic moment induced per atom μ_g for the ground state are given by:

$$\Delta W_g = -\Sigma_e |<g|g_J\mu_B\underline{B}.\underline{J}|e>|^2/(W_e - W_g), \quad \mu_g = -2\Delta W_g/B. \quad (25)$$

In the lanthanides these effects can be quite large because of the low energy of the excited states. However, even though no EPR is observed in such systems, they are not uninteresting for magnetic resonance for two reasons.

First, a good deal of work has been done on the EPR of Kramers ions introduced as probes into Van Vleck paramagnets. Exchange interaction between the probe and the host leads to line shifts, contributions to fine structure splittings for S-state probes, and indirect transferred hyperfine structure which leads to line broadening. From measurements of these effects, the exchange interaction between probe and host ions may be determined. A review of this field, together with references and recent developments, is given by Mehran.[55]

Secondly, nuclear magnetic resonance in Van Vleck paramagnets has a large electronic contribution both to the resonant frequency and to the transition probability; so it is arguable that it is appropriate to discuss it in an article on EPR. Off diagonal elements of the first two terms in equation (15), $<g|g_J\mu_B\underline{B}.\underline{J} + a\underline{I}.\underline{J}|e>$ produce second order terms in the spin hamiltonian for the lanthanide nucleus, giving extra contributions to the components of the g_N- and $\underline{P}$-matrices:

$$(\Delta g_N)_{ij} = \Sigma_e\{2g_J\mu_B a/(W_e-W_g)\} <g|J_i|e> <e|J_j|g> \quad (26)$$

$$(\Delta P)_{ij} = \Sigma_e\{a^2/(W_e-W_g)\} <g|J_i|e> <e|J_j|g> \quad (27)$$

This produces an "enhancement" of the nuclear magnetic and electric quadrupole interactions:

$$\underline{g}'_N = \underline{g}_N + \Delta\underline{g}_N, \text{ and } \underline{P}' = \underline{P} + \Delta\underline{P}. \quad (28)$$

The enhancement is anisotropic, and affects the interaction with both static and radio-frequency magnetic fields. This can both shift NMR to higher frequencies, and make transition probabilities much larger that would be expected for conventional NMR. Many measurements of this type have been made by Bleaney and his co-workers.[56]

The distinction drawn between non-Kramers' ions which exhibit EPR in type II doublets and those which don't, because the ground state is a singlet, is imprecise as it depends upon the microwave frequency f. EPR has been observed for one system of two neighbouring Tb^{3+} ions for which ion (1) has $\delta_1 <$ hf, and so exhibits EPR, while ion (2) has $\delta_2 >$ hf. However, as the ions are close neighbours, the transferred hyperfine interaction with the nucleus of ion (2) is resolved as well as the hyperfine structure of ion (1).[57]

5 Electron Nuclear Double Resonance (ENDOR)

ENDOR has been used for three different types of investigation in crystals containing lanthanide ions:
(a) self-ENDOR, ie. ENDOR of the lanthanide nuclei;
(b) ligand-ENDOR, which probes the delocalisation of the unpaired electrons on the lanthanide ions and the extent of covalent bonding;
(c) ENDOR of more distant nuclei, for which the interaction is purely dipole-dipole, owing to the compact nature of the 4f electrons, and so gives structural information about the neighbourhood of the lanthanide ion.

The combination of (b) and (c) enables one to construct a map of the nuclei surrounding the lanthanide ion.[58]

The advantages of ENDOR may be illustrated by considering the spin hamiltonian of a typical system with effective spin S and nuclear spins I_i:

$$H = \mu_B \underline{B}.\underline{g}.\underline{S} + \Sigma_i[\underline{S}.\underline{A}_i.\underline{I}_i + \underline{I}_i.\underline{P}_i.\underline{I}_i - \mu_N g'_{Ni}\underline{B}.\underline{I}_i]. \qquad (29)$$

Generally, from EPR with allowed transitions $\Delta M_S = \pm 1$, $\Delta M_I = 0$, one can find information only about the $\underline{g}$ and $\underline{A}$ matrices; and their accuracy is limited by the line width. Low accuracy information may be obtained about $\underline{P}$ and g'_N from forbidden transitions. ENDOR with allowed transitions $\Delta M_S = 0$, $\Delta M_I = \pm 1$ gives a direct measure of g'_N and $\underline{P}$, as well as of $\underline{A}$. g'_N identifies the nucleus involved. $\underline{P}$ gives a measure of the electric field gradient at the nucleus from all electrons, not just those which are unpaired; it may also have a contribution from the electric field gradient at the lanthanide ion from external charges, antishielded by the electrons of the lanthanide ion.[59,60] Because the line broadening mechanisms, which correspond to inhomogeneous variation of local magnetic field, produce an effect on nuclear moments which is about 10^3 smaller than the effect on electronic moments, the ENDOR lines can be 10^3 times narrower than EPR lines. This gives much greater precision in the measurement of $\underline{A}$, and allows the measurement of $\underline{A}$ in situations where the hyperfine structure is not resolved in EPR, which is often so for ligands of lanthanide ions. The symbol g'_N has been used deliberately rather than g_N, because the nuclear enhancement processes[51] [see section 4.6] can be significant for lanthanides. The spin hamiltonian, particularly for $S > 1/2$, may contain higher order terms, which are usually too small to affect EPR spectra, but which can be measured by ENDOR.[61] Interaction $\underline{I}_i.\underline{J}_{ij}.\underline{I}_j$ between nuclei has not been included in (29) because this is not commonly observed in lanthanide spectra.

For $S = 1/2$, one cannot generally obtain absolute signs of parameters in the spin hamiltonian, though relative signs are obtained. For ligands where the dipole-dipole interaction is dominant, its known sign may be used to infer absolute signs. Double ENDOR may be used to measure the relative signs of two $\underline{A}$ matrices.[62,63] For $S > 1/2$, when signs of the crystal field terms have been measured, it is possible to find for $I > 1/2$ the absolute signs from ENDOR transitions where $|M_S| > 1/2$. Even for $I = 1/2$, it may be possible to observe the effect of cross terms between $b_2{}^0(S)$ and $\underline{S}.\underline{A}.\underline{I}$ on the angular dependence of ENDOR, from which the relative signs of $b_2{}^0$ and $\underline{A}$ may be determined.[64]

5.1 Self ENDOR.- ENDOR of the nuclei of the lanthanide ion has been used to obtain accurate values of g_N. As diamagnetic compounds cannot be made, NMR cannot be used for such measurements. Values of g_N may be inferred from $\underline{A}$ parameters measured by EPR, through the relationship between $\underline{A}$ and the hyperfine parameter a in equation (15), because $a = N(2\mu_B\mu_N g_N N) <r^{-3}>$, where the numerical factor N is known for a particular electronic configuration.[65] However, the values of g_N deduced are limited by uncertainty in the value of $<r^{-3}>$ for 4f electrons. The only complication in the direct determination of g_N is the contribution of enhancement effects.

The large values of $\underline{A}$ for lanthanide nuclei can be determined to quite high precision, about 0.1 MHz, by EPR. Extra precision added by ENDOR is of little significance in the absolute value. However, it can be used comparatively in a number of ways.

The ratio of A values for two isotopes may be different from the ratio of g_N values because there is a contribution to the hyperfine structure from s-electrons, unpaired by exchange with the 4f sub-shell. The finite density of s-electron at the nucleus makes the hyperfine interaction sensitive to the distribution of nuclear magnetism, which is different for different isotopes.[66] The hyperfine structure anomaly $\Delta = \{[A^{(1)}/A^{(2)} x (g_N{}^{(2)}/g_N{}^{(1)}] - 1\}$ has been determined for a number of systems and interpreted in terms of the details of their electronic structure. The temperature and stress dependence of A may be interpreted in terms of vibronic interaction with the surrounding lattice.

The enhancement of the nuclear g factor may be used to determine the energy of an excited state using equation (26), provided that one excited state $|e>$ dominates the summation.[67]

On a pedagogical note, the fact that one can measure with precision a number of terms in the spin hamiltonian for a simple system in a site of high symmetry, allows a clear and quantitative demonstration of the relationship between terms in the spin hamiltonian and terms in the full hamiltonian which give rise to them in some order of perturbation theory.[38]

5.2 Ligand ENDOR.- Much of the early work on ENDOR of ligands of lanthanide ions was done in crystals with the fluorite structure, as was the first such measurement on CaF_2:[68] many references are cited in references [39] and [69]. For some such systems transferred hyperfine structure is partially resolved in the EPR,[70,71] but ENDOR is required for useful evaluation of the parameters. There is some evidence[48] that parameters determined by ENDOR are slightly different from those determined by EPR, but this work has not been followed up by a thorough investigation.

There have been several features of interest in this work. First, it has been possible to estimate and interpret the contribution of covalent bonding. This is not completely straightforward as dipole-dipole effects are large and not well known because of uncertainty about local distortion around the lanthanide impurity. Section 4.5 discusses models for finding the ligand distance R and for interpreting ligand interactions.

Secondly, the effects of temperature and stress upon ligand interactions have also been of interest. The effect of thermal expansion on dipolar contributions is straightforward, but temperature dependence of the non-dipolar contributions A_s and A_p is more complex and more interesting. Also, abrupt change of $\underline{A}$ may be produced by sudden ligand displacements at phase transitions. The interpretation of these changes assists in constructing models of the atomic structural changes [section 6.3]. Changes in $\underline{A}$ have also been measured for applied hydrostatic or uniaxial stress. In principle, the interpretation of these can give information about local compressibility.

Thirdly, ENDOR of the ligand shell has often given clear indication of the mechanism of charge compensation, such as through vacancies, substitutional ions or interstitials, and has sometimes revealed some rather unexpected local environments. For example, in some crystals with the fluorite structure, in which the ENDOR of ^{19}F ligands has not usually been difficult to observe, some lanthanides have shown ENDOR with no evidence of ligands, although more distant ions are evident. Furthermore, this was found for Gd^{3+} in one crystal of CdF_2,[72] yet other crystals gave easily detectable ligand ENDOR.[73] Similarly, missing ligand ENDOR has sometimes been found for 3d group dopants, eg. CdF_2:V^{3+} [74] and CdF_2:Mn^{2+}.[75] It seems probable in such systems that the F$^-$ ligands have been replaced by O^{2-}, for which the natural abundance of ^{17}O is too small for its ENDOR to be observed. This possibility is confirmed by the identification, by ^{17}O and ^{19}F ENDOR, of a trigonal site (T_1) for Yb^{3+} in CaF_2, which showed only one F$^-$ ion in the ligand shell, because the other seven were replaced by four O^{2-} ions.[76]

The early emphasis on fluorides was probably as much due to the ease of ENDOR for the large g_N of ^{19}F as to the interest in fluoride hosts. Since the early work many other ligands have been investigated: Cl, O, PO_4, OH_2. The tables of commonly investigated materials in sections 6.1 and 6.2 contain many materials in which the immediate ligand of Ln^{n+} is oxygen. Unfortunately the natural abundance of ^{17}O is too low for it generally to be possible to obtain ENDOR. In order to be able to make crystals with isotopically enriched ^{17}O, it is necessary to develop techniques which require relatively small quantities of oxygen bearing starting materials, to keep the expense within bounds. This has been possible for some materials, eg. MgO, ThO2, CeO$_2$, but not for many.[77]

5.3 ENDOR of distant nuclei.- By distant nuclei is meant those beyond the ligand shell. Because the 4f electrons are compact, the effects of covalent bonding are usually small for these nuclei. This can be assessed by factorising the **A**-matrix as in equation (19). A$_s$ can arise only from unpaired s-electron on the nucleus, so the magnitude of A$_s$ is a good indication of bonding contribution. The bonding contribution to **A**$_t$ for ligands has usually been found to be small compared with A$_s$, so that small A$_s$ for more distant nuclei has usually been taken to imply even smaller bonding contribution to **A**$_t$. Furthermore, **A**$_t$ can usually be transformed to the dipolar form of equation (21) by a judicious choice of z-direction, so that both the magnitude and direction of **R**, the position of the nucleus relative to the lanthanide ion, can be determined.

One of the nicest examples of this sort of measurement is in the work of Hutchison's group on ^{1}H and ^{2}D ENDOR in lanthanide nicotinate dihydrates (LnND) with formula [Ln(C$_5$H$_4$NCO$_2$)$_3$.2H$_2$O]$_2$.[78-80] The motivation for this work was the exploitation of ENDOR of distant nuclei in proteins, enzymes and other complex organic molecules which are capable of binding paramagnetic lanthanide ions at an active site, in order to determine the structure of the molecule near that site. To test this method, Hutchison's group used the nicotinates as a model system in which the Ln^{3+} ion is bonded at a low symmetry site to complex organic molecules, but where the crystal structure has been determined by X-ray crystallography. They determined the distances of the four nearest water protons at ~ 0.3 nm with a standard deviation of 0.6 pm. The differences from the distances determined by X-ray crystallography are about 6 pm. The only uncertain aspect was that for ENDOR it was necessary to use paramagnetic Ln^{3+} dilutely incorporated into a diamagnetic material LaND. For Nd^{3+} in LaND,[78] the local distortion is probably not large, but for Gd^{3+} in LaND[79,80] small distortion is less certain.

6 Materials investigated

Initially EPR was measured for a range of lanthanide ions in salts grown from aqueous solution, notably the ethyl sulphates, Ln(C$_2$H$_5$OH)$_3$.9H$_2$O (abbreviated LnES), and the double nitrates, Ln$_2$Mg$_3$(NO$_3$)$_{12}$.24H$_2$O. These were selected because (a) they are easy to grow from aqueous solution, (b) they form a long homologous series, and (c) the lanthanide site has high point symmetry. The advantage of (b) is that one can study many Ln^{3+} ions in the same crystal field; and (c) is helpful both in reducing the number of parameters needed to specify the crystal field [section 3], and in giving a clear relationship between the principal directions of the terms in the spin hamiltonian and the crystal structure [section 4.1]. Many of the chemically simpler salts of the lanthanides, which could be grown from aqueous solution, have lanthanide sites of low symmetry, and so were initially less attractive.

The possibility of growth from the melt was first exploited using the anhydrous chlorides, $LnCl_3$ (Ln = La to Gd); these also have high symmetry, but are hygroscopic and so difficult to handle. It was also found to be possible to incorporate lanthanide ions into several types of cubic crystal: alkaline earth fluorides grown from the melt (and also CeO_2 and ThO_2), where the cation site is eight-fold co-ordinated; and also into alkaline earth oxides, where the cation site is six-fold coordinated. When Ln^{n+} does not have the same charge as the cation it replaces, the charge compensating ion or vacancy takes a variety of forms which reduce the O_h point symmetry of the site, or it may be so remote that the departure from O_h symmetry is not observable. This leads to a large number of different sites.[69] By irradiation or electrolysis in the solid state, the lanthanide can often be reduced to the divalent state. The examples of this early work are summarised by Abragam and Bleaney,[2] and the measurements in alkaline earth fluorides by Anderson and Baker.[39,69]

The possibility of using lanthanide ions as the active species in solid state lasers prompted investigation of a wider range of materials, which can be grown of reasonable size, and are stable, robust and not water soluble. Here, symmetry is less important, indeed there are advantages in low symmetry in enhancing optical transition probabilities through configurational mixing by low symmetry components of the crystal field. Also leading to work on low symmetry sites is the possibility of using lanthanide probes, because of their more suitable ionic radius, to elucidate the local structure in either materials whose physical properties depend upon their low symmetry, such as ferro-electrics, ferro-elastics peizo-electrics; or others which are inherently of low symmetry, or interesting for other reasons, such as enzymes, proteins and other biological material.

These comments show that there are two broad classes of material which have been investigated: (a) lanthanide compounds which are diamagnetic, of La, Lu or Y (or of Van Vleck paramagnets), where a paramagnetic lanthanide substitutes for a host lanthanide ion; and (b) other compounds where a paramagnetic lanthanide substitutes for a different type of ion.

The last comprehensive review of EPR results, including lanthanides, was in 1959[1]. Many more results have been tabulated in more recent publications, but they do not represent a collection of all data measured. From 1971 to 1979, Buckmaster published an annual review of EPR under classified headings, so it is easy to find details of work published on lanthanides.[81] Since 1983, the *Rare Earth Bulletin* has catalogued the literature on EPR.

Although there are many cases of individual lanthanide ions in odd materials, most investigations have been made of particular materials containing many of the lanthanide series. Section 6.1 discusses some of the lanthanide compounds, both grown from aqueous solution, from the melt, from high temperature flux or by other means, which have been used as hosts for a range of Ln^{3+} ions. Section 6.2 discusses other materials in which a wide range of lanthanide ions have been incorporated, substituting Ln^{n+} for a cation of similar radius.

There is also a large number of materials in which one or two lanthanides have been incorporated as probes to study the symmetry of the site, or the change in symmetry as a function of some external parameter [see section 6.3 for example].

Sections 6.1 and 6.2 do not discuss a number of elemental metallic hosts in which interesting measurements have been made such as: cubic closed packed elements like Cu, Ag, Au; hexagonal close packed elements like Mg, Y. Nor do they discuss a wide range of intermetallic compounds which have been used as hosts. EPR in these metallic systems has been reviewed by Taylor.[82]

6.1 Measurements on lanthanide compounds.-Table 3 lists some of the lanthanide salts, both grown from aqueous solution, from the melt, from high temperature flux or by other means, which have been used as hosts for a range of Ln^{3+} ions. These compounds will not each be discussed in detail, but some comments will be made about a selection of them.

A large number of measurements have been made in $LaCl_3$ and isostructural materials. Although the crystal structure of $LnCl_3$ for Ln = Tb to Lu (and Y) is different from $LaCl_3$, it is possible to grow crystals of $LaCl_3$ containing the heavier lanthanides [see section 7 and table 5.8 of reference (2)]. The crystal field parameters for all Ln^{3+} ions in the simpler structure with C_{3h} point symmetry make interesting comparison with similar information for the ethyl sulphates: both crystals require four B_n^m to specify the crystal field.[2] The change in structure of $LnCl_3$ is an example of the reduction in size of the coordination sphere forcing a change of crystal structure at some point in the lanthanide series. $LaBr_3$ and $La(OH)_3$ have the same structure as $LaCl_3$, but the change to a different structure occurs earlier in the series for the larger Br⁻ ion; and for the smaller (OH)⁻ ion, $Ln(OH)_3$ forms a homologous series for all Ln^{3+}.

LaF_3 is an example of a structure where the symmetry of the La^{3+} site, as revealed by EPR of Ln^{3+}, was needed to provide confirmation of an ambiguously determined crystal structure.[83] The point symmetry is close to C_{3v}, but is lowered to C_{2v} by small displacement of the Ln^{3+} ion. LnF_3 has this structure, with 9-fold coordination with F⁻, for Ln = La to Sm, and is 8-fold coordinated with orthorhombic structure for Ln = Gd to Lu (and Y): another example of the change in size of the coordination sphere. However, it has been possible to incorporate the heavier Ln^{3+} into LaF_3.

Many measurements of EPR of Ln^{3+} doped into a diamagnetic salt of La, Lu or Y have been made as part of an investigation of interesting magnetic properties of the isomorphous Ln salt. EPR of the Ln salt itself may give wide lines [section 4.3], and may be impossible because of the effects of interactions, so may give limited information. EPR of the dilute salt gives much more

3: *EPR and ENDOR in the Lanthanides* 157

precise information about the ground state of the isolated Ln^{3+} ion, though it may be of limited value in discussion of the magnetic properties of the Ln salt because of differences in the Ln^{3+} site in the two materials.

Table 3 Lanthanide salts in which EPR has been measured for many lanthanide ions. n is the number of sites distinguishable by EPR.

host	space group	n	point symmetry
$Ln(C_2H_5OH)_3.9H_2O$	P63/m	1	C_{3h}(pseudo D_{3h})
$Ln_2Mg_3(NO_3)_{12}.24.H_2O$	R3	1	C_3(pseudo C_{3v})
$LaCl_3$, $LaBr_3$, $La(OH)_3$	C63/m	1	C_{3h}
LaF_3	C6/mcm	6	C_{2v}
$Ln_3Al_5O_{12}$	Ia3d	8	C_2
$LnVO_4$, $LnPO_4$, $LnAsO_4$	$I4_1$/amd	1	S_4
$LiYF_4$	$I4_1$/amd	1	S_4
$Cs_2NaLnCl_6$	Fm3m*	1	O_h*
$NH_4Ln(SO_4)_2.4H_2O$	$P2_1/b$**	2	C_1(pseudo C_4)
$Ln(BrO_3)_3.9H_2O$	$Cmc2_1$	6	σ_v
$LnCl_3.6H_2O$	P2/n	2	C_1
$[Ln(C_5H_4NCO_2)_3.2H_2O]_2$	$P2_1/c$	2	C_1
$LnBa_2Cu_3O_6$	P4/mmm	1	C_{4v}
$LnBa_2Cu_3O_7$	Pmmm	1	C_{2v}
LnB_6	Pm3m	1	O_h

* the symmetry is lower for some compounds (see text)
** symmetry sometimes lowered by phase transitions

The garnets ($Ln_3Al_5O_{12}$) and the zircons ($LnVO_4$, etc) are examples where there has been a great deal of interest in the magnetic properties of the Ln salt.[3] The zircons have tetragonal symmetry, the crystal field being specified by five parameters B_n^m, so they form the next simplest system after cubic and hexagonal symmetry. A lot of measurements have been made on these materials, and many of them have been discussed by Bleaney.[3] One limitation of materials like these where the Ln^{3+} is immediately coordinated with O^{2-} is that ENDOR of the ligands cannot be performed. In pure Ln compounds of these two materials there are examples of obtaining a reduced line width by the use of a very high magnetic field and a very high microwave frequency. Janssen has used fields up to 6T with resonance frequencies up to 1200GHz generated by a far infra-red laser for measurements on TmVO$_4$ [35] and Dy$_3$Al$_5$O$_{12}$ (DAG).[84] TmVO$_4$ is of interest, as it exhibits a cooperative Jahn-Teller effect with a transition at T$_D$ = 2.15K. The EPR lines observed were fitted to spin hamiltonian (22). The g-values were similar to those in a diluted sample. Some lines were observed where two coupled spins absorb a single photon, and others due to defect sites which slightly distort the symmetry of the Tm^{3+} site.[35] DAG was considered to be a well understood material, but the explanation of its magnetic properties by a simple Ising model depended upon the assumption that the crystals were free of imperfections. The measurements were made to test this, using a spherical sample at fields up to 7.3 T and frequencies up to 1629 GHz. In addition to the expected strong resonance from axially symmetric sites with g$_z$ = 18.02(7), a number of unexpected resonances were observed, some of which correspond to Dy^{3+} on octahedral crystallographic sites.[84]

Table 3 features a number of materials with cubic sites. In the epasolites (Cs$_2$NaLnCl$_6$) the Ln^{3+} site is 6-fold coordinated with Cl$^-$. EPR for some Ln^{3+} in Cs$_2$NaYCl$_6$ shows that the site is cubic,[85] and for others shows phase transitions at low temperature. EPR in undiluted compounds with Ln = Pr, Tb and Tm indicates a small tetragonal distortion of the Ln^{3+} site.[86]

The lanthanide bromates, Ln(BrO$_3$)$_3$.9H$_2$O, were expected to be similar to LnES, but there has been considerable controversy about their crystal structure. EPR has been used to distinguish between assignments of hexagonal and orthorhombic symmetry.[87,88] The actual space group is C$_{2v}$^{12}, the lanthanide site having σ_v symmetry. The structure is produced by a small distortion of a D$_{3h}$ structure at the rare earth site. The crystals are pseudo-hexagonal, with a tendency to twin about the pseudo-hexagonal axis, which has caused confusion in some measurements. The lower symmetry for the Ln^{3+} site was first confirmed by EPR of Nd^{3+} [87] and later for Gd^{3+}.[88]

The lanthanide ammonium sulphates have been studied because of their phase transitions [section6.3], and lanthanide nicotinates because of their low symmetry [section 6.4], because of their similarities to biological systems [section 5.3] and the interactions between the two Ln^{3+} ions in the dimer [section 7].

A number of electrically conducting systems have been studied. LnB$_6$ is an interesting cubic host material of this type. LaB$_6$ is metallic, SmB$_6$ is an intermediate mixed-valence metal where the configuration of Sm^{3+} fluctuates between 4f^6 and 4f^{5}5d, and YbB$_6$ is a semiconductor. A probe ion such as Er^{3+} exhibits EPR at a cubic site. Γ_6 is the ground state for YbB$_6$ and LaB$_6$, with a negative g-shift from the nominal value of g = 6, due to exchange with conduction electrons.[89] The exchange also leads to a temperature dependent line width. Er^{3+} in SmB$_6$ shows a Γ_8 ground state with a dynamic Jahn-Teller effect, indicative of enhanced electron-phonon coupling mediated by the valence fluctuations of the host.[90]

3: *EPR and ENDOR in the Lanthanides*

A class of lanthanide compound which has received a lot of recent attention, including EPR, is the high T_c superconductors based on $LnBa_2Cu_3O_7$ and Ln_2CuO_4. The material which exhibits superconductivity with the highest T_c is intermediate in oxygen concentration between the tetragonal insulating compound, $YBa_2Cu_3O_6$, and the orthorhombic compound, $YBa_2Cu_3O_7$. The most fruitful information has come from non-resonant, very low field microwave absorption, connected with weak Josephson links in the material; and from Cu^{3+} EPR, because the superconductivity is related to the Cu,O layers in the materials. The role of the lanthanide is minor, as shown by the insensitivity of the superconducting properties to substitution of different lanthanides.

Nevertheless, a number of measurements have been made of EPR of Gd^{3+} and other Ln^{3+} in both pure an undoped materials. Much of this work has been done on polycrystalline samples, because only recently have good single crystals become available, though sometimes the magnetic anisotropy has been used to produce alignment of crystallites in an applied magnetic field. Many measurements have been done on poorly characterised material, sometimes indicated by different results by different authors on nominally similar material.

No strong theme appears to run through the publications on lanthanide EPR in these systems, so a few instances are given to illustrate the results.

Some of the difficulties are illustrated by measurements of Er^{3+} in $Er_xY_{1-x}Ba_2Cu_3O_6$ for $0.02 < x < 1$. Interpretation of the broad anisotropic lines is complicated and susceptible to at least three different explanations.[91]

For Gd^{3+}, even at low concentrations in $YBa_2Cu_3O_7$, the lines tend to be wide, and only the central 1/2 to -1/2 transition is strong, other lines being very broad.[92] However, the structure has been fitted using fine structure terms up to b_6^m, including forbidden transitions.[93] This structure shows no major change for $YBa_2Cu_3O_y$ for $6 < y < 6.14$, but above 6.14 the lines other than the central line are split into up to five sub-bands, indicative of regular oxygen configurations of the four available O^{2-} sites closest to Gd^{3+} in the Cu(1) layers, with O^{2-} occupancy varying from zero to four.

Measurements of powder samples of $GdBa_2Cu_3O_y$ showed no distinct difference in EPR signal between the orthorhombic structure (superconducting below $T_c = 92K$) and the tetragonal structure (no superconducting transition).[94] The interaction between Gd^{3+} and Cu^{2+} is so weak that it does not suppress the superconductivity in the orthorhombic phase, indicating that the superconductivity is confined to the Cu layers. The line width for Gd^{3+} follows the temperature dependence of the magnetic susceptibility, and indicates an exchange interaction between nearest neighbour Gd^{3+} ions of about 0.1K. The EPR signal, strong for $T > T_c$, persists below T_c but becomes difficult to separate from a strong background microwave absorption which rapidly increases as T is lowered.

In $GdBa_2Cu_3O_7$, superconductivity and antiferromagnetism in the Gd sublattice co-exist below $T_N = 2.3$ K. The Gd ordering is three dimensional, showing that some coupling between Gd^{3+} ions must be transferred through the region responsible for the superconductivity. Measurements between 9 and 36 GHz show a change in line width which indicates that the exchange frequency f_{ex} lies between these two limits of Zeeman frequency.[95] From this, the correlation time and the exchange integral have been estimated, and shown to be in good agreement with that deduced from the Neel temperature. The exchange field is about three times the dipolar field. The g-value was found to increase with decreasing oxygen content. The line width, Neel temperature and exchange field were found to depend upon the nickel content of the material when Cu was replaced by Ni, whereas the effect of replacing of Cu by Co is small. As it is known that Ni replaces Cu in Cu(2) planes, while Co replaces Cu in Cu(1) chains, it is presumed that exchange between Gd^{3+} ions comes from superexchange via O^{2-} in the Cu(2) planes. The dependence of g-value upon oxygen content also suggests some interaction between the Gd spins and the superconducting carriers in the Cu(2) planes.

Measurements in $EuBa_2Cu_3O_{7-\delta}$[96] give a spectrum which is quite different from that found by the same group in the Gd salt,[97] in that it exhibits an extensive hyperfine structure. This is interpreted as arising from a small quantity of Eu substituting as Eu^{2+} for Ba^{2+} which occurs in pairs separated by 0.388 nm and connected by two Eu-O-Eu bridges. It has previously been shown that Eu in $LnBa_2Cu_3O_{7-\delta}$ can substitute either for Ln^{3+} or Ba^{2+}.[98] That this type of spectrum does not occur in the Gd salt is then simply accounted for, as Gd^{3+} does not substitute for Ba^{2+} in these compounds.

6.2 Insulating hosts which are not lanthanide salts.-Table 4 lists other materials in which a wide range of lanthanide ions have been incorporated, substituting Ln^{n+} for a cation of similar radius. The large amount of work on alkaline earth fluorides has been discussed above [section 6], together with mention of alkaline earth oxides. In the latter, Ln^{n+} is octahedrally coordinated with O^{2-}, but as very little work has been done on ^{17}O enriched samples, they have not provided data on ligand hyperfine interactions for comparison with the wealth of data for the 8-fold coordinated MF_2.

This sort of comparison is provided by the fluoride perovskites, MNF_3, where the divalent M^{2+} ion is 6-fold coordinated to F^-. Many lanthanides have been measured at cubic sites, including ENDOR determinations of the A-matrices of the ligands and more distant ions; and the theoretical techniques for interpretation of ligand hyperfine interaction discussed in section 4.5 have been applied. There are even isolated examples of cation neighbour ENDOR.[99] There has also been interest in comparing the crystal field, and fine structure, parameters with those in MF_2, particularly using the superposition model [section 3], and also in studying their changes under

3: EPR and ENDOR in the Lanthanides

applied stress [section 3.1]. Some lanthanides are believed to substitute at the N^+ site, which is 12-fold cordinated to F^-, when the M^{2+} site is small, e.g. Zn^{2+} or Mg^{2+}.[100] As with MF_2, there are many sites of lower than cubic symmetry, the replacement of F^- by O^{2-} being common. Several of the MNF_3 compounds undergo structural phase transitions, which have been studied by EPR and ENDOR of Gd^{3+} [section 6.3].

Table 4 Crystals in which EPR has been measured for many lanthanide ions. n is the number of sites distinguishable by EPR

host	space group	n	point symmetry
MF_2, CeO_2, ThO_2, TmH_2	O_h^5	1*	O_h*
MO	O_h^5	1*	O_h*
MNF_3, $MNCl_3$	O_h^5**	1*	O_h*
$KTaO_3$	O_h^5**	1**	O_h**
MWO_4	$I4_1/a$	1	S_4

* the symmetry is often lowered by local charge compensation
** the symmetry is lowered for some compounds by phase transitions

It has been found possible to substitute Eu^{2+} for the cation in many alkali halides with charge compensation achieved by the removal of a neighbouring cation.[101] Analysis of the crystal field parameters in terms of the superposition model [section 4.1] suggests that local distortions are significant. A model for calculating the local distortions has been proposed,[102] together with determination of the parameters $\bar{b}_2$, $\bar{b}_4$, t_2 and t_4, demonstrating that a satisfactory account of the measured parameters can be achieved by the combined superposition-distortion model. The spectroscopic properties of divalent lanthanides (Eu^{2+}, Sm^{2+} and Yb^{2+}) in alkali halides have been reviewed together with EPR data for Eu^{2+}.[103] Not only have pure alkali halides been studied, but also the effect on fine structure parameters b_n^m of the variation of x in mixed crystals like $Na_{1-x}K_xCl$ and $KCl_{1-x}Br_x$.

EPR has been observed for lanthanides incorporated into several semiconductors, such as GaSe, GaAs, CdS, CdTe, PbTe, SnTe, InP, SmS, SmSe, and SmTe. As an example of this work, Gd^{3+} has been used as a probe to investigate the structure of impurity centres in the layer semiconductors GaSe.[104] Six different Gd^{3+} centres were found with axial symmetry about the crystallographic c-axis, with different values of b_n^0. ENDOR was measured for ^{69}G and ^{71}Ga, and of ^{23}Na for one centre, and of ^{7}Li for three others. The transferred hyperfine parameters for the alkali metals were of purely dipolar form [section 4.5], and were therefore used to measure the distance between the alkali metal and Gd^{3+}. The multiplicity of sites is explained in terms of occupation of two different types of interlayer interstitial site v and w, at different distances from the Gd^{3+} site. The small Li^+ ion can go into either site, but the larger Na^+ can go only into w sites (K^+ was found to go into neither site). Two of the centres are uncompensated sites with v or w neighbours, so show no alkali ENDOR. The sixth site is accounted for by a small percentage of γ-polytype, where Gd^{3+} has one v and one w neighbour, the former occupied by Li^+. A consistent account of the magnitude of b_2^0 for the six sites has been given in terms of these models.

6.3 Structural phase transitions.-As the **g**-, **D**- and **A**-matrices of a paramagnetic species, or the fine structure parameters b_n^m for an S-state ion, are sensitive to the symmetry of its site, there is a considerable literature on the use of EPR to detect changes in such a site symmetry as a function of external parameters, such as might occur at a phase transition.[105] Much of the literature concerns radical species produced by irradiation, or 3d group probes, but lanthanide probes have been used as well, their larger ionic radius making them suitable for incorporation into a different set of materials from 3d group probes. If there is a good theoretical understanding of the interaction between the paramagnetic species and its environment, the quantitative change in the spin hamiltonian parameters may be interpreted in terms of specific structural distortion.

As most Ln^{3+} ions exhibit EPR only below about 20K, their use as a probe is limited (but see below); so Gd^{3+} and Eu^{2+} have been the most used probes. For these, the fine structure is the main indicator of changes in the environment. Gd^{3+} has the advantage over Eu^{2+} in not having an extensive hyperfine structure to confuse the spectrum: the hyperfine structure of ^{151}Eu and ^{153}Eu comprises 12 lines spread over about 18 mT. This can be an enormous advantage where a change of local point symmetry produces an increase in the number of symmetry related sites, so that each line in one phase may be split into several lines in the other phase.

The earliest use of a lanthanide ion in this type of measurement was in the lanthanide hexa-antipyrene iodides (LnHAPI) containing Gd^{3+} as a probe.[83] These salts form a homologous series throughout the whole lanthanide group, with hexagonal symmetry and one Ln^{3+} site at room temperature. However, EPR for most lanthanides, for which temperatures below 20K were required, showed more sites and lower symmetry. EPR of Gd^{3+} revealed a crystallographic transition at low temperature, the transition temperature increasing from LaHAPI (15K) to YbHAPI (130K), which lowers the symmetry of the site and increases the number of spectra from symmetry related but differently oriented sites. In these materials, unusually, the EPR spectrum of Ce^{3+} was observable up to 90K, so that its spectrum was also used to study the transition: this had the advantage that the spectrum has no fine or hyperfine structure, making it less complicated. The change in the **g**-matrix is indicative of the nature and magnitude of the distortion. Although less information is available than for the change in b_n^m parameters for Gd^{3+}, the difficulty of interpretation of the latter may not make that such a disadvantage.

The superposition model is ideal for interpreting the change in b_n^m in terms of local distortion, as the theory contains explicit dependence upon the ligand positions $\underline{R}_i$ [section 4.1]. Similar information may be obtained from ENDOR measurements of the ligand transferred hyperfine parameters [section 4.5].

Several of the fluoride perovskites, MNF_3, which have O_h^1 space group at high temperature, undergo structural phase transitions at low temperature. Extensive measurements of Gd^{3+} centres in $RbCaF_3$ and $RbCdF_3$, including Gd^{3+}-O^{2-} and Gd^{3+}-V_{Cd}, have been used as probes of the phase transition to D_{4h}^{18}.[106] It has also been possible to measure the shift of T_c under hydrostatic pressure.[107] Measurements have also been made for uniaxial stress in various crystallographic directions. EPR has also been used to study fluctuations near T_c[108] as well as dynamic clusters and disorder.[109] EPR in $RbCdF_3$ below T_c suggests that the nature of the transition is a rotation of adjacent CdF_6 octahedra in opposite directions about one of the cubic axes, so leading to domains for rotation about each of the three axes. ENDOR of the ^{19}F ligands confirms this model, in showing that both the ligands along the c-axis and those which have rotated about the c-axis, have nearly the same values of A_s and $\underline{A}_t$, and corresponding at 4.2K to a rotation of $\pm$ 9.9(2)°.[48]

A large number of measurements have been made of EPR of Gd^{3+} probes in $LnNH_4(SO_4)_2 4H_2O$.[110] The symmetry of the Gd^{3+} site is low, requiring a large number of parameters b_n^m to specify it. Gd^{3+} is nine-fold coordinated to O^{2-}, six in SO_4^{2-} tetrahedra and three in H_2O; the symmetry corresponding to a distorted mono-capped square antiprism {compare with the lanthanide nicotinates [section 6.4.1], and lanthanide glasses [section 6.4.2]}. The superposition model has been applied to parameters b_2^0 and b_4^0, but a complete analysis of all b_n^m has not been attempted because of incomplete crystal structure data, and uncertainty about the lattice distortion produced by the probe Gd^{3+} ion. The temperature dependence of b_n^0 shows abrupt discontinuities indicative of phase transitions, in some cases coupled with an increase in the number of symmetry related sites indicative of a change of space group.

$Pb_3(PO_4)_2$ undergoes a ferro-elastic transition near $T_c = 180°C$. Above T_c, the structure is rhombohedral, two different Pb^{2+} sites having D_{3d} (Pb_I) and C_{3v} (Pb_{II}) point symmetry. Below T_c, the structure is monoclinic, Pb_I changing to C_1 and Pb_{II} to C_2. Gd^{3+} probes appear to substitute only for Pb_{II}.[111] In both phases the g-matrix shows significant anisotropy, and b_2^m terms dominate (and have been expressed in D,E formalism); but significant b_4^m and b_6^m have been determined. There is an abrupt change in D and E at the transition, as well as a change in line width. The latter also broadens close to T_c, both for $T>T_c$ and $T<T_c$, because of crystal

field fluctuations induced by the soft mode. Small differences between the results for Gd^{3+} and Mn^{2+} as the probe indicate the presence of a compensating charge for Gd^{3+}.

Gd^{3+} has been used as a probe in $ThBr_4$ [112] and $ThCl_4$,[113] which undergo a normal to incommensurate phase transition at T_I = 95 and 70K respectively. In the incommensurate phase, where atomic positions suffer a modulation incommensurate with the crystal structure, the cation sites undergo a similar modulation of $b_n{}^m$ from site to site. This leads to a distribution of line shifts, like a restricted type of powder pattern, which gives EPR over a wide range of fields with singularities. Unlike a true powder pattern, the line shape depends upon the direction of $\underline{B}$ relative to the modulation axis. The amplitude of the line shift varies as $(T_I-T)^{0.35}$. The spectra for $T>T_I$ show that the symmetry of the Gd^{3+} site is lower than that of Th^{4+}. An application of the superposition model enabled the charge compensation mechanism to be identified: as a Cl^- vacancy in the nearest neighbour shell for $ThCl_4$, and a Br^- vacancy in the next-nearest neighbour shell for $ThBr_4$. In spite of the reduction of symmetry by charge compensation, the Gd^{3+} centre is a very sensitive probe of the local structure. Details of the incommensurate phase have been inferred using the superposition model. Near T_I, the line shape indicates the co-existence of the normal and incommensurate phases.

6.4 Sites of low symmetry.-Much early work concentrated upon lanthanide sites of relatively high symmetry: hexagonal, cubic or tetragonal. High symmetry is useful because only a few terms are needed to describe the crystal field, so the probability of obtaining a complete understanding of the system is greater.

Recently there has been increasing interest in low symmetry systems of a variety of types:
(a) biological or complex organo-metallic systems;
(b) optically active systems, which require spiral structures;
(c) laser systems, where the low symmetry increases the configurational mixing needed for strong optical transitions in the visible;
(d) glasses, where the sites have random distortions, because of their potential applications in optical devices, such as optical fibres, solid state lasers, or for up-conversion of IR radiation to visible fluorescence.[114]

The measurement and understanding of low symmetry sites, particularly if they are randomly distributed in a powder sample, is much more demanding because there are many more free parameters in the crystal field.

For non-Kramers ions at such sites one would expect a set of $(2J+1)$ widely spaced singlet states. It is therefore surprising that for some ions the ground state appears to be a doublet. This is quite common for Tm^{3+}, $4f^{12}$ (J = 6), in crystals of low symmetry, where the ground state often approximates to $|\pm 6>$, with a small zero field splitting.

6.4.1 Crystals of low symmetry.- The nicotinate dihydrates have already been mentioned [section 5.3] as sites of low symmetry. Yet for TmND, DyND, ErND and Tb^{3+} in LaND, the ground state is a doublet with an Ising-like g-matrix with $g_* = 2Jg_J$, and a small zero field splitting for the non-Kramers ions Tb^{3+} and Tm^{3+}.[34,57,115-7] This unexpected result suggests that there might be some hidden symmetry in the crystal field, and this may be so, as the eight oxygen ligands of Ln^{3+} form a distorted square antiprism. These four ions form an interesting set, because Tb^{3+} and Tm^{3+} both have J = 6 and Dy^{3+} and Er^{3+} both have J = 15/2. However, the effect of the same crystal field on the pair of ions with the same J will be different because they have different L and S [see table 1], so states $|J, M_J\rangle$ correspond to different admixtures of $|L, M_L, S, M_S\rangle$, and as the crystal field acts upon the orbital part of the states, its effect is different in the two ions. This is reflected in different magnitudes *and signs* of the Stevens parameters θ_n [section 3]. These are such that it is impossible to find a crystal field for which all four ions have ground states corresponding to predominantly $|M_J = \pm J\rangle$.

As information about the excited states in these systems is limited, it is impossible to derive the crystal field from experimental information. As an alternative, an attempt was made to predict the ground state by calculating the crystal field using the superposition model, as the crystal structure has been determined for many of the pure LnND, so the relative positions of the ligand oxygen atoms is known. This is a crude model as the oxygens are not equivalent: two are water oxygens, four are in the carboxylate bridges between the two Ln^{3+} components of the dimer, and two are chelated. Also, the distortion of the neighbourhood when an impurity is incorporated into LaND is not known. In the absence of knowledge about the excited states, the only way to assess the validity of this model is through the predicted components of the **g**-matrix, and δ values for non-Kramers' ions. These are in surprisingly good accord with experiment for some ions, but not for others.

6.4.2 Glasses.- Another type of system with low-symmetry sites is provided by various sorts of glass. The crystal field for lanthanides incorporated into glasses is likely to vary both in orientation and magnitude, reflecting the variations in coordination of the lanthanide ion.

Most EPR in glasses has involved Gd^{3+} or Eu^{2+} ions because their EPR is observable at room temperature. Because the symmetry of the sites is low, the fine structure is dominated by n = 2 terms, which may therefore be expressed in the D,E formalism. A distribution of fine structure parameters produces a spectrum with singularities at specific values of $g_{eff} = hf/\mu_B B$. This type of spectrum has been labelled the U-spectrum[118] in view of its ubiquitous occurrence in both vitreous and disordered polycrystalline materials. The spectra indicate both a broad distribution in magnitude of fine structure parameters and a random orientation. Griscom[119] hypothesised that large ions like Gd^{3+} in glasses tend to dictate their own environments, and this was confirmed by work in alkali-zinc borosulphate glasses[120] and oxy-halo borate glasses.[121] Their broad distribution of site symmetries contrasts markedly with iron group impurities in oxide, fluoride and borosulphate glasses. It has been shown for oxide glasses ($BaO-P_2O_5$, $MgO-P_2O_5$ and SiO_2-Al_2O_3),[122] using a range of spectrometers with different frequency between 2.7 and 10 GHz, that a plot of EPR intensity against g_e/g_{eff} shows a strong resemblance between spectra of different ions (Gd^{3+} or Eu^{2+}), or the same ion in different glassy matrices, at two different appropriately chosen frequencies f_1 and f_2. It is shown that this can be explained if the ratios of various terms in the fine structure spin hamiltonian is constant (equal to f_1/f_2) for Gd^{3+} and Eu^{2+} in the same glass, or for either ion in different glasses.

Examples of other lanthanides are provided by the fluorozirconate glasses based on ZrF_4, which are easily doped with lanthanides. The EPR spectrum of Tm^{3+} and Tb^{3+} in these glasses is characterised by sharp discontinuities in curvature of the first derivative, or as cusps in the second derivative.[123] The spectra are explained using a model where the ground state for both ions corresponds to $M_J = \pm 6$, with a small zero field splitting δ varying from zero with a distribution function $P(\delta)$. The axis of quantisation is assumed to be distributed randomly. Except for the distribution of values of δ, this situation is reminiscent of the ground state of these two ions in the low symmetry sites of the crystalline nicotinates [section 6.4.1] The predictions were found to be sensitive to the relative direction of microwave and steady magnetic fields. The theory, using equation (22) as spin hamiltonian, predicts discontinuities both at frequency independent fields $B_z = (A_z M_I/g_z \mu_B)$, and at $B_z = \pm (hf/g_z \mu_B) - (A_z M_I/g_z \mu_B)$, associated with sites for which $\delta = hf$ and 0 respectively. The occurrence of $|\pm 6>$ ground states indicates the occurrence of certain characteristic arrangements of fluorine ligand, rather than a random arrangement. Various types of fluorine coordinated sites were considered with coordination numbers between 8 and 9, which might give $|\pm 6>$ ground states. The superposition model was used to calculate the crystal fields for these models with intrinsic parameters taken from Er^{3+} in LaF^{3+}.[124] The results seem to indicate a preference for Tm^{3+} to occupy an 8-coordinated square antiprism site, and for the larger Tb^{3+} to occupy a 9-coordinated tricapped trigonal prism site.

Measurements have also been reported recently in phosphate glasses of Yb^{3+} [125] and Nd^{3+} [126] using electron spin echo methods to measure transferred hyperfine interactions with ^{31}P neighbours.

A matter of considerable interest in glasses, and in amorphous material, is the phonon density of states. Because spin-lattice relaxation depends upon this density of states, the relaxation of low concentrations of paramagnetic ions may be used as a probe of the density of phonon states. Yb^{3+} has been used in this way in silicate glasses.[127] In a previous study of Yb^{3+} in phosphate, silicate and borate glasses,[128] it was found that the SiO_4 tetrahedra cluster around Yb^{3+} in an approximately octahedral arrangement, giving a ground doublet about 330 cm^{-1} below the first excited state, which makes the Orbach process insignificant. The EPR is a broad featureless peak near g = 3. Pulse saturation-recovery measurements[127] showed relaxation rates dependent upon various parameters. For low power, short pulses and thin samples, the recovery was close to exponential. At higher concentrations of Yb^{3+} the relaxation was dominated by cross relaxation to exchange coupled Yb^{3+} pairs; but at low concentrations, the effects of cross relaxation are small, and the Raman process is found to be anomalously weak, varying as T^6 rather than the usual T^9 for crystalline solids. Several models for a modified density of phonon states are described, all involving a critical temperature or frequency at which vibrational excitations change character from extended to localised. The discussion is inconclusive, but suggests that the relaxation mechanism is associated with non-localised phonons. The temptation to draw parallels between the anomalous T dependence in these glasses with similar results for protein solutions is discouraged, even though crystalline protein samples also exhibit similar anomalous behaviour.

7 Interactions between paramagnetic ions

Measurements of interactions between neighbouring lanthanide ions go back almost to the earliest EPR measurements on lanthanides. The spectrum of neodymium ethyl sulphate for $\underline{B}$ parallel to the c-axis shows a partially resolved three line structure with intensities 1:2:1,[33] corresponding to the internal field set up by magnetic dipole-dipole interaction with two nearest neighbours along the c-axis.

Soon after this, interactions were measured with better resolution using semi-dilute crystals, with sufficiently low concentration of paramagnetic ions to avoid large spin-spin broadening, but with large enough concentration to have observable densities of paired ions.[128]

Exchange interaction between lanthanide ions is usually much smaller than between transition metal ions, and it usually does not have the simple Heisenberg form. It often does not overwhelm the other interaction mechanisms, so that it is sometimes possible to detect and measure the effect of magnetic dipole-dipole, electric quadrupole-quadrupole and virtual phonon exchange interactions. The investigation of interactions up to 1971 has been reviewed previously.[129]

A pair of interacting ions with $S_1 = S_2 = 1/2$ is described by the spin hamiltonian:

$$H = \mu_B \underline{S}_1 \cdot \underline{g}_1 \cdot \underline{B} + \mu_B \underline{S}_2 \underline{g}_2 \cdot \underline{B} + \underline{S}_1 \cdot \underline{J} \cdot \underline{S}_2. \tag{30}$$

For similar ions with $\underline{g}_1 = \underline{g}_2$ the interaction term may be separated into a scalar and a traceless term:

$$\underline{S}_1 \cdot \underline{J} \cdot \underline{S}_2 = J_o \underline{S}_1 \cdot \underline{S}_2 + \underline{S}_1 \cdot \underline{K} \cdot \underline{S}_2, \tag{31}$$

and the spins couple to form states of total spin $\underline{S} = \underline{S}_1 + \underline{S}_2$. EPR transitions are not allowed between states with $S = 1$ and $S = 0$, whose centres of gravity are separated by energy J_o, but only within the triplet of levels associated with $S = 1$ whose energies are affected by $\underline{K}$. Hence, for similar ions only the elements of $\underline{K}$ may be determined from pair spectra. For dissimilar ions, there is no such restriction, and all elements of $\underline{J}$ may be determined. That hyperfine interactions in lanthanide ions are typically large, and interactions $\underline{J}$ between ions are typically small, raises the possibility that neighbours with different isotopes, or even for the same isotope in different nuclear quantum states, the hyperfine interaction may make the ions sufficiently different for it to be possible to measure J_o as well as $\underline{K}$. In the context of the above formalism, this may be viewed in terms of matrix elements of the hyperfine interaction, say $\underline{S}_1.\underline{A}_1.\underline{I}_1$, admixing states corresponding to $S = 1$ with $S = 0$, and so allowing transitions between them which would otherwise be forbidden. The measurement of such pair spectra gives by far the most detailed type of information about interactions between neighbouring paramagnetic ions.

The early work on pair spectra for Ln^{n+} ions, reviewed in 1971,[129] was mainly in ethyl sulphates, $LaCl_3$ and $LaBr_3$, and Tm^{2+} pairs in some alkaline earth fluorides. Since then some measurements have been made in $La(OH)_3$ and a wider range of alkaline earth fluorides. But, considering the wide range of materials in which EPR has been observed for isolated ions, and also of materials in which the interactions are of interest to account for the bulk magnetic properties, the range of materials in which pair spectra have been measured is surprisingly small. There are several possible reasons for this: (a) the number of different types of pair, each giving its own spectrum, is much larger than the number of isolated ion sites, greatly increasing the complexity of the spectrum; (b) the inevitably larger line width makes the resolution of this complex spectrum more difficult, particularly the elucidation of its angular variation; (c) the great sensitivity of some interaction mechanisms to inter-atomic distance, and overlap with ligand wavefunctions, may mean that parameters derived from pair spectra may be of limited value in extrapolating to the different local structure appropriate to a pure lanthanide compound.

Even the interactions in the ethyl sulphates (LnES) have continued to be of interest. The pair spectrum of Nd^{3+} in LaES and YES, the first lanthanide system to be investigated in this way, has been measured under hydrostatic pressure up to 10 kbar,[130] using the hyperfine structure of one ion to make it dissimilar from the other, so that J_o could be measured as well as $\underline{K}$ in equation (31). In order to deduce the contributions from different mechanisms, corrections had to be made for local compressibility, the inter-ionic distance for a pair, and the energy spectrum of excited states for a pair of ions. It is concluded that the main contribution to the non-dipolar part of the interaction comes from isotropic ferromagnetic exchange, whose magnitude is the same for LaES and YES. It is suggested that this equality for the two hosts is because, in spite of the distance between the Nd nuclei being different, the length of the Nd-H$_2$O-H$_2$O-Nd path is the same in both hosts, indicating that the exchange path is indirect. Similar measurements have been made for Ce^{3+} pairs.[131] Theoretical interpretation of these measurements shows that virtual phonon exchange is the main non-dipolar mechanism, comparable to magnetic dipole-dipole interaction.[132]

Some of the well understood systems from older EPR and optical work have been used for more sophisticated measurements combining the two techniques. There are two ways in which optical excitation can be used to observe EPR: (a) by producing sufficient transient population of the excited state to allow conventional EPR, (b) by monitoring a change in the optical absorption or fluorescence as a result of population changes induced by EPR transitions (ODMR). An example is provided by the work of Hutchison's group on Nd^{3+} and Er^{3+} in LaCl$_3$ where both of these techniques have been used.[133-6] Sufficient population of Nd^{3+} in the excited states $^4I_{13/2}$ and $^4I_{15/2}$ for the observation of EPR and ENDOR was provided by efficient energy transfer from U^{3+} co-doped into the crystal and excited by broad band absorption of Ar$^+$ laser light. It was also possible in crystals containing 0.01 mole fraction of Nd^{3+} (and 0.003 m.f. of U^{3+}) to observe nn pair spectra for one component in the ground $^4I_{9/2}$ state and the other in the excited $^4I_{15/2}$ state.[137] There is, as usual, a slight shift in g-matrix from that for an isolated ion because of local distortion. There is even a very small change of the g-matrix of the ground state when a nn ion is excited into $^4I_{15/2}$. By assuming that the crystal field at the Nd^{3+} site is set up by the Cl$^-$ ligands, and using the superposition model [Section 3] with intrinsic parameters derived from optically measured crystal field parameters,[138] the g-shift for both ground and excited states could be accounted for by a displacement of the three Cl$^-$ ligands shared by the two nn Nd^{3+} ions by 3.5 pm towards the mid-point of the pair. This makes the Nd-Cl and Cl-Cl distances intermediate between those for LaCl$_3$ and NdCl$_3$. The g-shift appears to indicate no change in the Nd-Nd distance from that for La-La in LaCl$_3$. Because the g-matrices for ground and excited states are different, it was possible to measure all components of the interaction tensor between the ions. As for pairs with both ions in the ground state the contributions from EQQ and VPE interactions are estimated to be negligible, so that the non-dipolar contribution to the interaction is attributed to super-exchange.[139] There is a large change in the latter upon photo-excitation, and the super-exchange cannot be described by an isotropic exchange between real spins.

A novel mechanism of ODMR was produced by the heating effect of both optical and microwave absorption in a crystal of LaCl$_3$ containing Nd^{3+} ions (and also Er^{3+} ions) encapsulated in a fused silica envelope immersed in a bath of superfluid helium.[140] One Nd^{3+} transition is optically pumped with Ar$^+$ laser radiation, and the heating produced in radiationless transitions involved in the optical pumping cycle results in the populating of excited states to produce a large number (28) of fluorescent emissions that are not normally observed at liquid helium temperatures. EPR of the ground Kramers doublets of both Nd^{3+} and Er^{3+}, and the associated spin-lattice relaxation, produces a microwave heating in addition to that produced by optical pumping. This produces striking changes in the intensities of the 28 fluorescent emissions, providing a new mechanism for observing ODMR.

This work also revealed some novel, but unexplained, temporal oscillations of the intensities of the 28 fluorescent transitions, which occur under steady laser irradiation at zero applied magnetic field. These were also affected by applying a field to allow simultaneous excitation of EPR. These resonance and optical studies in the lanthanide chloride have been reviewed by Hutchison.[136]

Recently a series of measurements has been made in lanthanide nicotinate dihydrates (LnND) [see sections 5.3 and 6.4.1], where the nearest neighbour of any Ln^{3+} ion is the other Ln^{3+} ion in the dimer, about 0.4 nm distant, and more distant neighbours are considerably further away. This produces EPR spectra in which the nearest neighbour interaction produces resolved structure even in undiluted crystals. However, in a more dilute system, smaller line width leads to higher precision of measurement. For three of the heavier lanthanides, Er^{3+}, Dy^{3+} and Tm^{3+} the interaction was found to be purely dipolar, with an Ising-like form, $J_{zz}S_{1z}S_{2z}$.[34,57,107-9] Ho^{3+} and Tb^{3+} in nn pairs have singlet ground states:[141] it is interesting that Tb^{3+} with a La^{3+} neighbour has a doublet ground state. In contrast, the interaction for Ce^{3+} and Nd^{3+} has a non-dipolar component of similar magnitude to dipolar.[142,143] This is presumably due to super-exchange mediated via the four carboxylate bridges linking the Ln^{3+} ions 0.4 nm apart. This contribution for Nd^{3+} of about 0.1 cm^{-1} compares reasonably with a super-exchange contribution of 0.68 cm^{-1} measured for nearest neighbour Nd^{3+} ions in $LaCl_3$ distant 0.438 nm linked via three more direct Nd-Cl-Nd bridges.[144,145] The rather smaller super-exchange contribution for heavy ions in LaND is presumably because the radial distribution of the 4f electrons decreases more rapidly through the lanthanide series than the ionic radius [section 2], leading to smaller overlap of 4f wavefunctions with those of the ligands. In addition, for Nd^{3+} ions in LaND, there is also evidence that interactions with more distant ions have appreciable non-dipolar contribution: in this case the super-exchange path would have to go between dimers through hydrogen bonds.

Pair spectra have generally been measured for nearest neighbouring ions, though there have been some measurements for more distant neighbours. Also, most measurements have been for pairs of ions of the same species; though, in principle, it would be possible to use the same method to measure interactions between different species in crystals co-doped with the two lanthanides.

Another method has been used to measure interactions between different species by studying the magnetic resonance of one species as a probe ion introduced into an ordered magnetic salt of the other species.[146] The use of a magnetically ordered state avoids the large spin-spin broadening in a concentrated paramagnetic salt. The probe ion sees an effective field which is the sum of the externally applied field and the internal exchange and demagnetising fields, but if these are the same at all sites, the magnetic resonance lines can be narrow. Unlike pair spectra, the probe sees the internal field set up by all of its neighbours. In a two sub-lattice antiferromanget, this leads to separate resonances for each sub-lattice. In the antiferromagnetic phase there is small macroscopic magnetisation, so demagnetising fields are unimportant, and the condition for resonance for the two different sub-lattices is:

$$(hf/\mu_B)^2 = g_z^2(B\cos\theta \pm B_{int})^2 + (g_x B\sin\theta)^2. \tag{32}$$

By measuring the two resonant frequencies as a function of θ, the angle between $\underline{B}$ and the direction of sub-lattice magnetisation (z, which is along the crystallographic c-axis), one can find the exchange field B_{int} acting on the probe ion as well as the g-values of the probe. Such measurements have been made for Er^{3+} and Yb^{3+} in $DyPO_4$, which made it clear for $DyPO_4$ that the exchange interaction between Yb^{3+} and Dy^{3+}, and between Er^{3+} and Dy^{3+}, can not be represented by isotropic exchange between real spins.[147]

For an applied field in excess of B_{sf} (0.57 T for T = 0K in a needle shaped specimen of $DyPO_4$) spin-flop occurs. Here the magnetisation for both sub-lattices flops round to be almost parallel to the c-axis, because the anisotropy is so large. This changes the internal field and hence the field required for resonance. For external field applied at an angle θ to the c-axis, spin-flop occurs when $B > B_{sf}\sec\theta$, and the change in resonance condition is clearly shown by the results for Yb^{3+}.

Antiferromagnetic resonance also gives information about the exchange interaction between similar ions. It cannot be regarded as equivalent to the measurements described above with the probe species being the same as the host species, because the resonance is a cooperative excitation. The field for resonance when B is applied parallel to the axis is given by:

$$hf/g\mu_B = [B_A(2B_E + B_A) + (aB/2)^2]^{1/2} \pm B(1 - a/2), \tag{33}$$

where B_E and B_A are the exchange and anisotropy fields respectively and a is the ratio of parallel to perpendicular magnetic susceptibilities. The resonance has been observed for $GdVO_4$,[148] but the lack of precise agreement between various experimentally measured quantities which depend upon the parameters B_A, B_E and a indicates that the situation is more complex than the simple theories used assume.

Reasonably sharp resonance lines should also be observed for a probe in an ordered ferromagnet, or ferrimagnet, provided that a spherical specimen is used to eliminate line broadening due to a distribution of demagnetising fields, but only one line would be observed. Hence, measurement at two frequencies would be necessary to find the g-value and internal field.

A similar situation arises in a saturated paramagnet. An example of this is provided by TmND containing a low concentration of Gd^{3+} ions. The ground doublet (type II in the sense of section 4.6) of Tm^{3+} comprises mainly $M_J = \pm 6$, with a small zero field splitting δ. This is a system in which one can study the effect of paramagnetic saturation by rotating the crystal, because the principal g-value of Tm^{3+} is large: $g_z = 12$. In the ac plane, the g-values of the two symmetry related sites are equal, and vary between 11.4 and zero. At 35 GHz and 1.3 K, for $\underline{B}$ along the high g direction, at the field B of around 1 T required for EPR of Gd^{3+}, only 0.3% of the Tm^{3+} ions are in the $|->$ state, so the Tm^{3+} system is highly saturated, and the Gd^{3+} EPR lines are narrow; but for $\underline{B}$ perpendicular to this there is a random distribution of Tm^{3+} magnetic moments, and the Gd^{3+} lines are wide. Additionally, from the shift of the Gd^{3+} lines, one can measure the internal field due to exchange and dipolar interaction between Tm^{3+} and Gd^{3+} {compare with measurements in Van Vleck paramagnetics using a Gd^{3+} probe [section 4.6]}.

8 Conclusion

After nearly 50 years, the use of EPR to study lanthanide ions is a mature field. The current rate of publication is somewhat diminished from its peak, probably more so than for 3d ions. It has always formed part of a more general study of the properties of all of the crystal field levels of the lowest J manifold, involving magnetism, spin-lattice relaxation and optical spectroscopy. There are still many problems which have been only partially solved, because of their complexity, including the detailed theoretical understanding of the crystal field, ligand hyperfine structure, and exchange and other non-dipolar interactions between paramagnetic neighbours. The work on these topics continues, but does not seem to be converging on a simple solution. It seems unlikely that it would ever reach a stage where such properties could be deduced *ab initio* for a new substance of known structure. Most useful progress appears to be made by the use of simple empirical models, like the superposition model, for extrapolating from known parameters for one system to another unknown system.

The main future for EPR in lanthanides seems to be in device oriented materials involving lanthanide ions. There is increasing use of lanthanide containing materials, and EPR will continue to have a role in understanding the propereties of the Ln^{n+} ions in these materials. The use of Ln^{n+} probes to investigate local structure, and changes in that structure, also continues to be promising, as it gives access to a different range of materials than the use of 3d ions for this purpose.

References

1. J.W. Orton *Rep. Prog. Phys.* 1959 **22** 204.
2. A. Abragam and B. Bleaney *'Electron Paramagnetic resonance of transition Metal Ions'* OUP 1970.
3. B. Bleaney *'Handbook on the Physics and Chemistry of Rare Earths'*, Ed. K.A. Gschneider Jr. and L. Eyring, Elsevier, chapter 77 *"Magnetic Resonance Spectroscopy and Hyperfine Interactions"* 1988 p323.
4. A.A. Kaminskii *'Laser Crystals'*, Nauka, Moscow 1975 p256.
5. A.J. Freeman and R.E. Watson *Phys. Rev.* 1962 **127** 2058.
6. J.H. Van Vleck *'The Theory of Electric and Magnetic Susceptibilites'* OUP 1932.
7. *'Handbook on the Physics of the Rare earths'* ed. K.A. Gschneider Jr. and L. Eyring, Elsevier 1988, vol 11 *Atomic Theory and Optical Spectroscopy* B.R. Judd chapter 74 p81, *Highlights from the Exotic Phenomena of Lanthanide Magnetism* J.J. Rhine chapter 76 p293, *Magnetic Resonance Spectroscopy and Hyperfine Interactions* B. Bleaney chapter 77 p323.
8. K.W.H. Stevens *Proc. Phys. Soc. (London)* 1952 **A65** 209.
9. K.R. Lea, M.J.M.Leask and W.P. Wolf *Phys. Chem. Solids* 1962 **23** 1381.
10. M.I. Bradbury and D.J. Newman *Chem. Phys. Lett.* 1967 **1** 1917.
11. D.J. Newman *Adv. Phys.* 1971 **20** 197.
12. D.J.Newman and B.Ng *Rep.Prog. Phys.* 1989 **52** 699.
13. B.Z. Malkin *Crystal field and electron-phonon interaction in rare-earth ionic paramagnets*, in *'Spectroscopy of solids containing rare-earth ions'* Eds. A.A. Kaplyanskii and R.M. Macfarlane, Elsevier, Amsterdam 1987, p13.
14. G.E. Stedman and R.A. Minard *J. Phys. C: Solid State Phys.* 1981 **14** 4753.
15. C.D. Jeffries *'Dynamic Nuclear Orientation'* Interscience 1963.
16. B.Bleaney and R.P. Penrose *private communication* 1948.
17. D.J. Newman and W. Urban *Adv. Phys.* 1975 **24** 793.
18. W.B. Mims *'The linear electric field effect in paramagnetic resonance'* Clarendon Press, Oxford 1976.
19. I.N. Geifman, I.V. Kozlova, A.P. Levanyuk, and T.V. Son'ko *Sov. Phys. Solid State* 1988 **30** 407.
20. J.M. Baker and D. van Ormondt *J. Phys.C: Solid State Phys.* 1974 **7** 2060.
21. A. Abragam, J.F. Jaquinot, M. Chapellier and M. Goldman *J.Phys.C:Solid State Phys.* 1972 **5** 5629.
22. J.M. Baker and G. Currell *J. Phys. C: Solid State Phys.* 1976 **9** 3819.
23. R.Calvo, R.A. Isaacson and Z. Sroubek *Phys. Rev.* 1969 **177** 484.
24. J. Kuriata, T. Rewaj and N. Guscos *Acta Phys. Pol. A* 1991 **79** 531.
25. W. Pastusiak and J. Kuriata *Acta Phys. Pol. A* 1992 **81** 599.
26. I. Ivanenko and B.Z. Malkin *Sov. Phys. Solid State* 1970 **11** 1498.
27. see for example T.Rewaj and M.Krupski *Phys. Stat. Sol (b)* 1978 **88** K65 and 1980 **99** 285; T.Rewaj *Proc. Conf. RAMIS-85* 1985 p201.
28. T. Rewaj, M. Krupski, J. Kuriata and J.Y. Buzare *J. Phys. Condens. Matter* 1992 in press.
29. S.K.Misra and X. Li *Phys. Rev. B* 1992 **45** 2927.
30. see eg.C.A. Bates and H. Szymczak Z. *Phys.B* 1977 **28** 67; K.W.H. Stevens *Proc. Phys. Soc. (London)* 1952 **A65** 209.
31. R.P. Penrose *Nature, Lond.* 1949 **163** 992.
32. D.M.S. Bagguley and J.H.E. Griffiths *Proc. Roy. Soc.A* 1955 **228** 549.
33. B. Bleaney, R.J. Elliott and H.E.D. Scovil *Proc. Phys. Soc (London)* 1951 **A64** 933.

34. J.M. Baker, B. Bleaney, C.E. Davoust, C.A. Hutchison Jr., M.J.M. Leask, P.M. Martineau, M.G. Robinson, R.T. Weber and M.R. Wells *Proc. Roy. Soc. Lond. A* 1986 **403** 199.
35. I. de Wolf, P. Janssen and B. Bleaney *Phys. Letters* 1985 **108A** 221.
36. see eg B. Bleaney *'Handbook on the Physics and Chemistry of Rare Earths'*, Ed. K.A. Gschneider Jr. and L. Eyring, Elsevier, chapter 77 *"Magnetic Resonance Spectroscopy and Hyperfine Interactions"* 1988 p323 (section 7).
37. M.A.H. McCausland and I.S. MacKenzie in *'Nuclear Magnetic Resonance in Rare Earth Metals'*, Taylor and Francis, London 1980.
38. J.M. Baker, W.B.J. Blake and G.M. Copland *Proc. Roy. Soc. A.* 1969 **309** 119.
39. C.H. Anderson in *'Crystals with the fluorite structure'* ed: W. Hayes 1974 (chapter 5).
40. J. Owen and J.H.M. Thornley *Rep. Prog. Phys.* 1966 **29** 675.
41. J.D.Axe and G. Burns *Phys. Rev.* 1966 **152** 331.
42. B.R. McGarvey *J. Chem. Phys.* 1976 **65** 955.
43. B.R. McGarvey *'Electronic Magnetic Resonance of the Solid State'*, Ed. J.A. Weil, The Canadian Society for Chemistry, Ottawa 1987 p 83.
44. A.J. Freeman and R.E. Watson *Phys. Rev. Lett.* 1961 **6** 277.
45. J. Andriesson, D. van Ormondt, S.N. Ray and T.P. Das *J. Phys. B: At. Mol. Phys.* 1978 **11** 2601.
46. K. Baberschke *Phys. Lett.* 1971 **34A** 41.
47. J.M. Baker *J. Phys. C: Solid State Phys.* 1979 **12** 4039.
48. R.A. Allsopp, J.M. Baker, L.J.C. Bluck and J.Y Buzare *J. Phys. C: Solid State Phys.* 1987 **20** 451.
49. J.M. Baker and R.A. Allsopp *Rev. Roum. Phys.* 1988 **33** 353.
50. O.A. Anikeenok and M.V. Eremin *Sov. Phys. Solid State* 1981 **23** 1046; O.A. Anikeenok, M.V. Eremin, M.L. Falin and V.P. Meiklyar *J. Phys. C: Solid State Phys.* 1982 **15** 1557; O.A. Anikeenok, M.V. Eremin, M.L. Falin, A.L. Konkin and V.P. Mieklyar *J. Phys. C: Solid State Phys.* 1984 **17** 2813.
51. J.M. Baker and B.Bleaney *Proc. Roy. Soc. A* 1958 **245** 156.
52. F.I.B. Williams *Proc. Phys. Soc. (London)* 1967 **A91** 111.
53. J.W. Culvahouse, D.P. Schinke and D.L. Foster *Phys. Rev. Letters* 1967 **18** 117.
54. M.V. Eremin and M.L. Falin *Sov. Phys. Solid State* 1987 **29** 338.
55. F. Mehran *'Electron Magnetic Resonance of the Solid State'*, Ed J.A. Weil, Canadian Society for Chemistry, Ottawa 1987 p405; F. Mehran and K.W.H. Stevens *Phys. Rep.* 1982 **85** 124.
56. see eg B. Bleaney *'Handbook on the Physics and Chemistry of Rare Earths'*, Ed. K.A. Gschneider Jr. and L. Eyring, Elsevier, chapter 77 *"Magnetic Resonance Spectroscopy and Hyperfine Interactions"* 1988 p323 (section 8).
57. J.M. Baker, C.A. Hutchison Jr., M.J.M. Leask, P.M. Martineau, M.G. Robinson and M.R. Wells *Proc. Roy. Soc. Lond. A* 1987 **413** 515.
58. J.M.Baker, E.R.Davies and T.Rs. Reddy *Cont. Phys.* 1972 **13** 45.
59. R.M. Sternheimer *Phys. Rev.* 1951 **84** 244.
60. D.T. Edmonds *Phys. Rev. Lett.* 1963 **10** 129.
61. J.M. Baker and F.I.B. Williams *Proc. Roy. Soc. A* 1962 **267** 283.
62. R.J. Cook and D.H.Whiffen *Proc. Phys. Soc.* 1964 **84** 845.
63. J.M.Baker and W.B.J. Blake *Phys. Lett.* 1970 **31A** 61.
64. J.M. Baker and P.M.Martineau *J. Phys. C: Solid State Phys.* 1986 **19** 6253.
65. see eg B.R. Judd *'Operator Techniques in Atomic Spectroscopy'*, McGraw Hill, New York 1963 section 4-6.
66. A. Bohr and V.F. Weisskopf *Phys. Rev.* 1950 **77** 94.
67. J.M. Baker, W.B.J. Blake and G.M. Copland *Phys. Lett.* 1968 **26A** 504.
68. J.M. Baker and J.P. Hurrell *Proc. Phys. Soc.* 1963 **82** 742.

69. J.M. Baker in *'Crystals with the fluorite structure'* ed: W. Hayes, OUP, Oxford 1974 chapter 6.
70. U. Ranon and J.S. Hyde *Phys. Rev.* 1966 **141** 259.
71. J.Y. Buzare, M. Fayet-Bonnel and J.C. Fayet *J. Phys. C: Solid State Phys.* 1981 **14** 67.
72. L.J.C. Bluck *'Some Magnetic Resonance Experiments in Solids'*, D.Phil. thesis, Oxford University 1983.
73. J.M. Baker and T Christidis *J. Phys. C: Solid State Phys.* 1977 **10** 1059.
74. Yu.E. Mitrofanov, Yu.E.Poz'skii and M.L. Falin *Sov. Phys. Solid State* 1970 **11** 2976.
75. E.Y.S. Lee, W.J. Plant, N.S. Chung and R.L. Mieher *Phys. Rev. B* 1973 **7** 955.
76. T.Rs. Reddy, E.R. Davies, J.M. Baker, D.N. Chambers, R.C. Newman and B. Ozbay *Phys. Letters* 1971 **36A** 231.
77. J.M. Baker, T. Christidis, P.J. Walker and B.M. Wanklyn *J. Phys. C: Solid State Phys.* 1978 **11** 3071.
78. C.A. Hutchison Jr. and D.B. McKay *J. Chem. Phys.* 1977 **66** 3311.
79. C.A. Hutchison Jr. and T. E. Orlowski *J. Chem. Phys.* 1980 **73** 1.
80. R.A. Fields and C.A. Hutchison Jr. *J. Chem. Phys.* 1985 **82** 1711.
81. H.A. Buckmaster *Mag. Res. Rev.* 1973 **2** 273, 1974 **3** 127, 1978 **4** 63, 1979 **5** 25, 1980 **5** 121, 1980 **6** 85 & 139, 1983 **8** 283, 1986 **11** 81.
82. R.H. Taylor *'Magnetic Ions in Metals'*, Taylor and Francis, London 1977.
83. J.M. Baker and R.S. Rubins *Proc. Phys. Soc.* 1961 **78** 1353.
84. P. Janssen, M. Mahy and W.P. Wolf *Phys. Rev. B Condens. Matter* 1988 **37** 4851.
85. R.W. Schwartz and N.J. Hill *J. Chem. Soc. Faraday Trans. II* 1974 **70** 124; C.J. O'Connor, R.L. Carlin and R.W. Schwartz *J. Chem. Soc. Faraday Trans II* 1977 **73** 361.
86. B. Bleaney, A.J. Stephen, P.J. Walker and M.R. Wells *Proc. Roy. Soc. London A.* 1982 **381** 1.
87. D.R. Taylor and A.A.F. Sussums *J. Appl. Phys.* 1985 **57** 3736.
88. G. Bacquet, S.K. Misra, L.E. Misiak and F. Fabre *Solid State Comm.* 1990 **75** 369.
89. H. Luft, K. Baberschke and K. Winzer *Phys. Lett.* 1983 **95A** 186.
90. H. Sturn, B. Elschner, and K.-H. Hock *Phys. Rev. Lett.* 1985 **54** 1291.
91. M.X. Huang, J. Barak, S.M. Bhagat, L.C. Gupta, A.K. Rajavajan and R. Vijayraghavan *J. Appl. Phys.* 1991 **70** 5754.
92. A. Deville, B. Gaillard, L. Bejjit, O. Monnereau, H. Noel and M. Potel *Physica B* 1990 **165&166** 1319.
93. A. Rockenbauer, A. Janossy, L. Korecz and S. Pekker *J. Mag. Res.* 1992 **97** 540.
94. H. Kikuchi, Y. Ajiro, Y. Ueda, K. Kosuge, M. Takano, Y. Takeda and M. Sato *J. Phys. Soc. Japan* 1988 **57** 1887.
95. F. Nakamura, Y. Ochiai, H. Shimizu and Y. Narahara *Physica B* 1990 **165&166** 1315; F. Nakamura, Y. Ochiai, H. Shimizu and Y. Narahara *Phys. Rev. B* 1990 **42** 2558.
96. N. Guscos, M. Calamiotou, C.A. London, L. Likodimos, A. Koufoudakis, C. Mitros, H. Gamari-Seale and D. Niarchos *J. Phys. Chem. Solids* 1992 **53** 211.
97. N. Guscos, G.P. Tiberis, M. Calamiotou, Ch. Trikalinos, A. Koufoudakis, Ch. Mitros, H. Gamari-Searle and D. Niarchos *Phys. Stat. Sol. (b)* 1990 **162** 243.
98. N.E. Alekseevski, A.V. Mitin, V.I.Nizhankovskii, I.A. Garifullin, N.N. Garif'yanov, G.G. Khaliullin, E.P. Khylboy, B.I. Kochelaev and L.R. Taginov *J. Low. Temp. Phys.* 1989 **77** 87.
99. J.M. Baker and L.J.C. Bluck *J. Phys. C:Solid State Phys.* 1985 **18** 6051.
100. M.M. Abraham, L.A. Boatner, Y. Chen, L.A. Kolopus and R.W. Reynolds *Phys. Rev. B* 1971 **4** 2853; M. Arakawa, H. Ebisu, T. Yosida and K. Horai *J. Phys. Soc. Japan* 1979 **46** 2459.
101. J. Rubio, C. Ruiz-Mejia, V.U. Osguera and H.S. Murieto *J. Chem. Phys.* 1980 **73** 53 and 60.
102. Y.Y. Yeung and D.J. Newman *J. Phys. C: Solid State Phys.* 1988 **12** 537.

103. J. Rubio *J. Phys. Chem. Solids* 1991 **52** 101.
104. A.A. Klimov, V.G. Grachev, S.S. Ishchenko, Z.D. Kovalyuk, S.M. Okulov and V.V. Teslenko *Sov. Phys. Solid State* 1897 **29** 16.
105. F.J. Owens *Phase Transitions* 1985 **5** 81.
106. see eg J.Y. Buzare, M. Fayet-Bonnel and J.C. Fayet *Ferroelectrics* 1980 **26** 647.
107. K.A. Muller, W. Berlinger, J.Y. Buzare and C.J. Fayet *Phys. Rev. B* 1980 **21** 1763.
108. J.Y. Buzare and P. Simon *Ferroelectrics* 1984 **54** 115.
109. P. Simon, J.J. Rouseau and J.Y. Buzare *J. Phys. C; Solid State Phys.* 1982 **15** 5741.
110. S.K. Misra, X. Li, L.E. Misiak and C. Wang *Physica B* 1990 **167** 209.
111. M. Razeghi and B. Houlier *Phys. Stat. Sol (b)* 1978 **89** K135.
112. A.S. Hubert, J. Emery and J.C. Fayet *Sol. State. Comm.* 1985 **54** 1085.
113. N. Ait. Yakoub, J. Emery, J.C. Fayet and S. Hubert *J. Phys. C:Solid State Phys.* 1988 **21** 3001.
114. P.W. France, S.F. Carter, C.R. Day and M.W. Moore *Fluoride Glasses (Critical Reports on Applied Chemistry* **27** ed A.E. Comyns (Chichester: Wiley) 1989 pp87-122.
115. J.M. Baker, C.A. Hutchison Jr. and P.M. Martineau *Proc. Roy. Soc. Lond. A* 1986 **403** 221.
116. J.M. Baker, M.I. Cook, C.A. Hutchison Jr., M.J.M. Leask, M.G. Robinson, A.L. Tronconi and M.R. Wells *Proc. Roy. Soc. Lond. A* 1991 **434** 695 .
117. J.M. Baker, M.I. Cook, C.A. Hutchison Jr., P.M. Martineau, A.L. Tronconi and R.T. Weber *Proc. Roy. Soc. Lond. A* 1991 **434** 707.
118. L.E. Iton and J. Turkevich *J. Phys. Chem.* 1983 **81** 435.
119. D.L. Griscom *J. Non-Cryst. Solids* 1980 **40** 211.
120. B. Sreedhar, J. Lakshmana Rao, G.L. Narendra and S.V.J. Lakshman *Phys. Chem. Solids* 1992 **53** 67.
121. M.V. Ramana, P.S. Lakshmi, K. Sivakumar, S.G. Sathyanarayan and G. Sastry *Phys. Stat. Sol (a)* 1991 **126** K181.
122. L. Cugunov, A Mednis and J Kliava *J. Phys.: Condens. Matter* 1991 **3** 8017.
123. E.A. Harris and D. Furniss *J. Phys.: Condens. Matter* 1991 **3** 1889.
124. Y.Y. Yeung and D.J. Newman *J. Chem. Phys.* 1985 **82** 3747.
125. A.A. Antipin, S.B. Orlinskii and V.I. Shlenkii *Sov. Phys. Solid State* 1987 **29** 913.
126. A.A. Antipin, S.B. Orlinskii, Ya.K. Fedorov and V.I. Shlenkii *Sov. Phys. Solid State* 1988 **30** 323.
127. S.R. Stevens and H.J. Stapleton *Phys. Rev. B* 1990 **42** 9794.
128. J.M. Baker and B. Bleaney *Conference de Physique des basses temperatures*, Paris 1955 83 .
129. J.M. Baker *Rep. Prog. Phys.* 1971 **34** 109.
130. I.M. Krygin and A.D. Prokhorov *Sov. Phys. Solid State* 1987 **29** 368.
131. I.M. Krygin and A.D. Prokhorov *Sov. Phy. JETP* 1984 **59** 344.
132. M.V. Eremin, A.Yu. Zavidonov and B.I. Kochelaev *Sov.Phys. Solid State* 1987 **29** 1364.
133. C.A. Hutchison Jr., J.M. Clemens, J.P. Hessler and E.D. Liu *Semicond. Insulators* 1978 **3** 61.
134. C.A. Hutchison Jr., and E.D. Liu *J. Lumin.* 1978 **12/13** 665.
135. R. Furrer and C.A. Hutchison Jr. *Phys. Rev.* 1983 **B27** 5270.
136. C.A. Hutchison Jr. *'Electronic Magnetic Resonance in the Solid State'*, Ed. J.A. Weil, Canadian Society for Chemistry, Ottawa 1987 p383.
137. J.M. Clemens and C.A. Hutchison Jr *Phys. Rev.* 1983 **B28** 50.
138. J.C. Eisenstein *J. Chem. Phys.* 1963 **39** 2134.
139. J.D. Riley, J.M. Baker and R.J. Birgeneau *Proc. Roy. Soc. Lond. A.* 1970 **320** 369.
140. C.A. Hutchison Jr., G.G. Utterback and P.M. Martineau *Phys. Rev* 1989 **B39** 4051.
141. M.G. Robinson *High resolution optical studies of some magnetic compounds*, D.Phil. thesis, Oxford 1986.

142. A.L. Tronconi *Magnetic resonance in crystalline solids*, D.Phil. thesis, Oxford 1989.
143. A.A. Jenkins *private communication* 1992.
144. J.M. Baker, J.D. Riley and R.G. Shore *Phys. Rev.* 1966 **150** 198.
145. K.L. Brower, H.J. Stepleton and E.O. Brower *Phys. Rev.* 1966 **146** 233.
146. B. Bleaney *Proc. Roy. Soc. Lond. A* 1991 **433** 461.
147. M.M. Abraham, J.M. Baker, B. Bleaney, A.A. Jenkins, P.M. Martineau and J.Z. Pfeffer *Proc. Roy. Soc. Lond. A* 1991 **435** 605.
148. M.M. Abraham, J.M. Baker, B. Bleaney, J.Z. Pfeffer and M.R. Wells *Proc. Roy. Soc. Lond. A* 1992 **4** 5443.

4
Simulation and Analysis of ESR Powder Patterns

BY PHILIP H. RIEGER

1 Introduction

Randomly oriented rigid samples—powders, glasses and frozen solutions—have long presented a challenge to ESR spectroscopists. Information is lost in the spectrum of a powder which may be obtainable from an dilute single crystal, and rigid samples often give less well-resolved spectra than liquid solutions. The ESR purist disdains powder spectra and insists on a dilute single crystal and the non-specialist user of an ESR spectrometer is often baffled by the complexity of powder spectra and so sticks with liquids. There are many cases, however, where powder spectra can be sufficiently well resolved to give information rivalling that from dilute single crystal spectra and vastly richer in detail than that from isotropic spectra.

The simulation and analysis of magnetic resonance powder spectra has been reviewed by Swalen and Gladney[1] and by Taylor et al.[2] A recent compilation of ESR computer software[3] lists more than 30 programs for simulation and/or analysis of powder spectra, and there are probably at least that many which are not included in the list. This suggests a large number of approaches to the problem, but in fact most programs are derived from a rather small number of distinct simulation or analysis methods. This review will present an overview of the problem and discuss several variations on a general simulation approach and some methods for the extraction of parameters from experimental spectra.

2 The Anatomy of a Powder Pattern

Consider the case of an $S = 1/2$ system with a rhombic **g**-matrix (three distinct principal values) and no hyperfine coupling. The resonant field is given by eq (1) where ν_0 is the microwave frequency and g is a function of the

$$B = \frac{h\nu_0}{g\mu_B} \qquad (1)$$

polar and azimuthal angles, θ and φ, describing the orientation of the principal axes of the **g**-matrix in the magnetic field, eq (2).

$$g = \left[\left(g_x{}^2\cos^2\phi + g_y{}^2\sin^2\phi\right)\sin^2\theta + g_z{}^2\cos^2\theta\right]^{1/2} \quad (2)$$

One way of viewing the problem is in terms of a three-dimensional plot of resonant field as a function of $\cos\theta$ and ϕ as shown in Figure 1. Clearly there are minimum and maximum values of B beyond which no resonant absorption occurs. Between these limits, absorption occurs at all values of the field, but the probability is not uniformly distributed with field. In the simple case of Figure 1, orientation of the x- or z-axes along B correspond to maximum and minimum resonant fields, respectively. In both cases, the resonant field changes rapidly with angle so that only the orientations close to these axes contribute to the total absorption at B_x or B_z. Orientation of the y-axis along B corresponds to a saddle point and there are many orientations which have B close to B_y. Thus the probability of finding molecules with resonant fields near the saddle point value is considerably greater than that near the extrema, and the probability curve is sharply peaked as shown in Figure 1. To a first approximation, magnetic resonance absorption is directly proportional to the probability function, and first derivative presentation then will result in a positive-going peak at the minimum field, a negative-going peak at the maximum field, and a baseline-crossing feature at the intermediate saddle-point field. Although the details may be much more complex in practice, experimental spectra usually can be understood in terms of these three kinds of features.

2.1 "Extra" Features in Powder Spectra. The situation becomes somewhat more complicated with the addition of nuclear hyperfine coupling. When the odd electron is coupled to one magnetic nucleus, the resonant field is given to first order by eq (3), where we have assumed coincident **g**- and **A**-principal axes, g is given by eq (2) and K by eq (4).

$$B = \frac{h\nu_0}{g\mu_B} - \frac{Km}{g\mu_B} \quad (3)$$

$$gK = \left[\left(g_x{}^2A_x{}^2\cos^2\phi + g_y{}^2A_y{}^2\sin^2\phi\right)\sin^2\theta + g_z{}^2A_z{}^2\cos^2\theta\right]^{1/2} \quad (4)$$

180   Electron Spin Resonance

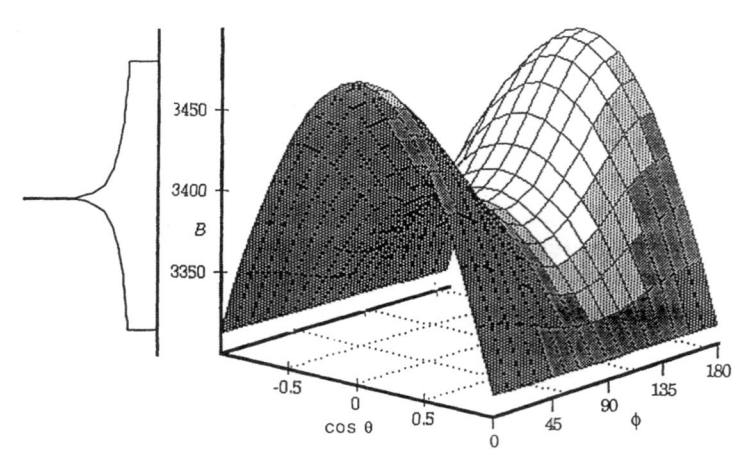

Figure 1. Resonant field surface for ν = 9.5 GHz, **g** = (1.950, 2.000, 2.005) and resulting powder absorption spectrum.

Features in the first-derivative spectrum correspond to discontinuities in the absorption spectrum which arise when $\partial B/\partial\theta$ or $\partial B/\partial\phi$ is zero. In general, discontinuities are expected when one of the principal axes is oriented along the magnetic field, but because of the extra terms from the hyperfine contribution, there may be other orientations which zero one of the derivatives. Ovchinnikov and Konstantinov[4] have shown that an "extra" feature is expected corresponding to an orientation of B in the ij-plane when the inequalities of eq (5) are satisfied. Thus there could be as

$$\left(2A_i^2 - \frac{h\nu_0 A_i}{m}\right) < \left(\frac{g_i^2 A_i^2 - g_j^2 A_j^2}{g_i^2 - g_j^2}\right) < \left(2A_j^2 - \frac{h\nu_0 A_j}{m}\right) \quad (5)$$

many as three off-axis extra features in a spectrum. A hypothetical example, shown in Figure 2, meets the criterion of eq (5) in all three planes. For **g** = (1.998, 2.005, 2.012), **A** = (0, 10, 20) × 10^{-4} cm^{-1}, I = 1, m = -1, orientation of the z-axis along B results in the global minimum, but the x- and y-axes also give local minima. The maximum field occurs in the xz-plane (θ = 63°), and saddle points are found in the yz-plane (θ = 58°) and in the xy-plane (ϕ = 27°). In practice, no feature is seen corresponding to the x-axis minimum;

as seen in Figure 2, this minimum is very sharp with a very small overall probability.

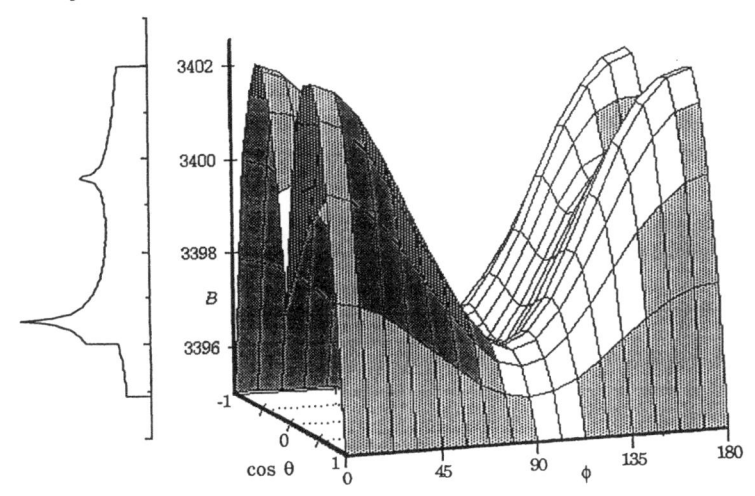

Figure 2. Resonant field surface for ν = 9.5 GHz, **g** = (1.998, 2.008, 2.012), **A** = (0, 10, 20) × 10⁻⁴ cm⁻¹, $I = 1$, $m = -1$ and resulting powder absorption spectrum.

2.2 Non-coincident Principal Axes.

For many years, it was assumed that powder pattern spectra carried no information about the relative orientations of the principal axes of the **g** and **A** matrices. While this conventional wisdom has proven incorrect in general,[5] there are many cases where axis orientation information is indeed lost in a randomly oriented sample. Consider a spin Hamiltonian with a single nuclear hyperfine term for which **g** and **A** share the y-axis, but the x- and z- axes differ by the Euler angle β.

The resonant field is given (to first order) by eq (3) and g is given by eq (2). If we consider resonances in the xz-plane (ϕ = 0) and assume that the g-anisotropy is not too big, K is given approximately by eq (6). The derivative

$$K \approx \left[A_x^2 \sin^2(\theta - \beta) + A_z^2 \cos^2(\theta - \beta)\right]^{1/2} \quad (6)$$

$\partial B/\partial\theta$ is zeroed for values of θ given by eq (7) where the parameter F is given by eq (8).

$$\tan 2\theta = \frac{\sin 2\beta}{F + \cos 2\beta} \qquad (7)$$

$$F = \left(\frac{g\mu_B B}{mK}\right)\left(\frac{K^2}{A_x^2 - A_z^2}\right)\left(\frac{g_x^2 - g_z^2}{g^2}\right) \qquad (8)$$

When the axes are coincident ($\beta = 0$) or when the g-anisotropy is much larger than that of **A** so that F is large, $\tan 2\theta \approx 0$. The observed features then correspond to $\theta = 0$ and $90°$, i.e., to orientation of the **g** principal axes along the field. When the **A**-anisotropy is much larger than that of **g**, F is small and $\tan 2\theta \approx \tan 2\beta$, i.e., the features correspond to orientation of the **A** principal axes along the field. In these cases, the spectrum would contain no information about the relative orientation of the axes.

When the anisotropies of **g** and **A** are comparable such that $F \sim 1$, powder pattern features correspond to other angles and the field positions are functions of g_x, g_z, A_x, A_z, and β. Furthermore, the effect is scaled by the quantum number m, so that the various features of a hyperfine multiplet correspond to different orientations and thus to different combinations of the spin Hamiltonian parameters. For a spin-1/2 nucleus, of course, the single line spacing would yield no information, so that again orientation information is lost, but for $I \geq 1$, variations in the spacings of spectral features can give a precise measure of the principal values of **g** and **A** and the relative orientations of the principal axes.

Consider, for example, the case of [CpMn(CO)(PMe$_3$)$_2$]$^+$.[6] The spectrum is characterized by **g** = (2.141, 2.032, 1.997) and **A**^{Mn} = (16.0, 16.5, 119.0) $\times$ 10^{-4} cm^{-1} (there is also nearly isotropic coupling to the two ^{31}P nuclei which we will ignore here). The y-axes apparently are coincident, and $\beta = 55°$. A plot of resonant field in the xz-plane is shown in Figure 3a; the dependence of the extrema on m is readily apparent. For $m = \pm 1/2$, the extrema approach $\cos \theta = 0$ and ± 1, for $m = \pm 5/2$, the extrema approach $\cos \theta = -0.573$ and 0.819 ($\theta = 90 - \beta$ and $180 - \beta$), and for $m = \pm 3/2$, the extrema are intermediate between these limits. The six component powder absorption lines are shown in Figure 3b. Notice that the six nearly evenly spaced

4: Simulation and Analysis of ESR Powder Patterns

divergences correspond to orientations along the common y-axis. Resonances in the xz-plane, on the other hand, spread over about 730 G (5 $a_{||}$ = 638 G) and seven apparent parallel features are seen, two of which correspond to m = -1/2. These seven outer features are well resolved in the experimental spectrum[6] and the positions yield accurate values of g_x, g_z, A_x, A_z, and β. Only a few of the inner features are resolved, but there is sufficient information to determine all parameters.

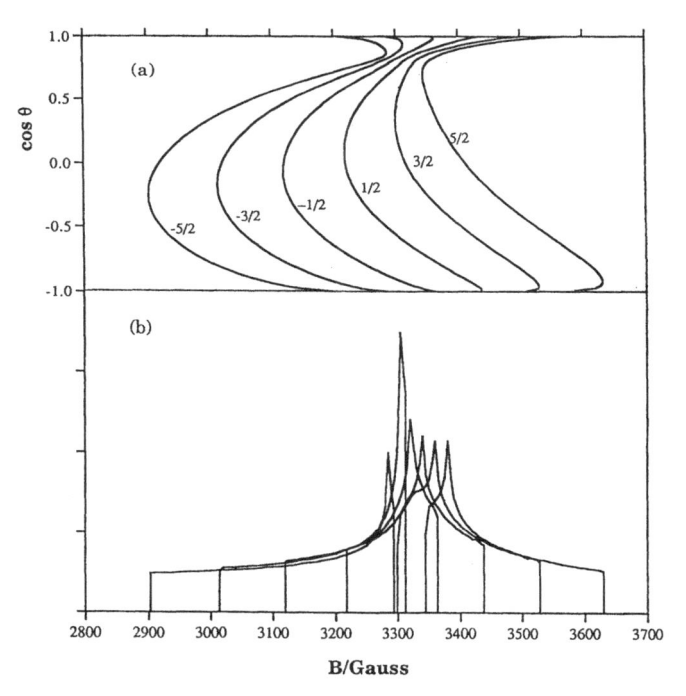

Figure 3. (a) Resonant field in the xz-plane for ν = 9.5 GHz, **g** = (2.141, 2.032, 1.997), **A** = (16.0, 16.5, 119.0) × 10^{-4} cm^{-1}, I = 5/2; (b) component powder absorption spectra averaging over all angles.

2.3 Powder Pattern Component Line Shapes and Linewidths.

Isotropic ESR spectra generally have Lorentzian line shapes with widths resulting from exponential relaxation processes. While Lorentzian component shapes are sometimes seen in powder spectra, Gaussian or composite Lorentzian/Gaussian shapes are much more common. Gaussian lines arise from inhomogeneous broadening which may be instrumental in origin. In the present context, the Gaussian line shape arises from unresolved hyperfine coupling or from variation in the local environment of the paramagnetic species. For radicals or paramagnetic complexes in frozen solution, site variation may arise from variable solvation or ion-pairing; there may also be variations in radical-radical dipolar or spin exchange interactions resulting from a distribution of radical-radical distances. For radicals in glasses, there may be small structural differences among paramagnetic sites which give rise to a similar distribution.

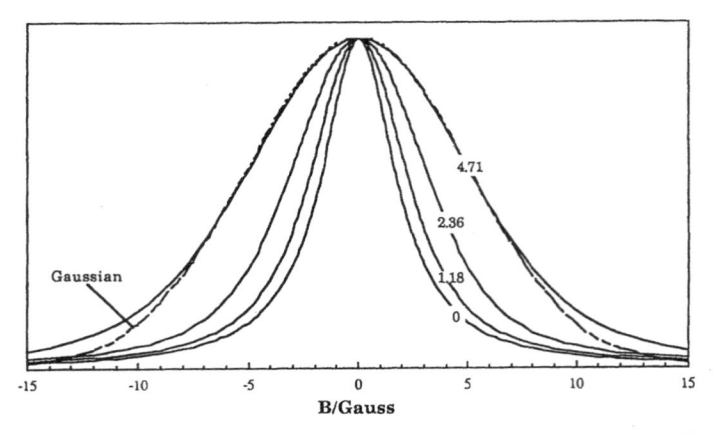

Figure 4. Computed line shapes for a Lorentzian absorption line with width at half-height of 4 G with a Gaussian distributions of line centers with widths at half-height of 0, 2.35, 4.71 and 9.42 G. For comparison, a Gaussian line with 10.0 G width is also shown.

Local environmental differences may lead to a distribution in one or more of the **g**- or **A**-matrix components; this gives rise to a distribution in resonant field for a given orientation. If the distribution of "line" centers follows the normal error curve with a width greater than the intrinsic Lorentzian width, the resulting line shape will be Gaussian. Mixed Gaussian/Lorentzian line shapes are expected when the width of the distribution of centers is comparable to the intrinsic Lorentzian width. This effect is shown in Figure 4. If the environmental effects are different for different components of **g**, the Gaussian widths will be orientation-dependent. If components of **A** are affected, the Gaussian widths may be both orientation-dependent and m^2-dependent.

When a solution of a paramagnetic species is frozen to form an ideal glass, the orientations are random, but there is also a random distribution of radical-radical distances. Thus dipolar coupling or spin exchange may lead to a distribution of component widths. Curiously enough, however, a Gaussian distribution of Lorentzian widths does *not* lead to Gaussian line shapes. For a distribution of widths as in eq (9), the resulting line shape is

$$P(w) = \sqrt{2/\pi s^2} \exp[(w - w_0)^2/2s^2], \quad w \geq w_0 \tag{9}$$

nearly Lorentzian but with a width greater than w_0. If the component width distribution is symmetrical about w_0, the resulting line shape has a sharper peak and broader wings than a Lorentzian.

In studies of radiation-damaged glasses, where hyperfine coupling patterns are often relatively simple and much of the information content of the spectrum resides in the distribution of widths, considerable attention has been paid to the analysis of linewidth anisotropies.[7] Although orientation- and m^2-dependence of linewidths is fairly common in the spectra of frozen solutions of radicals or paramagnetic complexes, there has been relatively little work on the analysis and interpretation of such effects. Accordingly, there has been a systematic loss of possibly valuable information.

One of the few cases where linewidths of powder spectra have been interpreted in detail is the study by Kevan and Schlick[8] of motional averaging in spectra of peroxy radicals trapped in solid matrices. In this case, the modified Bloch equations were used to predict Lorentzian linewidth contributions which were then incorporated in computer simulations of the spectra.

3 Computer Simulation of Powder Patterns

The absorption intensity F for a paramagnetic center is a function of the magnetic field strength B, the orientation of the field in a molecule-based coordinate system, g and D, the g- and electron spin-spin interaction matrices, $\{A, Q\}$, the set of nuclear hyperfine and quadrupole interaction matrices, and $\{m\}$, the set of nuclear spin quantum numbers. For a randomly oriented sample, the absorption must be averaged over all orientations and summed over all quantum numbers. The resulting average, $\langle F(B) \rangle$, is given by eq (10). In general, the integration limits need not

$$\langle F(B) \rangle = \sum_{\{m\}} \frac{1}{4\pi} \int_0^{2\pi} d\phi \int_0^{\pi} \sin\theta \, d\theta \, F\big(B, \theta, \phi, g, D, \{A, Q\}, \{m\}\big) \quad (10)$$

extend over the entire sphere; for example, when the principal axes of all matrices are coincident, the integration may be limited to an octant ($0 \leq \theta, \phi \leq \pi/2$) and integration over ϕ is unnecessary for an axial system.

3.1 Analytical Solutions. In the simple case of a rhombic g-matrix and no fine structure or hyperfine coupling, the shape of the absorption curve can be computed analytically.[9] If the resonant fields corresponding to principal values of g are $B_x > B_y > B_z$, the shape of the absorption vs. field curve is given by eq (11),[10] where $\langle F(B) \rangle = 0$ for $B < B_z$ or $B > B_x$; $K(k)$ is the complete elliptical integral of the first kind[11] with k given by eq (12).

$$\langle F(B) \rangle = \frac{2}{\pi} \frac{B_x B_y B_z}{\left(B_x^2 - B_y^2\right) B^2 \left(B^2 - B_z^2\right)} K(1/k), \quad B_x > B > B_y$$

$$= \frac{2}{\pi} \frac{B_x B_y B_z}{\left(B_y^2 - B_z^2\right) B^2 \left(B_x^2 - B^2\right)} K(k), \quad B_y > B > B_z \quad (11)$$

$$k = \left[\frac{\left(B_x^2 - B_y^2\right)\left(B^2 - B_z^2\right)}{\left(B_y^2 - B_z^2\right)\left(B_x^2 - B^2\right)}\right]^{1/2} \quad (12)$$

The analytical approach is cumbersome at best and cannot be extended to more general cases with fine structure or hyperfine coupling. However, eq (11) is a good approximation to the shape of the absorption curve between singularities even when the principal fields arise from the combination of Zeeman, fine structure and nuclear hyperfine terms. Thus if the turning points for a transition—B_{min}, B_{sp}, and B_{max}—can be determined from the spin Hamiltonian parameters, these can be approximately equated to B_x, B_y and B_z and eq (11) used as an approximation to the line shape. Lindsay[12] has used this approach in cases where the hyperfine coupling is too large to permit a perturbation theory solution to the spin Hamiltonian and the spectrum is so wide that point-by-point matrix diagonalization is impractical. Kreilick[3,13] uses this method as a fast alternative in a powder pattern simulation program.

3.2 Numerical Methods Since only the simplest cases can be dealt with analytically, numerical methods are almost always required to compute the absorption function. In practice, this means that the orientation space is divided into discrete increments, usually as a uniform grid in $\cos\theta$-ϕ space. Thus eq (10) can be written as a sum. The problem then is reduced to computing the absorption intensity at a large number of orientations and summing these in accordance with eq (13).

$$\langle F(B) \rangle = \sum_{\{m\}} \frac{1}{4\pi} \sum_{\phi = 0}^{2\pi} \Delta\phi \sum_{\cos\theta = -1}^{1} \Delta(\cos\theta)\, F\big(B, \cos\theta, \phi, g, D, \{A, Q\}, \{m\}\big) \quad (13)$$

There are two general approaches to powder pattern simulation: (1) When the electron Zeeman interaction is larger than the other terms of the spin Hamiltonian (fine structure, hyperfine coupling, quadrupole coupling, etc.), perturbation theory can be used to obtain analytical expressions for the energies of the various levels as functions of the spin Hamiltonian parameters, orientation, and field—the high-field approximation.[14,15,16] These expressions then can be rearranged to compute the resonant field for a particular orientation and set of nuclear spins. (2) When perturbation theory cannot be used—most commonly because of large fine structure or quadrupole coupling terms—the Hamiltonian matrix must be computed and diagonalized for each orientation, quantum number set and magnetic field. This approach is discussed by Swalen and Gladney[1] and will not be pursued here.

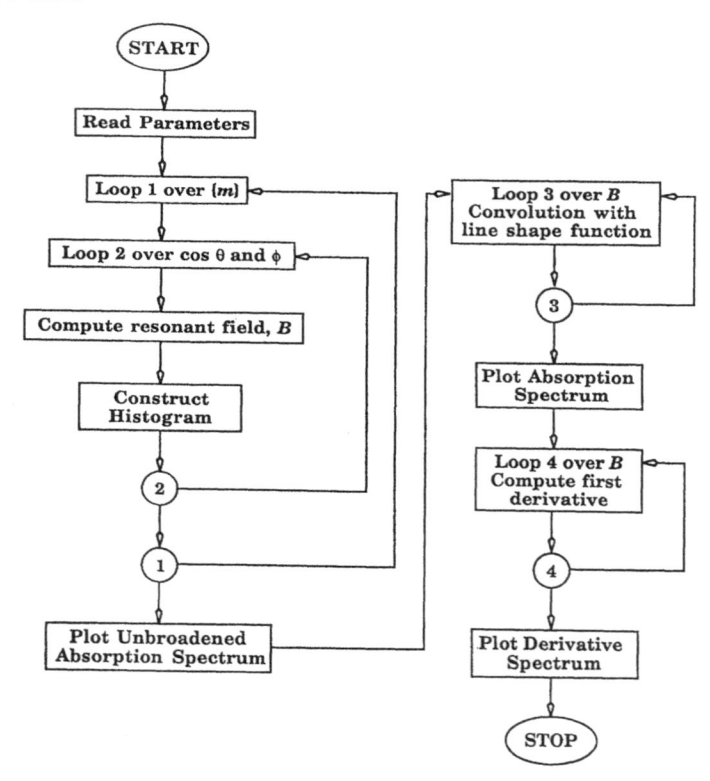

Figure 5. Computer flow chart for the Taylor powder pattern simulation algorithm.

In the following sections, we will examine one simulation strategy in detail, looking in particular at limitations which are attacked by other strategies.

3.3 The Taylor Method. Probably the most straightforward powder spectrum simulation method is that developed by Taylor and Bray[17] in the 1960's. The Taylor program has been used extensively by the solid-state physics community and, largely as a result of the accident of location, has been used in all of our work.[5,6] Although there are many simulation methods in use, almost all developed independently of Taylor, most other methods can be thought of as derived from Taylor's algorithm and it thus provides a convenient starting point for discussion.

The Taylor algorithm is shown schematically in Figure 5. The algorithm can be divided into two parts: (1) The first (and most time-consuming) part uses a set of two nested loops. The outer loop runs over all sets of nuclear spin quantum numbers and the inner loop covers a uniform grid in cos θ-ɸ space. For each quantum number set and orientation, the resonant field for each transition (Δm_S = +1, Δm_I = 0) is computed using an appropriate perturbation theory solution to the spin Hamiltonian. The contributions of the calculated resonant fields are accumulated to form the unbroadened powder absorption spectrum. The absorption spectrum is best thought of as a histogram: The field axis is divided into equal increments, and, assuming uniform transition probabilities, the absorption contribution is simply the number of times a resonant field is found within a field increment.

(2) The unbroadened powder absorption spectrum then is numerically convoluted with a Lorentzian or Gaussian shape function, $S(B_i)$, eq (14).

$$F_c(B_n) = \sum_{i=-k}^{k} S(B_i) F(B_{n+i}) \qquad (14)$$

The limits ±k are chosen to extend 3 - 5 times the component linewidth. Composite line shapes can be produced by successive convolution with Lorentzian and Gaussian functions. The broadened powder absorption spectrum then is numerically differentiated. Since the field increments are uniform, differentiation is very simple, consisting of subtraction of adjacent values of the absorption intensity, eq (15).

$$\left(\frac{dF}{dB}\right)_{n-1/2} = \frac{F_c(B_n) - F_c(B_{n-1})}{\Delta B} \qquad (15)$$

This approach has several advantages. First of all, it is easy to modify for any kind of powder spectrum simply by replacing the subroutine which uses the perturbation theory expression to compute the resonant field(s) from the quantum numbers, orientation, and spin Hamiltonian parameters. The unbroadened powder absorption spectrum can be plotted separately or saved for later convolution with other shape functions or width parameters. Hyperfine components can be simulated separately by limiting the outer loop to a single set of quantum numbers.

3.4 Limitations and Improvements. There are some important disadvantages and limitations inherent in the simple algorithm. Variations introduced by other authors can correct one or more of these limitations.

3.4.1 Number of orientations. In a typical simulation, ca. 10^5 orientations are sampled, requiring considerable computer time (we use the rule of thumb that each orientation and quantum number set requires 100-200 μs on a mainframe computer). The very large number of orientations is required primarily to give a good representation of the absorption spectrum near the singularities (minima, maxima and saddle points). The field axis typically is divided into 10^3 increments. Although there are an average of 100 units of absorption probability per field division, the probability is much greater at or near a saddle point divergence and much smaller numbers may accumulate near the minimum and maximum fields. When a hyperfine component is spread over a wide range of field, this may lead to excessive "simulation noise" near the extrema. The noise can be suppressed by using a larger number of orientations, but computer time goes up accordingly.

Most of the orientations sampled are in the uninteresting regions between the singularities. Thus some time can be saved by using more closely spaced points near the singularities. A simple application of this strategy is used in Pilbrow's version of the simulation algorithm:[18] Equal increments in θ are used up to θ = 43° (with the sin θ weighting factor included) and equal increments in cos θ are used for θ > 43°. In programs by Collison and Mabbs,[3,19] cos θ - φ space is divided into equal areas, Δ(cos θ)Δφ, but with variable increments in cos θ and φ. Thus for the first increment along the z-axis, Δφ = π/2 and Δ(cos θ) is small, the next increment has Δφ = π/4 and Δ(cos θ) twice as big, and so on until, in the xy-plane, Δ(cos θ) is large and Δφ is small. This method of subdividing the surface of a sphere is a bit more efficient than the use of constant Δ(cos θ) and Δφ increments, and for the same number of points, gives better definition near the x-, y-, and z-axes. However, off-axis features may be less well defined with this method. A really "smart" program would know the location of singularities in advance and adjust the increment size accordingly; to my knowledge, there has been no attempt to implement such an approach.

Interpolation offers another general method for reduction of the number of orientations. In effect, the $\cos\theta$–ϕ–B surface is approximated by a series of flat tiles similar to the surfaces shown in Figures 1 and 2. Belford[20] has described a three-point Gauss point approach which does essentially this.

3.4.2 Variable transition probabilities. In the simplest version of the Taylor algorithm, transition probabilities are equal for all allowed transitions ($\Delta m_S = +1$, $\Delta m_I = 0$), but there are cases where probabilities vary. Many years ago, Bleaney[21] showed that systems with large g-matrix anisotropy exhibit variations in transition probability, and in principle, corrections should be made in a simulation program. For systems where $g_{max} - g_{min} \leq 0.2$, such corrections are of marginal importance, but for larger anisotropies, the effects can be quite important. The transitions in spectra of $S > 1/2$ systems have nonuniform transition probabilities which must be accounted for in a simulation.

The Taylor algorithm is easily modified to deal with these problems simply by computing the transition probability along with the resonant field and adding this value rather than unity to the absorption spectrum in constructing the histogram. Pilbrow's programs,[15,18] for example, incorporate intensity corrections by essentially this method.

For $S > 1/2$ systems or those with quadrupole coupling, "forbidden" lines may be important and the intensities of these must be accounted for. A perturbation theory treatment of quadrupole coupling in $S = 1/2$ systems has been presented which is easily incorporated in the Taylor algorithm to account for forbidden lines.[22] However, perturbation theory usually fails when forbidden transitions are important and matrix diagonalization techniques are then required.

3.4.3 Variable component widths. When the powder absorption spectrum is completely assembled before convolution with a line shape function, there is no way to incorporate orientation- or quantum number-dependent linewidths. When such effects are required, it is necessary to construct a series of absorption spectra, covering a range of "site" parameters, which are then added together to give the desired spectrum. While this approach may accurately mimic the origins of the width variations, it can be extremely tedious.

Taylor[17] approached this problem by a set of subroutines which "mollify" the computed powder absorption spectrum to incorporate small variations in one or more of the parameters without completely recomputing the entire spectrum. Thus, for example, if it assumed that g_x in Figure 1 is distributed over a range of values, the absorption spectrum is altered for a particular value of g_x simply by moving the maximum field value and renormalizing the envelope.

Another quite different approach was used by Lefebvre and Maruani.[23] Instead of first constructing a histogram or "stick spectrum" and then convoluting it with a shape function, the shape function (Lorentzian or Gaussian, absorption or derivative) is incorporated within the main computation loop. That is, when the resonant field for a particular orientation and set of nuclear spin quantum numbers is computed, the broadened line is added to the spectrum. Thus linewidths can be made dependent on orientation or quantum number. The disadvantage of this approach is that it is functionally equivalent to doing a convolution integration for each of the ca. 10^5 orientations instead of ca. 10^3 field points; computer time is considerably greater. Changing a linewidth parameter of course requires a completely new simulation.

4 Analysis of Powder Spectra

There are two basic techniques for the analysis of anisotropic spectra to obtain spin Hamiltonian parameters: (1) Comparison of a simulation based on guessed parameters with the experimental spectrum and parameter refinement based on the discrepancies; and (2) Calculation of spin Hamiltonian parameters from the field positions of certain features of the experimental spectrum. Both techniques have been commonly used at various levels of sophistication. Here we will examine the requirements and limitations of the two methods.

4.1 Comparison of Experimental and Simulated Spectra.

The ideal analysis method would use a simulation method capable of exactly reproducing the experimental spectrum, given the right spin Hamiltonian and linewidth parameters. A simulation based on approximate parameters would then be compared with the experimental spectrum, point by point, and a least-squares refinement algorithm used to adjust the parameters. The comparison procedure would be repeated until the simulated and experimental spectra match according to some statistical criterion. Unfortunately, there has been no attempt to achieve this ideal in a general program.

In most cases, spectroscopists have fitted experimental spectra by trial and error, starting with estimates of spin Hamiltonian and linewidth parameters, derived perhaps from fitting of spectral features (see below). Parameters are adjusted until the simulation visually matches the experimental spectrum. In such cases, errors in the fitted parameters usually are unknown, and there are probably many cases in the literature of apparent fits to parameters which in reality are essentially unknown. This situation is both unfortunate and preventable. Taylor et al.[2] discuss several examples of parameter error estimation which can be followed in general. The procedure is simple, if a bit tedious: If N parameters are to be determined, N + 1 simulations are done; in each simulation, one of the parameters is stepped by a small amount and the effect on the simulation is assessed visually. Parameters which have little effect clearly have large uncertainties. In its simplest application, this method can easily miss cross-correlated parameters, but it will give at least a rough indication of parameter reliability.

More recently, there have been several attempts to approach the ideal analysis method, at least for specific systems. Thus Fajer and co-workers[24] have reported the use of a simplex algorithm to compare simulated and experimental spectra of nitroxide spin labels. In this case, there were six spin Hamiltonian parameters (the components of g and A^N); four linewidth parameters were also used. Starting with parameter estimates, a digital simulation was compared with the experimental spectrum and χ^2 computed as a measure of goodness-of-fit. Each of the ten parameters was then stepped and the comparison repeated to generate a χ^2 surface in ten dimensions. The parameters were then adjusted, following the gradient in the χ^2 surface, and the process repeated. In practice, ca. 200 iterations were required to find the best fit.

Fajer's methods can probably be extended to other relatively simple cases. However, there are only a few cases where component shapes and widths are sufficiently well understood at present to make the method work. Furthermore, computer time increases roughly as the square of the number of fitted parameters so that there is a practical limit to the complexity of spectra which can be dealt with in this way.

4.2 Fitting of Spectral Features. The second method of analysis focuses on resolved features of the experimental spectrum. In well-resolved spectra, maximum, minimum, and saddle point features can be identified and field positions can be measured accurately. As shown in Figure 6, the maximum and minimum fields correspond to peaks in the derivative spectrum, and the saddle point field corresponds to the baseline crossing of the central feature. The situation is somewhat more ambiguous in axial spectra. As shown in Figure 7, the parallel feature appears as an easily measured minimum field peak (for $g_\parallel > g_\perp$), but because of the assymmetry of the perpendicular feature, the proper field position lies somewhere between the baseline crossing and the outer peak When the principal axes of **g** and **A** are coincident, these features often will correspond to orientation of one of the principal axes along the field. The measured positions then can be fitted to a perturbation theory solution to the spin Hamiltonian in much the same way as in the analysis of an isotropic spectrum.

For example, in the spectrum of an axially symmetric mononuclear transition metal complex which exhibits metal hyperfine coupling (^{51}V, ^{55}Mn, ^{59}Co, ^{63}Cu, etc.) and no ligand couplings, the field positions of the parallel features (minima or maxima) and perpendicular features (divergences) are given by eqs (16) to second order in perturbation theory.

$$B_\parallel = \frac{h\nu}{g_\parallel \mu_B} - \frac{A_\parallel m}{g_\parallel \mu_B} - \frac{1}{2B}\left(\frac{A_\perp}{g_\perp \mu_B}\right)^2 [I(I+1) - m^2] \quad (16a)$$

$$B_\perp = \frac{h\nu}{g_\perp \mu_B} - \frac{A_\perp m}{g_\perp \mu_B} - \frac{1}{4B}\left[\left(\frac{A_\parallel}{g_\parallel \mu_B}\right)^2 + \left(\frac{A_\perp}{g_\perp \mu_B}\right)^2\right][I(I+1) - m^2] \quad (16b)$$

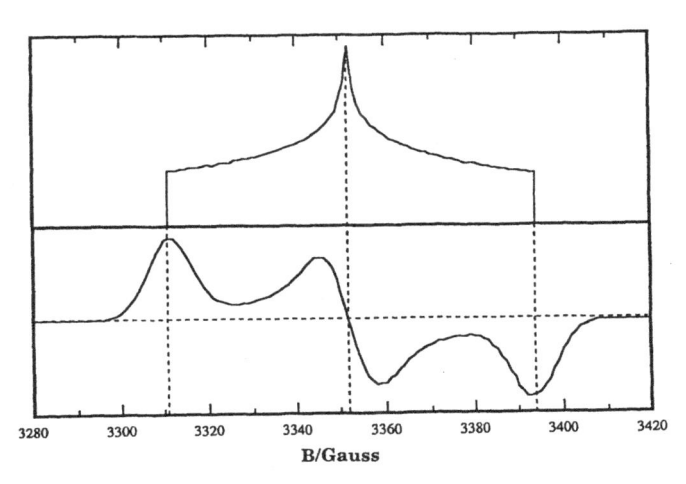

4: *Simulation and Analysis of ESR Powder Patterns* 195

Figure 6. Unbroadened powder absorption spectrum and first-derivative curve (5 G wide Gaussian components) for ν = 9.5 GHz, g_x = 2.000, g_y = 2.025, g_z = 2.05.

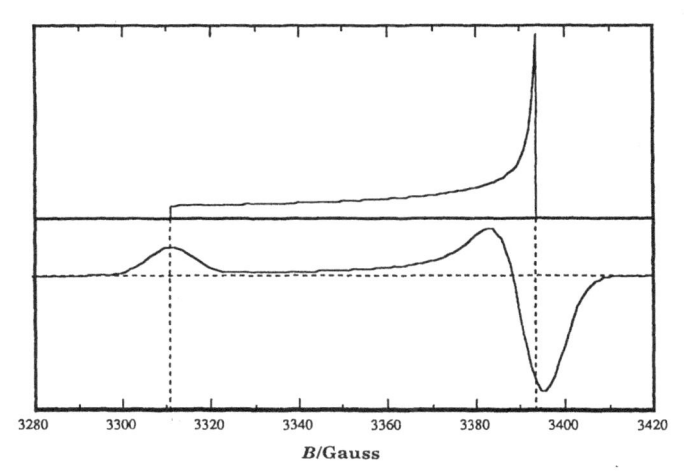

Figure 7. Unbroadened powder absorption spectrum and first-derivative curve (5 G wide Gaussian components) for ν = 9.5 GHz, $g_\perp$ = 2.000, $g_\parallel$ = 2.05.

Thus well-resolved spectra show two clearly recognizable sequences which can be least-squares fitted to eqs (16) using estimates of $A_\parallel$ and $A_\perp$ in the second-order terms. When off-axis resonances occur (as in most spectra of copper complexes), they can be ignored in this approach, provided they are accurately identified.

When the principal axes of **g** and **A** are non-coincident or when off-axis resonances cannot be ignored, analytical expressions analogous to eqs (16) are too complex and involve too many parameters to permit a linear least-squares fit of measured field positions, but the general approach is still possible. We have made use of a simple algorithm to search for minima, maxima, and saddle points in the xz, yz, and xy-planes.[25] (Restriction to these planes is possible if at least one of the principal axes is common to all matrices.)

The algorithm is shown schematically in Figure 8. The resonant field is first computed for orientation of one of the axes (e.g., the x-axis) with the field and then for four displacements (e.g., for $\cos\theta = \pm 0.01$, $\phi = 0$ and for $\cos\theta = 0$, $\phi = \pm 0.01$). The approximate derivatives, $\partial B/\partial(\cos\theta)$ or $\partial B/\partial\phi$, are then computed and examined to determine whether the orientation is a minimum or maximum (B increases or decreases for all displacements) or a saddle point (B increases for displacements in one plane and decreases for displacements in the other). This procedure is repeated for the other two axes. The derivatives obtained for each orientation are then compared to see whether other features are expected. For example, if $B_x > B_y$ and $(\partial B/\partial\phi)_x < 0$, then $(\partial B/\partial\phi)_y < 0$ suggests that no other features lie in the xy-plane; $(\partial B/\partial\phi)_y > 0$ suggests a minimum in the xy-plane. Similarly if $(\partial B/\partial\phi)_x > 0$, $(\partial B/\partial\phi)_y < 0$ suggests a maximum in the xy-plane and $(\partial B/\partial\phi)_y > 0$ suggests both a minimum and a maximum. If additional features are expected in a plane, the plane is sampled using search grids of decreasing size until the extremum is found; it is then checked to see whether it is a true extremum or a saddle point.

With an algorithm to locate off-axis resonances, a non-linear least-squares routine can be used to analyse the positions of features from experimental spectra and to refine parameter estimates. Using estimates of the spin Hamiltonian parameters, the fields of observable features are computed and compared with experimental values to obtain $\delta B = B_{expt} - $

4: Simulation and Analysis of ESR Powder Patterns

B_{calc}. These deviations are related to the errors in the parameters, ΔP_i, by Taylor series expansion, eq (17). For each predicted feature k, estimates of

$$\delta B_k = \sum_{i=1}^{N} \left(\frac{\partial B}{\partial P_i}\right)_k \Delta P_i \qquad (17)$$

the derivatives, $(\partial B/\partial P_i)_k$, are obtained by stepping the parameters, and a linear least-squares fit of the δB_k to the $(\partial B/\partial P_i)_k$ is performed to obtain ΔP_i's which are used to adjust the parameters. The procedure is repeated until the parameter corrections approach zero. The scheme is summarized in the computer flow chart of Figure 9.

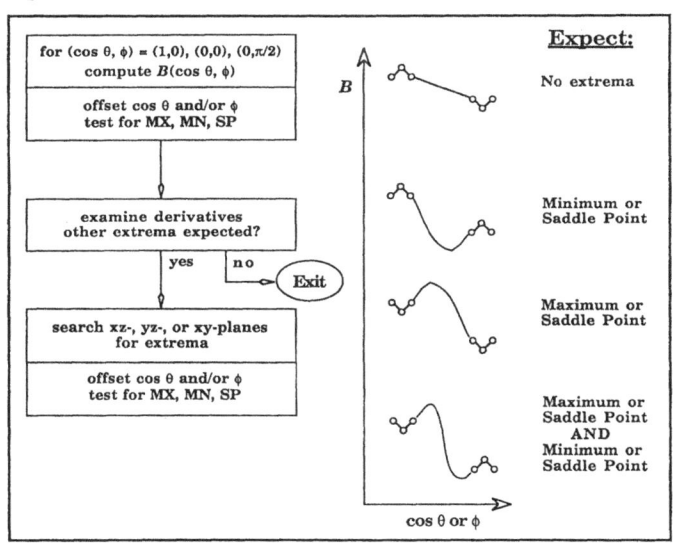

Figure 8. Computer algorithm for location of off-axis features resonances.

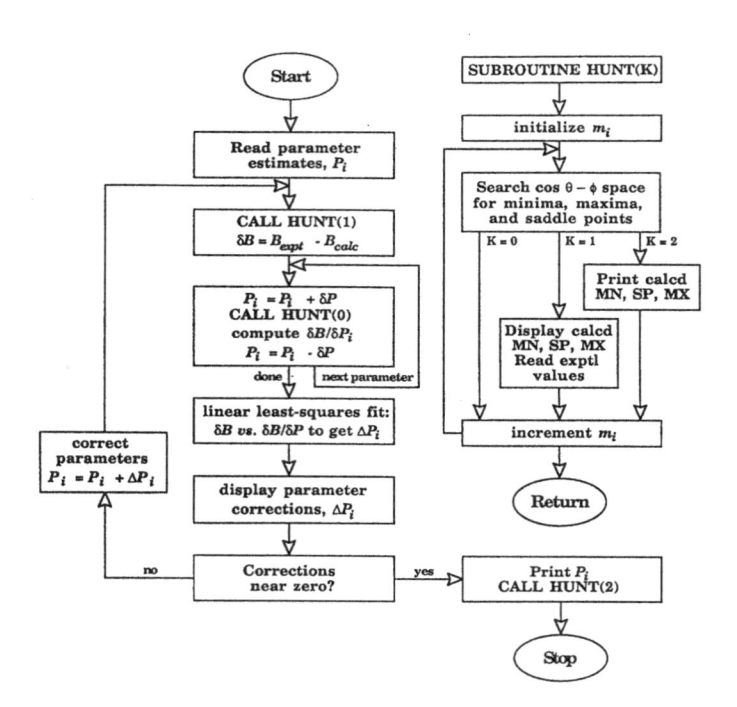

Figure 9. Computer flow chart for non-linear least-squares fitting of spin Hamiltonian parameters to observed spectral features; the search algorithm of Figure 8 is used to locate spectral features in cos θ - φ space.

References:

[1] J. D. Swalen and H. M. Gladney, *IBM J. Res.*, 1964, **8**, 515.
[2] P. C. Taylor, J. F. Baugher, and H. M. Kriz, *Chem. Rev.*, 1975, **75**, 203.
[3] R. Cammack, *Computers in ESR Spectroscopy Software Database* (University of London, 1992).
[4] I. V. Ovchinnikov and V. N. Konstantinov, *J. Mag. Reson.*, 1978, **32**, 179.

[5] B. M. Peake, P. H Rieger, B. H. Robinson, and J. Simpson, *J. Am. Chem. Soc.*, 1980, **102**, 156; *Inorg. Chem.*, 1981, **20**, 2540; W. E. Geiger, P. H. Rieger, B. Tulyathan, and M. D. Rausch, *J. Am. Chem. Soc.*, 1984, **106**, 7000.

[6] R. D. Pike, A. L. Rieger, and P. H. Rieger, *J. Chem. Soc., Faraday I*, 1989, **85**, 3913.

[7] D. L. Griscom, *J. Non-Cryst. Sol.*, 1978, **31**, 241; *ibid.*, 1980, **40**, 211; J. Kliava, *Phys. Stat. Sol. B*, 1986, **134**, 411.

[8] L. Kevan and S. Schlick, *J. Phys. Chem.*, 1986, **90**, 1988.

[9] The analytical approach has been extended to several other relatively simple cases by Y. Siderer and Z. Luz, *J. Mag. Reson.*, 1980, **37**, 449.

[10] J. E. Harriman, *Theoretical Foundations of Electron Spin Resonance* (Academic Press, New York: 1978), pp 158-162.

[11] See, for example, I. S. Gradshteyn and I. M. Ryzhik, *Table of Integrals, Series, and Products*, 4th ed (Academic Press, New York: 1980), pp 904-5.

[12] K. Kernisant, G. A. Thompson, and D. M. Lindsay, *J. Chem. Phys.*, 1985, **82**, 4739; D. M. Lindsay, G. A. Thompson, and Y. Wang, *J. Phys. Chem.*, 1987, **91**, 2630.

[13] R. Kreilick, private communication.

[14] A. Rockenbauer and P. Simon, *J. Mag. Reson.*, 1973, **11**, 217.

[15] P. D. W. Boyd, A. D. Toy, T. D. Smith and J. R. Pilbrow, *J. Chem. Soc., Dalton Trans.*, 1973, 1549.

[16] P. H. Rieger, *J. Mag. Reson.*, 1982, **50**, 485.

[17] P. C. Taylor and P. J. Bray, *Lineshape Program Manual*, (Brown University, 1968); *J. Mag. Reson.*, 1970, **2**, 305.

[18] A. D. Toy, S. H. H. Chaston, J. R. Pilbrow, and T. D. Smith, *Inorg. Chem.*, 1971, **10**, 2219; J. R. Pilbrow and M. E. Winfield, *Mol. Phys.*, 1973, **25**, 1073.

[19] D. Collison, private communication.

[20] L. K. White and R. L. Belford, *J. Am. Chem. Soc.*, 1976, **98**, 4428.

[21] B. Bleaney, *Proc. Phys. Soc., A*, 1960, **75**, 621.

[22] J. A. DeGray, P. H. Rieger, N. G. Connelly, and G. Garcia Herbosa, *J. Magn. Reson.*, 1990, **88**, 376.

[23] R. Lefebvre and J. Maruani, *J. Chem. Phys.*, 1965, **42**, 1480.

[24] P. G. Fajer, R. L. H. Bennett, C. F. Polnaszek, E. A. Fajer, and D. D. Thomas, *J. Mag. Reson.*, 1990, **88**, 111.

[25] J. A. DeGray and P. H. Rieger, *Bull. Mag. Reson.*, 1987, **8**, 95.

5
Inorganic and Organometallic Radicals

BY MARTYN C. R. SYMONS

1 Introduction

As in past years, this report, following on from 12B, covers the past two years, roughly from July 1989 to August 1992. There seems to be less to report than on previous occasions, although the field of solid-state materials science still proves to be very fruitful for esr spectroscopists. Also some of the newly developing techniques have been brought to bear on some 'old' problems, with interesting results.

Classification follows the same lines that Peter Atkins and I established many years ago.[1] Starting with electrons, solvated and trapped (2), I move on through atoms and monoatomic ions (3) A–B· molecules and ions (4) ·AB$_2$ radicals (5) ·AB$_3$ radicals (6), ·AB$_4$ and ·AB$_5$ species (7) and other molecular or ionic radicals (8) that are less readily classifiable. Then comes the important section on inorganic materials which, as usual, is not done justice herein because it is such a large specialist area. [It is really worthy of a separate chapter in this series]. This is followed by an equally brief section on spin-trapped inorganic radicals (10), and finally a section on metal carbonyls and related species (11). I treat the classifications in (3)–(7) liberally, bringing in related species for convenience and for comparative purposes.

This is the last "Inorganic Radical" report that I shall write. I would like to express deep gratitude to many contributors to this field who have supplied me with abundant reprints, helpful advice, and great encouragement—thank you.

1.1 Books and Reviews

Several extremely useful books have appeared. That by Pillbrow[2] is of general use to esr spectroscopists although largely confined to transition metal complexes, as is the interesting treatise on 'exchange coupled systems' by Bencini and Gatteschi.[3]

The Annual Report (C) by Almond and on matrix isolated species includes a lot of information on radicals, including esr studies thereof.[4] The book edited by Lund and Shiotani on Radical Ionic Systems[5] contains many useful chapters, especially those by Byberg on Anion Radicals in Inorganic Crystals, and by Knight on inorganic radical cations in rare-gas matrices.[6] In the former, Byberg concentrates on those centres that he has so thoroughly explored over the years. This work is largely concerned with primary and secondary radiation products for oxysalts, especially KClO$_4$, and the overview is extremely useful for understanding structure and mechanism. Knight's review includes a detailed description of the equipment used and of methods of radical-cation generation. There must be counter anions in these systems, but often there seems to be no clear evidence thereof. The Review cites recent evidence for a range of different counter anions including OH$^-$, ·CO$_2^-$, and sometimes ·CH$_2^-$, F$_2^-$, Cl$_2^-$, or hal$^{-(7\ 10)}$. Some of the cations detected and studied in recent years are summarised in Table 1.

Chanon and his coworkers have written a very thorough review on radical cations and anions, entitled "One Electron More, One Electron Less. What does it change? Activations Induced by Electron Transfer. The Electron, and Activating Messenger".[11] This is an

5: Inorganic and Organometallic Radicals

Table 1. Some Inorganic Radical Cations studied in Rare–Gas Matrices by Esr spectroscopy (discussed by Knight in Ref. 5, p73).

C_2^+	H_2O^+	Ag_n^+
CO^+	CH_2^+	AgH^+
BF^+	Mg^+_{3-6}	Xe_2^+
N_2^+	SiH_2^+	PH_3^+
AlH^+	PdH_2^+	NH_3^+
Mg_2^+	NH_3^+	P_4^+
AlF^+	N_4^+	SnH_4^+
SiO^+	O_4^+	CCl_4^+
Mn_2^+	N_2CO^+	S_8^+
CuH^+	$C_2O_2^+$	$S_4N_4^+$
Cu_2^+	H_2CO^+	$N_2O_4^+$
Ag_2^+	F_2CO^+	$S_3N_2^+$
Au_2^+	CH_4^+	$Se_3N_2^+$
GaP^+	CH_3CHO^+	$N_2H_4^+$
$GaAs^+$	$P_2H_6^+$	F_3N^+
Si_2^+		

exciting and challenging review that fully lives up to its title. It is especially useful to esr spectroscopists because it covers all techniques, and tends to put esr in its place. It is, of course, concerned with all types of compound, inorganic and organic, but there is little on transition metal complexes. There are very useful tables giving ionization potentials, electron affinities and E° values.

Spaeth has written a series of extremely useful reviews relating to radicals in solid-state systems. One is concerned with ENDOR, both in its normal form but also with ENDOR induced esr spectroscopy and with "double ENDOR".[12] The application of these techniques to a wide range of defects such as F-centres and hole-centres, hydrogen atoms, and various radiation induced defects shows how much extra information is made available by these methods, which help to establish the precise locations of the defects.

In the second Review, ODMR techniques are discussed in the same context. This includes the application of ENDOR and the use of magnetic circular dichroism.[13] In the third,[14] Spaeth describes the way in which ENDOR can be used to determine the atomic coordinates of point defects in solids with high accuracy.

As a bridge between this and the next section, I mention Ikeya's review of what he calls esr "microscopy" as applied to solids.[15] This is an imaging technique which uses an X-band spectrometer fitted with field gradient coils that can achieve very large gradients (as large as 20mT/mm). Applications to the study of paramagnetic centres in fossils, diamond and other materials are described. Reference to this technique to the study of diamonds can be found in section 9, below.

1.2 Techniques

As mentioned above a particularly intriguing method of achieving very small images, developed by Ikeya and his coworkers is described as 'near-field scanning esr microscopy'.[15] Samples are scanned over an aperture (*ca.* 1mm) in the microwave cavity, and a spatial resolution of 0.2mm can be obtained.

There are now many examples of the application of nmr techniques in the esr field. Freed and his coworkers have shown that two-dimensional FT esr methods can be extended into the slow-motion limits in the 10^{-6}–10^{-3} range, despite the large line-widths involved.[16,17] This method complements the 2D-ELDOR technique for studying motions in the very slow tumbling limit.[18] Astashkin and Schweiger have shown how what is termed 'transient mutation' can be used as an important alternative to pulsed esr spectroscopy, and can be particularly useful in unravelling data from overlapping features in solid-state studies.[19] Loop-gap resonators are still being developed extensively, and Schweiger and coworkers have described the use of a pulsed ENDOR probehead involving a bridged loop-gap resonator for X-band studies.[20]

For most workers, 4K is considered to be the lowest temperature needed in esr studies. However, this is still quite hot, and certain studies require much lower temperatures. A normal device for cooling to ca. 0.4K for X-band spectroscopy has now been described, which should greatly simplify the task of reaching such low temperatures.[21]

Finally I refer to quite a different type of "technique", namely a chemical rather than a physical one. This has been developed by Roberts and his coworkers, and is described as 'polarity reversal catalysis'.[22] This involves a catalytic cycle in which an electrophilic radical (such as RO) is converted into a nucleophilic radical (in this case an amine-alkylboryl radical) which then undergoes highly specific reaction with the substrate of interest. This is symbolised in reactions (1)–(4).

$$\text{El} \cdot + \text{H–Nuc} \rightarrow \text{H–El} + \text{Nuc} \cdot \quad (1)$$
$$\text{Nuc} \cdot + \text{H–El} \rightarrow \text{h–Nuc} + \text{El} \cdot \quad (2)$$
$$\text{El}^1 \cdot + \text{H–El}^2 \rightarrow \text{H–El}^1 + \text{El}^2 \cdot \quad (3)$$
$$\text{Nuc}^1 \cdot + \text{H–Nuc}^2 \rightarrow \text{H–Nuc}^1 + \text{Nuc}^2 \cdot \quad (4)$$

In addition to X-band imaging there has been considerable activity in the field of low frequency imaging.[23–25] The reason for moving to low frequencies (usually in the 250–1000 MHz range) is to minimise losses from water caused by dielectric relaxation. By working at, say 300 MHz, one can use a 500ml beaker full of water as the solvent.[26] Hence whole body imaging becomes a possibility. Of course, sensitivity is being sacrificed, but is of the essence, if useful imaging is to be performed. These spectrometers are not available commercially, so I find myself back in full circle to the days when we had to construct our own X, K, or Q band spectrometers.[27,28]

The sad fact is that most of the "natural" radicals one would like to be able to detect by esr give such broad lines that they are undetectable in fluid solution. For example, $\cdot$OH, $\cdot$O$_2^-$, $\cdot$NO, RO$\cdot$ and RS$\cdot$ are all of considerable biological importance, but because of orbital degeneracy and spin-orbit coupling, Δg can be very large and averaging gives undetectably broad lines. So the rather indirect method of spin-trapping has to be used.

2 Trapped and Solvated Electrons

2.1 Introduction

Most of the work in the area of solvated electrons has been either theoretical, especially involving molecular dynamics calculations, or studies at very short time intervals mainly using optical detection. Extensive work has been done on gas-phase clusters of solvent molecules associated with one electron, and there are interesting discussions as to when, and if, the electron moves from an outer, Rydberg-type orbital, into an inner, localised orbital of the type often envisaged for electrons in solution. There has been little direct esr study of these systems—on the other hand, applications of modern sophisticated spin-resonance techniques to electrons trapped in solids has been extensive. There have been several important Reviews in this area. For example, a book entitled "Excess Electrons in Dielectric Media", edited by Ferradine and Jay-Gerin gives a thorough coverage of these fields.[29] Of particular interest are chapters by Sanche (Primary Interactions of Low Energy Electrons in Condensed Matter), by Schiffer (Excess Electron Live History by Dielectric Relaxation) by Schmidt (Electronic Energy Levels in Non-polar Dielectric Liquids) by Jay-Gerin and Ferradini (Electron Solvation in Polar Liquids) and by Ogasawara (Excess Electrons in Polar Matrices). However there is little reference to the use of esr spectroscopy in these Reviews.

2.2 Electrons in Fluids

Work with non-aqueous solvents has concentrated on species such as Na^-, formed, for example, in fluid hexamethylphosphoramide (HMPA), rather than the solvated electrons which are minority species. Interest still centres on electrons in water, but again, esr spectroscopy is not usually used. The studies by Bartels and his coworkers on electron diffusion,[30] and of Abramezyk and Kroh on the near infrared absorption spectra of electrons in glassy alcohols and water are highly pertinent.[31] They failed to observe the NIR band in aqueous H_2O glasses even at 6K, whereas when D_2O was used this band was stable. The nature of the species responsible for this initial band, which is highly controversial, is discussed, but esr spectroscopy has not been used to help solve the problem.

As mentioned in my previous Review, Jeevarajan and Fessenden have shown that aquated electrons formed by normal photolysis of SO_3^{2-} in water lead to good steady state esr spectra for e^-_{aq}.[32] This work has been extended to explore polarisation effects of the electron esr spectra produced by photoionisation of various sources.[33] Thus photoionisation of phenoxide gave a signal in absorption whilst photolysis of thiophenoxide gave an emissive signal.[27] However, in radiolyses, both solutions gave electron signals in emission [Fig. 1].

2.3 Electrons in Solids

This is a much more extensively studied area, with a large increase in interest since my last report, largely because of the application of new techniques and the study of new materials. Much of this work has been done by Spaeth and his group.[34] Some examples of their work are now described.

When F-centres, generated in KCl doped with cyanide ions, are optically bleached at 220K, the electrons migrate to sites adjacent to the CN^- ions, on the [110] axis.[35] There is said to be a strong "coupling between the F-centres and the CN^- ions". ENDOR studies show hyperfine coupling to ^{13}C, but no ^{14}N coupling was detected. Thus delocalisation into the π^*-levels of CN^- is not involved but a σ^*-interaction is probable (Fig. 2).

The salt BaFBr has been extensively studied, primarily because, when doped with europium (Eu^{2+}) it is an important storage phosphor. The parent crystal has a matlockite structure, with layers in the sequence $F^- - Ba^{2+} - Br^- - Br^- - Ba^{2+} - F^-$. Trapped electron (F) centres are formed on irradiation, and these are involved in image formation. Electrons are trapped both at F^- vacancies, $F(F^-)$, and at Br^- vacancies $F(Br^-)$. These F centres have been fully characterised both optically and magnetically.[36,37] It seems clear that the special properties of this phosphor depend not only on the presence of Eu^{2+}, but also on the presence of O^{2-}, which is very difficult to remove. The importance of this ion was established using ^{17}O.[38] Charge compensation leads to bromide ion vacancies which accept the radiation generated electrons. There is considerable spatial correlation between the

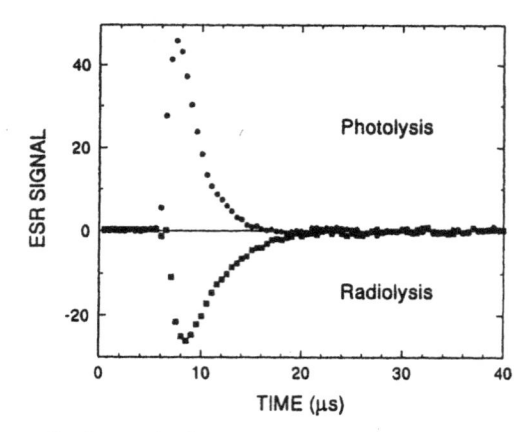

Figure 1 ESR time profiles for e_{aq}^- in photolysis and radiolysis experiments. Positive values represent absorption. The solutions contained 10 mM phenol at pH 12.5 and 20 mM phenol at pH 12.3. The signal channel in the radiolysis experiment was gated off during the 0.5-μs radiolysis pulse (at about 6 μs) and for 1 μs after. The growth of the ESR signal in both cases is governed by the magnitude of the microwave magnetic field (about 15 mG), and the spin relaxation time of e_{aq}^- and does not necessarily reflect growth of the polarisation. The decay depends mainly on spin relaxation (photolysis) and the rate of reaction of e_{aq}^- [taken from Ref. 5].

Eu^{2+}, F-centres and holes (Br_2^- V_k centres) and this is thought to be of major importance in the photostimulated emission of this phosphor. As well as helping to generate F (Br^-) centres, the oxide ions also trap holes (to give $O \cdot^-$ or possibly $(O-Br)^{2-}$ centres?).

A somewhat similar situation to the F-centre CN^- ion pairing occurs for KBr crystals doped with hydroxide ions (symbolised $F_H(OH^-)$).[39] However, the OH^- ions lie in the 4th shell around the F-centre, which is curious. A thermal bistability is observed which is connected with the orientation of the OH^- ions relative to the F-centres.

A general study of F-centres in a range of alkali halide crystals using ODMR and MCP methods has been reported.[40] An ENDOR study of calcium fluoride doped with lithium, containing F_{2A}^+ centres has also been described.[41] Hyperfine coupling to ^7Li as well as ^{19}F was observed.

Finally the interesting solids known as sodides can form trapped electron centres. These salts essentially comprising Na^+ cations and Na^- anions, can form anion vacancies in which electrons are trapped as in conventional salts. The g-values, hyperfine splittings and relaxation times are discussed.[42]

3 Atoms and Monatomic Ions

3.1 Introduction

Although primarily concerned with the electronic structures of paramagnetic atoms or ions, generally in condensed media, their interactions with the matrix and their reactivities are also considered. This remains an extrememly active field. Of various related studies I mention the work of Gurtler et al.[43] who studied the vacuum UV spectra for chlorine atoms in neon matrices. These showed the presence of the different sites which on careful annealing

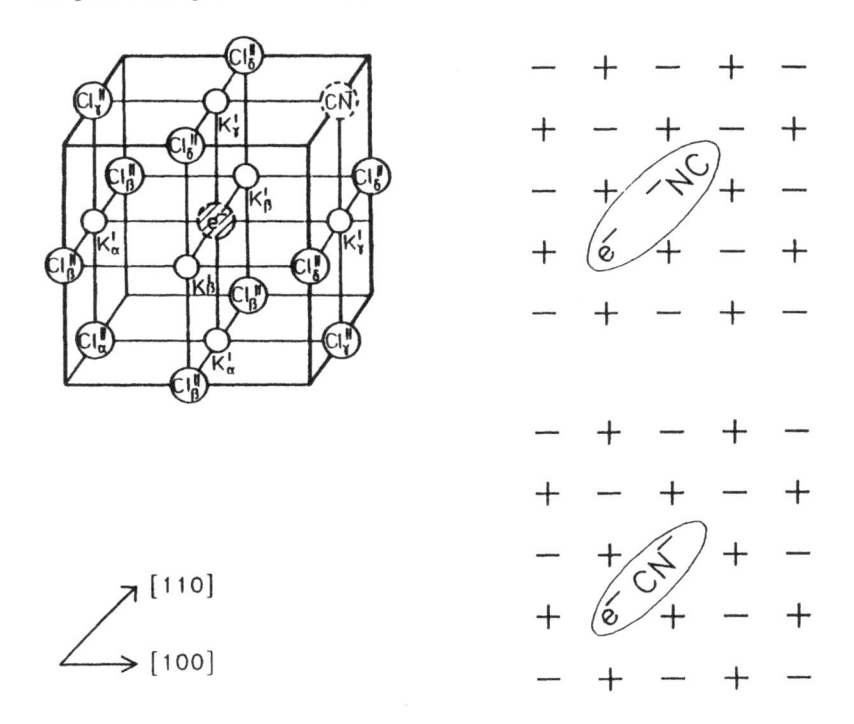

Figure 2 Model for $F_H(CN^-)$ centres in KCl (left). Orientations of the CN^- dipole along a (110) crystal axis (right) [taken from Ref. 34].

formed only a single preferred site. The results are very close to those for the gas-phase, showing that Cl–Ne interactions are not important.

3.2 Hydrogen Atoms

The hydrogen atom adduct of a chloride ion, $HCl\cdot^-$ has been treated theoretically.[44] H· atom adducts of F^-, Cl^-, Br^- and I^- have been detected by esr spectroscopy and apart from the fluoride ion species, which has virtually complete hydrogen atom character, they all show considerable delocalisation, increasing from Cl^- to I^-. We described these species as σ^* radicals, $H \dotdiv hal^-$.[45] Since the interaction in $H \dotdiv Cl^-$ is weak, the problem is an extremely subtle one, and difficult to handle theoretically.

Hydrogen atoms are readily generated and trapped in irradiated solid hydrogen, and have normal esr spectra. On going from H_2 matrices to HD or D_2, there is a large change in linewidth for H centres, which has been interpreted in terms of two alternative trapping sites.[46] It has always puzzled me that when H is surrounded by H_2 molecules there seems to be no tendency to form $H_3\cdot$ molecules. This is a 3-electron molecule with two bonding electrons and one non-bonding electron and so is expected to have some stability. Possibly the explanation lies in the fact that there are many H_2 molecules surrounding each atom which then is more stable when 'poised' in the centre. This problem also arises for atoms in rare-gas matrices, where superhyperfine structure clearly establishes that the central position is favoured. There remains the possibility of quantum-mechanical tunnelling between the sites. The trend in linewidths, shown in Fig. 3, is certainly one that goes against any expectation in terms of normal matrix effects.

A different type of isotope effect has been observed in the well established and extremely fundamental reaction (5).

$$e_{aq}^- + H_2O \rightleftharpoons H_{aq} \cdot + OH^- \tag{5}$$

As well as being one of the simplest and yet least known reactions, this is important in showing how competitive e_{aq}^- and hydrogen atoms really are. This reaction can only be studied reversibly in alkaline solutions or glasses. In low temperature ice, hydrogen atoms are probably formed from "H_3O^+" cations by reaction with dry electrons (6),

$$e^- + H_3O^+ \rightarrow H \cdot + H_2O \tag{6}$$

and it turns out that this process shows a marked $^1H/^2H$ isotope effect.[47] When the hydrogen content of H_2O/D_2O ices falls below 3%, electrons are detectably trapped at 4.2K. They are preferentially converted into $H_t \cdot$ rather than $D_t \cdot$ with $k_H/k_D > 10^3$. This preference falls rapidly as the total hydrogen content increases.

Reaction (1) using a range of relative concentrations of H and D solvents has been studied by Han and Bartels.[48] They have given precise values for the apparent Arrhenius parameters for this reaction, with clear but quite small differences between the A and E_a parameters for $H + OH^-$ and $D + OD^-$. Some weight is put on these in an extensive attempt to probe structure and mechanism for this basic reaction.

Some of the problems discussed above, which try to distinguish between localisation of H · as in H ∸ Cl⁻, and delocalisation in a matrix site are partially resolved by the interesting study of Baker and his coworkers.[49] At 300K H atoms in X-irradiated Li_2O show hyperfine coupling to eight ^7Li nuclei, thus clearly occupying a cubic site in the crystal. However, on cooling to 4K, coupling is to only four ^7Li nuclei. This is expected in terms of preferential trapping at one face of the cube, close to the four cations. On warming tunnelling between the six equivalent sites of the cube occurs to give the room temperature results. One wonders if, at even lower temperatures, there might be a tendency to move towards two, or even one of the cations?

3.3 Copper, Silver and Gold

Two papers treat these "coinage" metals collectively. In one, the metal atoms are trapped from the gas-phase in vapour deposited alkali-metal chlorides (M^+Cl^-) using a rotating cryostat.[50] Curiously, this method of isolation gives much reduced pure metal atom character than is found when precursor cations grown into M^+Cl^- crystals are converted into atoms by radiolysis—and also for gas-phase metal atoms. For example, for silver (^{109}Ag) treating the gas-phase hyperfine coupling as 100% that in the reduced crystal centre is ca. 90–100% whilst that for the deposited atoms is 66–72% depending on the alkali-metal ions.

It is suggested that these drastic reductions in hyperfine splittings arise because, in the rotating-cryostat work, the atoms become trapped in asymmetric sites at interfaces and are thus more like surface trapped sites. There are two major reasons for loss of hyperfine coupling—one is delocalisation of spin-density onto ligands, whilst the other is asymmetric coordination resulting in s-p hybridisation of the unpaired electron (these are not, of course,

5: *Inorganic and Organometallic Radicals* 207

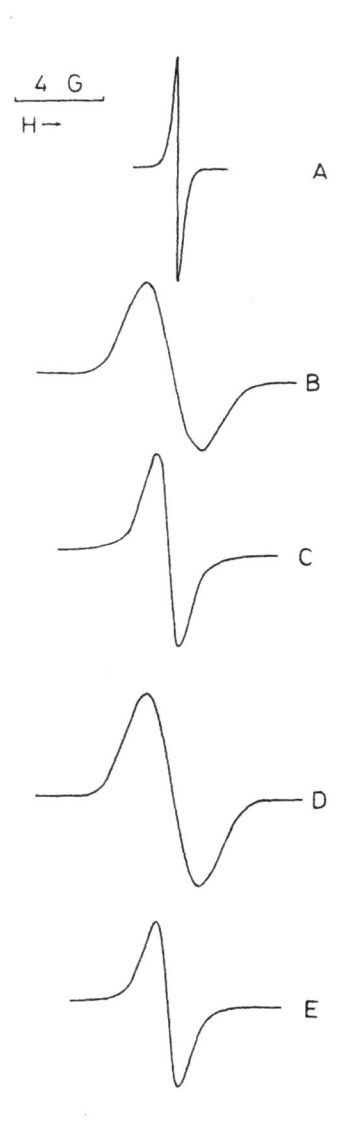

Figure 3 Low field features of H(D) atoms produced in γ-irradiated hydrogen at 4.2K. (A) H in H_2, (B) H in HD, (C) H in D_2-H_2 (1mol%), (D) D in HD, (E(D in D_2 [taken from Ref. 46].

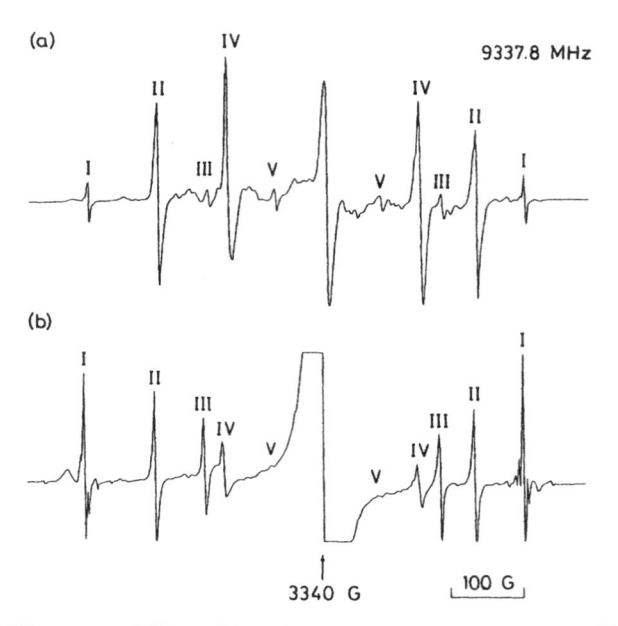

Figure 4 (a) ESR spectrum of ^{107}Ag + SiO + adamantane at 77K; (b) ESR spectrum of ^{107}Ag + SiO + adamantane at 77K with a higher degree of SiO polymerisation (G = 10^{-4}T) [taken from Ref. 55].

independent effects). Since the anisotropic contribution to the coupling from outer p-orbital occupancy is relatively small, the observed effect is just a reduction in the large isotropic splitting. However, there should also be a significant negative g-shift, and these are indeed quite large (ca. −0.01) which supports the concept of trapping at asymmetric sites.

Much of the work of Howard, Mile and their coworkers is directed towards the reactions of these matrix atoms with a range of simple substrates rather than on the atoms themselves. For example, reactions of Cu, Ag and Au atoms have been allowed to react with a range of acyclic and cyclic alkenes and unconjugated dienes, at 77K.[51] Reactions follow the pattern found for ethene and propene, with some exceptions. Copper and gold atoms give both mono- and di-liganded complexes, but silver atoms give only the mono-liganded derivatives, together with cluster compounds. Allyl radicals are also formed in these systems.

Copper atoms react with P(Me)$_3$, P(OMe)$_3$ and PF$_3$ to give a range of complexes.[52,53] For P(Me)$_3$ and P(OMe)$_3$ the major products are the tricoordinate, 17 electron species Cu[PMe$_3$]$_3$ and Cu[P(OMe$_3$)]$_3$. The results indicate planar coordination at copper with the SOMO largely the non-bonding $4p_z$ orbital on copper. However, for PF$_3$, the mono-liganded species, Cu–PF$_3$ is the main product, as is also the case for silver atoms.[54] The copper complex is clearly not a phosphoryl radical, but can be thought of as σ* species with ca. 70% spin density in the 4s orbital, ca. 8% in the phosphorus 3s orbital with small delocalisation onto fluorine. The remaining spin density is probably in the $4p_z$ (Cu) orbital giving the expected σ^* structure, Cu $\dotdiv$ PF$_3$. A similar structure is found for Ag $\dotdiv$ PF$_3$.

The same group have studied the reactions of silver atoms and SiO molecules in hydrocarbon matrices at 77K.[55] A range of radical products include Ag(SiO), Ag(SiO)$_2$, Ag(SiO)$_3$ and Ag(SiO)$_n$. The relative intensities of these products depends on the degree of polymerisation of the SiO prior to reaction with silver, as indicated in Fig.4. The strong central line is due to occluded radicals in the SiO clusters., and the complexes are complexes of dimers, trimers etc. of SiO, rather than being Ag^0 with a range of (SiO) ligands. The structures of these interesting species are discussed.

There is an extensive literature on metal-atom clusters in zeolite cages, which lend themselves to the stabilisation of these species. Kevan and coworkers have extended their studies to silver atoms and Ag$_2^+$ dimers and their interaction with water molecules in synthetic fluorohectorite and beidellite.[56]

3.4 Aluminium, Gallium and Thallium
Not content with Cu, Ag and Au, Howard and Mile have also explored the trapping and reactions of aluminium and gallium atoms in hydrocarbon matrices at 77K. Aluminium atoms react with ammonia in adamantane at 77K to give three distinct derivatives, the diliganded centre, Al(NH$_3$)$_2$, the tetraligated centre Al(NH$_3$)$_4$, and the rearranged hydride, HAlNH$_2$.[57] For Al(NH$_3$)$_4$ most of the spin density is on aluminium, with ca. 20% 3s population indicating a strongly hybridised orbital. Delocalisation onto the ligands is small. The diliganded species Al(NH$_3$)$_2$ was less well defined, with the unpaired electron thought to be in a non-bonding $3p$ orbital normal to the plane of the bent molecule as in (I).The rearranged species HAlNH$_2$ has a beautifully defined spectra showing a doublet of sextets, each feature being split into further narrow components. The hyperfine parameters are very similar to those for the previously studied species HAlOH.

I   II

Reaction of aluminium atoms with Et$_2$O,[58] and with a range of ethers,[59] has also been described by these workers. The results for Et$_2$O are well defined, leading to the suggested rearranged hydride derivative CH$_3$CH – Al – H(OEt) or possibly CH$_3$CH$_2$OCH$_2$CH$_2$ – Al – H. Other possible assignments are discussed. This work has recently been extended to cover other R$_2$O derivatives, and several novel insertion compounds have been detected.[59]

In extending these studies to gallium atoms, these workers have explored reactions with ethene,[60] and with benzene.[52] For the former, the expected Ga[C$_2$H$_4$] π-complex was detected, with a SOMO comprising largely the Ga $4p_x$ orbital. Some evidence for the cyclic species II was also adduced. With benzene, the species is essentially a gallium atom weakly perturbed by complexing with benzene molecules in some unspecified way.[61] The results contrast with the Al(C$_6$H$_6$) complex which is well defined.

In a different key, there are two studies of thallium centres—one in NaCl and KCl crystals,[62] and one in K$_2$SeO$_4$ crystals.[63] Tl^0 centres are, in first order, s^{2}p^1 atoms, and a Tl^0(1) centre has been studied extensively in alkali halide crystals, using esr and optical methods. Two new Tl^0 centres have now been discovered having orthorhombic symmetry. Careful analysis suggests that the thallium is now at anion centres, having switched from the original cation sites of the parent Tl$^+$ ions.

The centre in K$_2$SeO$_4$ is Tl^{2+} rather than Tl^0.[54] This is formally an s^1 ion and hence the hyperfine splitting is very large. This splitting changes with temperature, and there is thought to be long-range correlations between centres induced by the "phason" mode between the defects. A theoretical explanation of this phenomenon is given.

3.5 Lead and Bismuth

Lead centres (both Pb$^+$) have been studied in alkaline-earth fluorides by Spaeth *et al.*[64] and in SrCl$_2$ by Schoemaker and coworkers.[65]

In alkaline earth fluorides, Pb$^+$ is in a cation site adjacent to a fluoride vacancy. In contrast with Pb$^+$ doped KMgF$_3$, these crystals could not be induced to lase. This is shown to be associated with F$_A$-centres in these crystals.[64] X-irradiation of SrCl$_2$ doped with Pb^{2+} ions gave several types of Pb$^+$-centres at 80K. The dominant species with trigonal symmetry is attributed to Pb$^+$ at a cation site with a nearest-neighbour anion vacancy. The other centres were also Pb$^+$ but their local environments were not determined. For heavily doped samples, a Pb$^+$(1) defect perturbed by a Pb^{2+} ion was detected.[64]

A Bi^0 (6p^3) centre has been studied in KCl crystals.[65] The centre is formed from Bi^{3+} impurity ions, by X-irradiation above 220K. This is quite unusual since it involves three electron addition events. In my own experience, it is difficult to induce a second electron to add in radiation studies, but to induce a third addition is remarkable. The results show that cation or anion vacancies play a vital role in this conversion. The identification as Bi^0 (6p^3) is compelling, but requires that the normal ground-state, ^{4}S$_{3/2}$, must be split by the crystal field into 3/2 and 1/2 states in order to explain the S = 1/2 spectrum. Note that for nitrogen atoms, with much smaller spin-orbit coupling, the quartet ^{4}S state is formed, all three transitions being readily detected,[66] the zero-field splitting being quite small. Crystal-field calculations of the g- and quadrupole parameters also support the identification, as does the hyperfine coupling.

3.6 Oxygen and Sulphur Centres

Centres described as O$_F^-$ have been detected in BaFBr, which were referred to above. This centre is formed by reaction between Br$\cdot^-$ centres and oxide impurities (7).

$$Br_2^{\cdot -} + O^{2-} \rightarrow 2Br^- + O^{\cdot -} \tag{7}$$

A detailed ENDOR study[67,68] suggests that the centre is O$^-$ trapped in an F$^-$ site. Thus there is no tendency to form the σ* species, O $\doteq$ F^{2-}, which might have been expected since it is isoelectronic with F$_2^-$.

Finally, the radical S·$^-$ has been mentioned amongst a range of other sulphur centred species such as HS·, SO$_2^-$, HS–SH$^-$, HSS·$^{2-}$ and HSS·$^-$.[69] It is best in this particular case to consider all these species together. The S·$^-$ centre was studied many years ago in alkali halide crystals, but not in glassy matrices as in the study by Sevilla and coworkers, who used

8M NaClO$_4$, 12M LiCl or alkali metal hydroxide glasses at 77K. Primary glass radicals (O·$^-$, Cl$_2$·$^-$) and e$^-$ react with H$_2$S, HS$^-$ or S^{2-} to give HS· and S·$^-$ radicals. In the alkaline glass, O·$^-$, which is a primary product, reacts to give S·$^-$ which has a very strongly shifted g value, (2.153). This is turn reacts with oxygen to give the well known ·SO$_2^-$ radical. In the chloride glass, S·$^-$ is formed from S^{2-} and Cl$_2$·$^-$, and again I find it a surprise that S $\doteq$ Cl^{2-} σ* centres were not observed. The results for HSS· (2.071, 2.025 and 2.002) are of interest in view of the controversy that still centres around RSS· and RSSR$_2$· centres formed in organic systems.

4 Diatomic and Related Radical Systems

4.1 Introduction

Classification is always difficult. On this occasion, in view of the information available, I have decided to discuss 4.2, hydrides, 4.3, σ-radicals, where the SOMO is a σ-orbital, 4.4, π-radicals, where the SOMO is π, and 4.5, transition metal dimers.

4.2 Hydrides

I start by considering reports on ·OH and ·SeH radicals, and then go on to various metal hydrides. One of the most 'popular' radicals for spin-trapping studies is the hydroxyl radical. This is because, in fluid aqueous solution, although hydrogen-bonding must strongly lift the orbital degeneracy, as it does in ice, this is a variable and fluctuating effect that must contribute strongly to the width of the isotropic feature, which has never been detected. Although the first stage of aquation for ·OH, the species (H$_2$O·OH) has not been detected. Kim and coworkers have, however, carried out a very complete theoretical study of this species.[70] There are at least three possible structures, namely the σ* species H$_2$O $\doteq$ OH (i), the species HOH $\cdots$ ȮH (ii), with water acting as H-bond donor, or the species ·OH $\cdots$ OH$_2$ (iii), with ·OH acting as the donor. The species explored in this paper is (iii), and the interaction is comparable with that in the water dimer.

The isostructural ·SeH radical has been less widely studied, except in the gas-phase. A laser magnetic resonance study by Ashworth and Brown[71] has probed the rotational structure for the $v = 0$ level of the $2_π$ ground-state of this species, and they were even able to detect features for ·^{77}SeH in natural abundance (0.9%).

Many metal hydrides have been studied in the gas-phase and in low temperature matrices (see previous reports). A useful theoretical review of some of these has been published,[72] which links with several other theoretical studies on metal hydrides by Goddard and his coworkers. In the present study, the species HfH·$^+$ to HgH·$^+$, together with BaH·$^+$ and LaH$^+$, have been studied. Factors controlling bond energies and characters and excitation energies have been probed. These hydrides are found to be significantly more stable than the corresponding first or second row hydrides.

Finally, the MnH molecule has been studied in the gas-phase by Varberg *et al*[73] and the results compared with Weltner's results for MnH in rare-gas matrices. This molecule is high spin ($^7\Sigma+$), and ^{55}Mn and ^{1}H hyperfine splittings were well defined in both studies. An interesting decrease in the magnitude of the ^{55}Mn Fermi contact interaction for gas *versus* solid state is discussed.

4.3 σ-A–B Radicals

Inevitably, this is an arbitrary classification for some high-spin species, but in other cases it is clear-cut. For the molecules NaMg, NaCa, NaSr, KMg and KCa, formed by codeposition in an argon matrix, the $^2\Sigma$ ground state is indeed clear.[74] However, the unpaired electron is quite strongly biased towards the alkali-metal atoms (ca. 75%). These are three electron systems so one expects that the σ-bonding electrons will favour the alkaline earth-metal and the σ^* electron the alkali-metal, as observed.

A Pb_2^{3+} dimer has been detected in NaCl doped with Pb^{2+}.[75] Mention was made in section 3 about mononuclear lead centres in ionic crystals. In this case, for heavily doped material, one of the centres formed on X-irradiation at 77K is best described as Pb_2^{3+}. This species could not be detected in KCl matrices, implying that its formation has little to do with strong bonding between the lead atoms. From studies of $^{207}Pb^{2+}$ doped crystals it is shown that the two lead atoms are effectively equivalent.

In terms of normal valence-electron arguments, this is a 5 electron unit and the SOMO could be σ or π. The nearly isotropic coupling to ^{207}Pb shows that it must be σ, but the large g-shifts show that the vacant π-levels are very close to this orbital. The spectra were simulated using the newly developed "simulated anneal" method, based on the Metropolis Monte Carlo algorithm.[76]

Knight and his coworkers have described their matrix-isolation studies of the radical cations Si_2^+ and Ge_2^+, using pulsed laser vaporisation.[77] These centres have two extra valence electrons (7e), three of which are unpaired, giving $^4\Sigma$ ground states. These results also nicely illustrate the proximities of the $\sigma-$ and $\pi-$ molecular orbitals for such A−B radicals. However, in these cases, the atomic s-contributions to the σ-orbitals are relatively small. A theoretical calculation for $^{29}Si_2^+$ gave good agreement with experiment. The zero-field splitting is very large, and the spectra resemble those for high-spin transition metal complexes, as shown in Fig. 5 for $Si_2^+ \cdot$.

I have found an example of a 9-electron σ^*-radical, namely $\cdot BO$.[78] This species is isoelectronic with the CN radical, which has its unpaired electron quite strongly localised on carbon, which is again indicative of a σ^* orbital (in this case, largely non-bonding). All isotopic forms of BO have been isolated in neon, the new results being in excellent agreement with gas-phase data, as is usually the case for neutral centres in rare-gas matrices. Using the simple method of comparing experimental isotropic and anisotropic hyperfine couplings with calculated values for 100% orbital population[1] the SOMO for the $\cdot BO$ radical is found to have ca. 40% $2s$ and 41% $2p$ (σ) character on boron, and from the ^{17}O parameters, very low density on oxygen. This is in line with the simplest model for $\cdot BO$, the SOMO being the non-bonding s-p hybrid orbital on boron. It's satisfying that simple theory often works well!

In the same publication there is a study of the radical $\cdot BNH$, which I discuss here because it is the protonated version of $\cdot BN_4^-$, which is isoelectronic with $\cdot BO$ and $\cdot CN$. As expected, the unpaired electron is again largely centred in a boron centred hybrid orbital, and the hyperfine parameters are almost indistinguishable. The nitrogen spin density is again very small, and there is a small proton coupling corresponding to a spin density of ca. 2%. The spectrum for BNH is beautifully defined, as shown in Fig.6.

Finally, moving up to 15 electron systems, the $\cdot F_2^-$ σ^* centre (V_k centre), which has been studied extensively by esr spectroscopy over many years, has been treated theoretically by Feller.[79] The calculated esr parameters agree remarkably well with experiment.

4.4 π-Radicals

The $\cdot NO$ radical is an 11-electron species with a $(\pi^*)^1$ configuration which means that there is strong angular momentum around the z-axis via spin-orbit coupling. This in turn means that liquid and even solid state spectra are undetectably broad. This is most unfortunate since

5: *Inorganic and Organometallic Radicals* 213

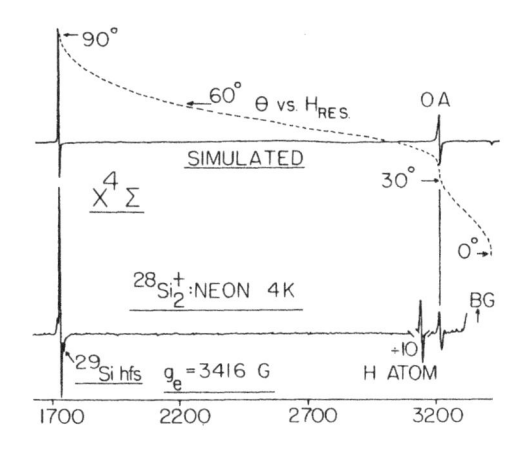

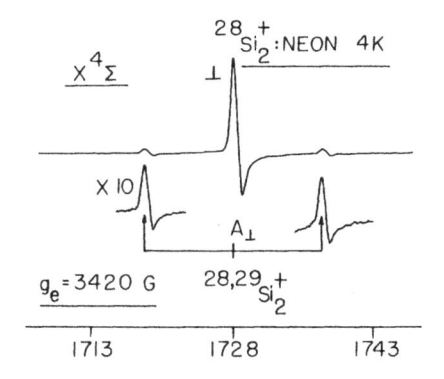

Figure 5 Upper Figure: First derivative ESR spectra for $^{28}Si_2^+$ in neon at 4K (lower trace) and the simulated spectrum (upper trace). The line marked OA is the "off-angle", extra line which arises because of the mode of evolution of the θ *versus* H_{res} curve (....). Lower Figure: Hyperfine features from $^{28,29}Si_2^+$ in the low field (perpendicular) region [taken from Ref. 77].

·NO is now known to be of great biological significance. It is, of course, one of the very few stable inorganic radicals, which is also bad news for esr spectroscopists since it means that conventional spin-traps are not effective (see section 10, below).

However, ·NO is very efficiently scavenged by deoxyhaemoglobin or deoxymyoglobin, and the detection of these complexes by esr is an important method of studying biological ·NO. A most interesting alternative to this has been described by Mordvinţcev *et al.*,[80] who recommend the use of ferrous diethyldithiocarbamate as a trap. This gives a mononitrosyl

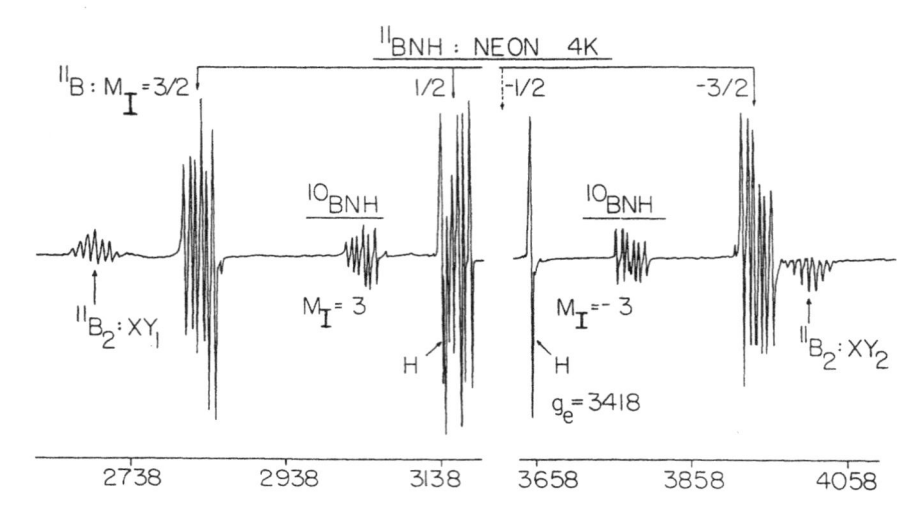

Figure 6 The isotropic ESR spectrum of $^{11}B^{14}NH$ consisting of a widely spaced $^{11}B(I=3/2)$ quartet further split into smaller $^{14}N(I=1)$ triplets and H doublets is shown for a neon matrix sample at 4K. The lowest- and highest-field groups of the analogous $^{10}B(I=3)$ septet are labelled $M_I = \pm 3$. The H atom doublet lines and the perpendicular (XY) lines of B_2 in its $X^3\Sigma$ ground state are also indicated. Each fine structure transition of B_2 is split into a septet hyperfine patern for two equivalent $^{11}B(I=3/2)$ nuclei [taken from Ref. 78].

derivative that is characterised by an isotropic triplet at $g = 2.03$ in the liquid phase. Although the lines are broad, it is claimed that a lower limit of detection of only 0.5 nmol can be obtained. The trap is actually dissolved in yeast cell membranes so the ·NO is trapped within the membrane rather than in aqueous solution. The trapping efficiency is as high as 60% in the absence of oxygen, but falls in aerobic solution in part because of the rapid reaction between O_2 and ·NO.

Another stable paramagnetic molecule that can normally only be studied directly in the gas-phase is triplet oxygen. Matsumoto and Kuwata have used esr spectroscopy on a supersonic beam of oxygen gas. Under these conditions only the lower levels are strongly populated, and, using a specially constructed cavity, good esr spectra have been obtained.[81]

The $^1\Delta$ state of dioxygen is thought to be of considerable importance, especially in biological systems. The use of spin-traps to detect $^1\Delta O_2$ is discussed in section 10, below. However, these molecules have orbital magnetism, and can be detected in the gas-phase by esr methods. The isoelectronic molecule NF has been studied in the gas phase in its $^1\Delta$ form by Bradburn and Lilenfield.[82] These workers also detected $^1\Delta NH$ by the technique.

As usual, there are several reports on the superoxide radical-anion, another of the few stable inorganic radicals. Again, the species is not readily detectable in the liquid phase, but is often detectable in solids, provided there is a well defined "crystal-field" lifting the orbital degeneracy. [Hydrogen bonding achieves this most effectively]. The $O_2 \cdot^-$ centre often occurs in oxide crystals. It has been studied recently on MgO surfaces by esr spectroscopy.[83] It was formed from oxygen *via* reactions with various "slightly acidic" X–H molecules such as propene or benzene.

Protonated superoxide, $HO_2 \cdot$, is readily formed from hydrogen peroxide or by protonation of $O_2 \cdot^-$. It is more reactive, but more readily detected by esr spectroscopy because the hydrogen lifts the orbital degeneracy extensively. These radicals have been formed by photolysis of aqueous glasses of hydrogen peroxide and the spectra studied in the $g = 2$ and $g = 4$ regions. In the former, quite well defined spectra for $HO_2 \cdot$ radicals were detected, with no trapped OH radicals and no side-features characteristic of pairwise trapped species. However, there was a well defined feature at $g = 4$, which proves that radical-pairs must be present.[84] This paradox is explained in terms of the nature of the glassy medium, which comprises $H_2O + H_2O_2$ molecules randomly disposed, but strongly hydrogen-bonded. Photolysis gives two ·OH radicals, but these fly apart with enough energy for each to attack near-neighbour H_2O_2 molecules, giving two trapped $HO_2 \cdot$ radicals. These will be separated over a range of distances such that, for large distances spin-spin dipolar coupling will simply broaden the normal $HO_2 \cdot$ spectra at $g = 2$, whilst those closer together will give large zero-field splitting, but this will vary over a wide range so that the expected parallel and perpendicular features will be smeared out. However, this will only affect the intensities of the half-field, $M_s = \pm 1$ feature, which is almost isotropic. Hence this is readily detected. The importance of this result is that pairwise trapping in glasses may be completely missed if only the central spectral region is studied.

Finally, the isostructural anion, $\cdot S_2^-$, has been studied in NaBr crystals.[85] The crystals were doped with sulphide ions and exposed to X-rays at room temperature. The esr parameters (g-values) are reasonable for $S_2^- \cdot$ but differ considerably from those previously found for this centre.

4.5 Diatomics Containing One Transition Metal

This section links back with the section on transition-metal hydrides (4.2). The molecules PdB and PdAl in neon matrices have recently been studied.[86] Both radicals have a $^2\Sigma$ ground-state. Well defined hyperfine coupling to all nuclei was detected and the g-shifts were relatively small by comparison with results for ^{105}PdH. The unpaired electron is estimated to be *ca.* 60% on Pd and *ca.* 40% on B or Al. The orbital on Pd comprises a $d_{z^2} - s$ hybrid, the 5s orbital population being *ca.* 0.5. *Ab initio* calculations support these arguments.

Moving from boron and aluminium to carbon and silicon as 'ligands', the molecules VC, NbC, VSi and NbSi have been studied in several rare-gas matrices.[87] These also have doublet ground-states, but they are probably 2Δ, and hence there are large negative values for $\Delta g_\parallel$ ($= g_\parallel - g_e$), although some orbital quenching must occur. This, together with a marked variation in the parameters as the rare-gas matrix is changed, suggests that there is weak coordination of rare-gas molecules to the metal ion, although no extra hyperfine coupling was observed. The spectral interpretation and analyses are difficult, and careful explanations are given together with detailed considerations of alternative ground-states for these interesting molecules.

4.6 Diatomics Containing Two Equal Transition Metals

These include the cations Co_2^+, Ir_2^+ and Mn_2^+, and the molecule Mn_2. The Co_2^+ and Ir_2^+ cations were detected by Weltner and his coworkers in their study of Co, Rh and Ir trimers, discussed in section 5.[88] These both have $^6\Sigma$ ground-states, and give clearly defined hyperfine splitting, also the spectra are very weak. These species have not previously been studied.

The cation Mn_2^+ has also been studied, together with Mn_2 and CrMn mentioned in Section 4.7.[89] The molecule shows antiferromagnetic exchange coupling between the two sets of $3d^5$ electrons. The bonding is no more than Van der Waals in nature, and a range of spin states can be thermally populated. Loss of one electron to give Mn_2^+ yields a weak σ bond, but the high-spin structure remains, with 11 unpaired electrons giving a $^{12}\Sigma$ ground state.

4.7 Diatomics with Different Transition-metal Atoms

The molecule CrMn· is isoelectronic with Mn_2^+ discussed above. However, bonding is now much stronger, and the ground-state is $^4\Sigma$ with multiple bonding.[90] This set of three species, Mn_2, Mn^{2+} and CrMn, which are so very different from each other, present a real challenge to theory.

The ScCr molecule has a $^6\Sigma$ ground-state, in contrast with the isoelectronic molecule TiV which is $^4\Sigma$. Both ^{45}Sc and ^{53}Cr hyperfine features were resolved in solid argon at 4K.[91] There is a large zero-field splitting, but the g-shifts are quite small.

Weltner and coworkers have also studied Cu, Ag and Au derivatives of Mo and W [MoCu, MoAg, MoAu, WCu, WAg and WAu], which all have $^6\Sigma$ ground-states with zero-field splitting parameters, D, greater than $2cm^{-1}$. For the ^{197}Au derivatives both hyperfine and nuclear quadrupole splittings were detected.[92]

5 Triatomic · AB_2 Radicals

5.1 Introduction

As in the last section, I start with species having hydrogen atoms as ligands, since only σ-bonds need to be considered. For classificatory purposes I include alkyl derivatives since in a formal sense, π-bonding is still not important, although hyperconjugation may be very significant, especially for radical cations. I have then used the usual classification in terms of numbers of valence electrons, ranging from three to twenty one!

5.2 Hydrogen and Alkyl Derivatives (·AH_2, ·AH(R), ·AR_2)

The simplest triatomic molecule, $H_3\cdot$, was discussed in section (3.2) when considering hydrogen atoms trapped in a hydrogen matrix. Under those conditions there is no clear evidence for H_3 formation. However, $H_3\cdot$ has a relatively long-lived state in the gas-phase,[93] and hence there remains the chance that in the right matrix, it might be stabilised sufficiently for esr studies. Possible alternative structures for $H_3\cdot$ have been discussed by Peng et al.[94]

A range of aluminium centred radicals have been matrix isolated by Knight and his coworkers.[95] These include ·AlH_2, discussed here, and also HÁl(OH) discussed in section 5.3 and ·Al(OH)$_2$ discussed in section 5.6, below. The radicals ·AlH_2, ·AlHD and ·AlD_2 were isolated in neon at 4K. These radicals have not been previously studied by esr spectroscopy, although their UV spectra are known from gas-phase studies. They were formed during photolysis of a system containing aluminium atoms and H_2O or D_2O in very low concentrations. The mechanism of this reaction is not clear. The spectra for AlH_2 shows well defined sets of triplets from the two equivalent hydrogen atoms ($A_{iso}(^1H) = 128MHz$) and a range of hyperfine features for aluminium. ($A_{iso}(^{27}Al) = 834MHz$). The "experimental" orbital populations, deduced in the usual way,[1] are given in ref. 93. The sum of these populations adds to a satisfactory 98%. Also the p:s ratio of 2.81 is close to expectation for this simple structure. Theoretical calculations for ·AlH$_2$ give results in good agreement with experiment. Thus these species hold no surprises.

The ·NH$_2$ radical, which has a π-structure rather than the σ-structure of ·AlH$_2$ (or ·CH$_2$$^+$), has been very extensively studied. Theoretical calculations do not usually produce good values for the ^{1}H hyperfine coupling constants for such radicals. However, a new treatment for ·NH$_2$, ·NHD and ·ND$_2$ has given very good agreement with experiment.[96]

The radical H$_2$O·$^+$, which is isoelectronic with ·NH$_2$, has been less well studied because it is a strong acid and is rapidly converted into the OH radical in the presence of a proton acceptor. The situation that arises when H$_2$O·$^+$ is formed in femtosecond photoionisation of pure water has been probed by studying the decay of optical bands (460, 410nm) assigned to H$_2$O·$^+$, which is found to decay in less than 100fs.[97] Hence, although H$_2$O·$^+$ is readily detectable by esr spectroscopy under the right conditions,[98] it cannot be studied in liquid water by this technique.

As part of their extensive studies of small radical-cations in rare-gas matrices, Knight and his coworkers have managed to isolate CH$_3$ÖH$^+$ radicals.[99] Attempts to prepare such species in freon matrices have failed because methanol forms cyclic aggregates even at extremely low concentrations because of the marked cooperativity in hydrogen bonding.[100] These radicals have the same formal structure as H$_2$O$^+$ and CH$_3$ṄH. However, because of the unit positive charge, hyperconjugation is large, and the methyl proton hyperfine coupling (229MHz) reflects this.

Although radical-cations R$_2$O·$^+$ and R$_2$S·$^+$ have been studied extensively, R$_2$Te·$^+$ cations have not previously been described. They are formed by radiolysis of dilute solutions of R$_2$Te molecules in CFCl$_3$, a technique known to give the parent radical-cations. As expected, the SOMO is largely the 5p_z orbital on tellurium, normal to the radical plane. The (positive) g-shifts are large because of the large spin-orbit coupling constant for Te, and a linear correlation between the maximum g-shifts for R$_2$O·$^+$, R$_2$S·$^+$, R$_2$Se·$^+$ and R$_2$Te·$^+$ and the corresponding spin-orbit coupling constants was demonstrated. Also, the ^{1}H coupling constants for the methyl protons in this series (R = CH$_3$) were shown to fall linearly with the decrease in ionisation potentials. Using relatively concentrated solutions, Me$_2$Te·$^+$ radicals reacted to give the novel σ* species (Me$_2$Te − TeMe$_2$)$^+$ on annealing.[101]

5.3 HAB radicals

The species discussed in this section are HÄlOH radicals, the well known HĊO radical, and a series of RO$_2$· radicals. All the hyperfine splittings for HÄl^{17}OH have been measured.[102] These radicals were generated in the system used to generate ·AlH$_2$ and ·Al(OH)$_2$.[93] Cis and trans conformers were detected with slightly different parameters. *Ab initio* calculations agree well with the experimental results for this species, which are considered further in section 5.4, below.

The radical HĊO is closely related to HÄlOH, being isoelectronic with HḂO$^-$, and indeed, the ^{1}H hyperfine coupling constants are quite similar. Using laser magnetic resonance in the far infra-red, Brown *et al.* have produced a complete set of structural and rotational parameters for this radical.[101]

In quite a different area, Sevilla and his coworkers have probed the esr spectra for a range of peroxyl radicals, ROÖ, using ^{17}O to help distinguish between results for different R groups.[102] These radicals are familiar to all radiation chemists, and are often unwelcome guests when oxygen has not been completely removed from the system. They are of great

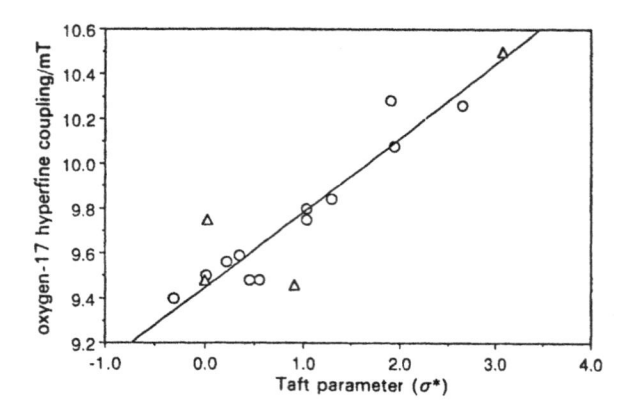

Figure 7. The relationship between the Taft substituent parameter (σ^*) and oxygen-17 hyperfine coupling constant on the terminal oxygen of peroxyl radicals. $A(1) = 9.41 + 0.33\sigma^*$, $r_2 = 0.912$. The regression line was calculated using only those radicals for which reliable Taft parameters could be calculated (o). Four radicals for which Taft parameters could only be estimated are shown (Δ), but not inluded in the regression fit [taken from Rev. 102].

importance in their role in chain autooxidation reactions. However most of us are under the impression that all $RO_2\cdot$ radicals give the same spectrum, dominated by the large shift in g_z, where z is approximately the O–O direction. This remains true for ^{16}O derivatives, but there are clear differences in the ^{17}O hyperfine data which reflect shifts in spin-density between the two oxygen atoms. These changes are nicely displayed in Fig.7, taken from ref. 102, which shows how the reactivity of $RO_2\cdot$ radicals increases as the spin-density on the outer oxygen increases. It is satisfying that this reaction-rate correlation does not hold for $RSOO\cdot$ radicals. Indeed, it is surprising that these radicals have g-components so similar to normal $RO_2\cdot$ radicals, since they are in a sense more closely related to radicals such as $\cdot O_3^-$ or $\cdot S_3^-$.

5.4 Nine Electron Radicals

The radicals $\cdot B_3$, $\cdot Al_3$ have been studied by Weltner's group,[103] and the related radical $\cdot Ga_3$ by Howard et al.[104] The former species was studied in rare-gas and methane matrices between 3 and 30K, whilst the latter was studied in adamantane, using a rotating cryostat at 77K.

In all cases, it seems probable that the ground-states are doublets and that the three nuclei are equivalent, so that these unprecedented molecules must exist as equilateral triangles with bond-angles of 60°. Equally unusual is the fact that at 4K, the hyperfine coupling to ^{11}B is isotropic. This gives a *total* $2s$ population of only *ca.* 15%, and 85% p-character should give rise to a large anisotropy, which is not observed. The only reasonable explanation is that these triangular molecules are rapidly rotating even at 4K, not just about the symmetry axis, but quite freely. Cooling to 2.5K in a krypton matrix gave a marked line-width variation across the spectrum (broad-narrow-broad), just as would be expected as the tumbling rate is reduced. Clearly 2.5K is still 'hot' for these species, which have a remarkably compact structure and are very keen to rotate—or could there be some other explanation for this previously unknown type of molecule?

Even the heavier molecule, ·Ga₃ gives an almost isotropic spectrum, indicating considerable averaging of the expected anisotropy.[104] The conclusion is that the SOMO is σ in type, using s-p hybrid atomic orbitals, but the expected anisotropy is lost as a result of an unprecedented desire to rotate. These seem to me to be amongst the most fascinating species discussed herein.

5.5 11, 13, and 15 Electron Radicals

In contrast with ·B₃ etc., these species are more conventional in their structure and spectra. Here I consider the 11-electron species BNB, the 13 electron species BCO and the 15-electron species NNO⁺. The BNB molecule (¹¹B¹⁴N¹¹B and ¹⁰B¹⁴N¹¹B) has been prepared in rare-gas matrices for vaporised boron nitride.[105] The molecule has a ²Σ ground state, with 'normal' parameters that are well reproduced by theory. The SOMO comprises mainly sp orbitals on the two boron atoms with small spin density on nitrogen. The BNB· radical is isoelectronic with CBC· and C₃⁺. The results for BNB suggest non-linearity, whilst C₃⁺ is probably linear. There seems to be very little resistance to bending. So the σ-frame is still being utilised rather than the π system. Hence there seems to be no particular constraint requiring non-linearity. As usual, the esr spectra for these species are beautifully defined. It is noteworthy that there is no resolved ¹⁴N splitting on the parallel lines, but the perpendicular lines are well defined, 20G, triplets. This very unusual behaviour requires a negative value for A_{iso}, the estimated populations on ¹⁴N being −1% 2s and +11% 2p_z. This accords with a structure having a nodal system through the central nitrogen so that the s character arises only from spin-polarisation.

The BCO molecule now has two more electrons to accommodate. It was prepared by reacting boron atoms with CO in rare-gases at 4K.[106] In this case the ground state is ⁴Σ with a relatively small zero-field splitting, D, of 0.074cm⁻¹. All three electrons seem to be largely confined to boron, so the CO group can be thought of as a ligand. Thus the three electrons occupy the 3(2p) boron orbitals almost equally. Again a surprising result.

Finally, though not esr spectroscopy, I mention the interesting infrared study of Jacox on N₂O⁺, a 15 electron centre.[107] So far as I know, this centre has yet to be studied by esr spectroscopists.

5.6 17 Electron Radicals

We are now on familiar ground, with many well known radicals in this class, the most important being ·NO₂. One of the most studied of these species is ·CO₂⁻, since this is an important metabolic product, and is formed in the radiolysis of carbonates, which in various forms constitute a range of important minerals. Also carbonate is a component in bones and shells. So ·CO₂⁻ trapped in such matrices may act as a method of estimating exposure to ambient radiation, and hence to dating, or it may be usable to show whether or not bone- or shell-containing food has been irradiated.

Moens et al. have exposed apatites to ¹³CO₂ molecules, which presumably are confined to surface sites, and then exposed these to ionising radiation. Both ¹³CO₃⁻ and ¹³CO₂⁻ radicals were clearly detected, but the ¹²C centres which were expected to be formed from internal carbonate ions were not detected. In addition to ·CO₂⁻ radicals, the species ·CO⁻ was also tentatively identified in these systems.[108]

A range of four different types of ·CO₂⁻ centres have been described in γ- and X-irradiated synthetic calcite.[109] Presumably they all occupy somewhat different sites, and possibly some differences in the parameters arise because of variable libratory motions. *Ab*

initio calculations are always helpful for esr spectroscopists. These have been done both for $\cdot CO_2^-$ and also for the ion-paired form, $M^+ \cdot CO_2^-$, by Ramondo *et al.*[110]
Derivatives of $\cdot CO_2^-$, formed by the addition of X_3Si radicals to CO_2, have been studied by Johnson and Roberts.[111] [$X = CH_3$ in most studies]. These radicals have the expected parameters for species RO–$\dot{C}$O behaving as perturbed $\cdot CO_2^-$ radicals.

Finally in this section I return again to the work of Knight *et al.* on AlX_2 species (X = H or OH).[93] Turning to $Al(OH)_2$ this can be thought of as the diprotonated form of $\cdot AlO_2^{2-}$ which is isostructural with $\cdot CO_2^-$. The esr properties accord well with this analogy. In this paper, results for a range of small aluminium radicals have been compared, and are reproduced in Table 2.

The order is that of increasing magnitude of A_{iso} for aluminium, which increases from 75MHz in $\cdot Al(CO)_2$ to 2819MHz in $AlF^+ \cdot$. This trend follows the extent to which Al can be thought of as Al^{2+} with weakly bonded ligands [so AlF^+ is close to $Al^{2+} \cdots F^-$]. The greater the Al^{2+} character, the greater the $3s^1$ contribution and the greater A_{iso} (^{27}Al) becomes.

5.7 19 Electron Radicals

The 19 electron radical $\cdot FO_2$, isostructural with the very stable $\cdot ClO_2$ molecule, has been studied in the gas-phase by laser magnetic resonance in the far infrared.[112] The results are in accord with expectation. Much interest centres around the regions of avoided level-crossings where useful transitions occur, just as in studies of muons.

5.8 Transition Metal Centres

These include the hydrides and the dioxides of Co, Rh and Ir.[113,114] The dihydrides and dideuterides were prepared in argon and krypton at 4K. They all have $S = 1/2$ ground-states and hence are non-linear. The spin densities on the hydrogen ligands are relatively small (ranging from 4 to 20%) and the g-shifts increase extensively going through the series, as expected. Clearly the ground-states are $^2A_1$ which accords with some theoretical calculations but not with others.

The $\cdot MO_2$ radicals, $\cdot CoO_2$, $\cdot RhO_2$ and $\cdot IrO_2$ were formed by reacting the laser-vaporised metal atoms with dioxygen and trapping in argon at *ca.* 2K. They all have $^2\Sigma$ ground-states. Hyperfine coupling from ^{59}Co, ^{103}Rh, ^{191} and ^{193}Ir and ^{17}O were all quite well defined, giving a detailed map of the SOMO for these dioxides. As expected, the oxygens act as if they were oxide ligands, with the electron spin largely confined to the metal, with *ca.* 50% s–σ character.[114]

6 Tetraatomic AB_3 Radicals

6.1 Introduction

Classification for these examples is most convenient in terms of the central atom, A. Before describing the results for a range of such radicals, I refer to a theoretical calculation on the $BH_3 \cdot^+$ radical.[115] So far as I know this species has not so far been detected by esr spectroscopy. The missing electron must come from the σ-frame, and a Jahn – Teller distortion from the D_{3h} symmetry of BH_3 is expected. The interesting question is, what is this distortion likely to be? Esr spectroscopy could give some answers for comparison with this, and many other, theoretical calculations.

6.2 Carbon Centred $\cdot AB_3$ Radicals

The simplest centre, $\cdot CH_3$, has been studied using a newly designed 4K ENDOR system.[116] The radicals were generated from CH_3I and Cs atoms in an argon matrix. Analysis of the ^{13}C

TABLE 2
A Comparison of the Observed Magnetic Parameters for Several Small Aluminum Radicals (MHz)

	g_x	g_y	g_z	Al: A_x	A_y	A_z	A_{iso}	H: A_x	A_y	A_z
AlH$_2$	2.0026(6)	1.9973(6)	2.0012(8)	784(2)	786(2)	932(2)	834(2)	123(1)	130(1)	130(1)
HAlOH	2.0015(4)	1.9975(4)	2.0015(4)	888(1)	870(1)	993(1)	917(2)	286(1)	290(1)	288(1)
Al(OH)$_2$	1.998 (1)		2.000 (1)		1182(3)	1295(4)	1220(4)		<30 MHz	
AlH$^+$	1.9996(3)		2.0018(3)		1537(2)	1685(2)	1586(2)		440(2)	447(2)
AlF$^+$	2.0000(5)		2.0015(5)		2782(6)	2893(8)	2819(7)	^{19}F:90(5)		473(5)
AlO	2.0004(3)		2.0015(3)		713(1)	872(1)	766(1)	---	---	---
HAlCH$_3$	2.002(1)	2.002(1)	2.000(1)	712	723	880	772	154	146	157
PdAl	2.0343(5)		2.010(1)		84(1)	182(2)	117(2)	---	---	---
Al(CO)$_2$	2.0021(3)	1.9990(3)	2.0043(3)	146(1)	39(1)	39(1)	75(1)			

ENDOR spectrum for $^{13}CH_3$ showed that the lines are inhomogeneously broadened because of a distribution of coupling constants. Another well studied radical $\cdot CO_3^{3-}$, formed by electron addition to carbonate has been studied in irradiated tooth enamel.[117] The use of human tooth enamel in esr dosimetry is becoming one of the major methods of estimating accidental exposure to ionising radiation, and $\cdot CO_3^{3-}$ is a major product. In this study, response curves for a range of photon energies are reported.

Various halogen substituted radicals have been studied. An esr spectrum for the radical $\cdot CFCl_2$ has been computed theoretically by Shida and his coworkers.[118] The important point underlying this study is that the powder spectrum must be extremely broad and poorly defined relative to most esr spectra for radical cations, which explains why, when $CFCl_3$ is used as a matrix, the matrix radical signal does not interfere (See Fig.8). Whilst this has always been assumed by workers in this field, it is helpful to have such an explicit demonstration. In fact, Roncin and coworkers have given reasons why the matrix radical is $(\cdot CFCl_3)^-$ rather than $\cdot CFCl_2$,[119] although there are also cogent reasons for favouring $\cdot CFCl_2$,[120] one of which is the reaction of Cl^- discussed below.

Exposure of chloral hydrate crystals to γ-rays gave a dominant radical thought to be $Cl_2\dot{C}CH(OH)_2$, presumably formed by electron capture and loss of Cl^-. If this is correct then one would expect an electron loss centre also, which could be $Cl_3\dot{C}C(OH)_2$. This could well be present, but the spectra are very complicated.

Another dichloro carbon radical, $Cl_2\dot{C}CO_2^-$, was detected during visible light photolysis of reduced ferredoxin in the presence of $Cl_3CCO_2^-$ as an electron acceptor.[121] Well defined powder spectra were detected for $Cl_2\dot{C}CO_2^-$. The computed spectra for this radical based on the assumption that it is nearly planar at the radical centre fits well with the experimental spectrum, suggesting that it is, in fact, close to planarity.

6.3 Nitrogen Centred $\cdot AB_3$ Radicals

Again, the simplest derivative, $\cdot NH_3^+$, has been studied (together with $\cdot^{13}CH_3$ mentioned previously) using ENDOR spectroscopy at 4K.[116] This radical and $\cdot^{15}NH_3$ were generated by radiolysis of NH_3 in argon or neon at 4K. Again inhomogeneous broadening of the nitrogen hyperfine features indicated a range of slightly different trapping sites.

When ammonium or alkylammonium halides are irradiated at 77K, one of the centres formed is the $R_3N\cdot^+$ radical cation.[123] These cations have well established esr spectra, but our aim in this work was to see if there was any tendency to bind to adjacent halide ions to give the σ^* species, $R_3N \doteq hal$.[124] The results were positive, and indeed, in several systems these radicals dominated. The significance of these results is discussed in section 8, below.

When nitrates are irradiated at low temperatures, simple electron transfer usually ensues, to give $NO_3\cdot$ radicals, where the SOMO is on the oxygen ligands, and $\cdot NO_3^{2-}$, when it is largely confined to an s-p hybrid on nitrogen. Bannov et al.[125] have detected these species in irradiated $NH_4^+NO_3^-$ crystals, together with a radical identified as $O\dot{N}OO$, presumably formed from $NO_3\cdot$. The peroxy radical seems to be considerably more stable than $NO_3\cdot$. On further warming this also decayed, and signals from $\cdot NO_2$, and surprisingly, $\cdot OH$ radicals were detected.

6.4 Sulphur and Selenium $\cdot AB_3$ Radicals

The $\cdot SO_3^-$ radical, formed in radiolysis of sulphates has, fortuitously, an extremely narrow almost isotropic feature and hence is readily detected in irradiated solids. It seems that yields of $\cdot SO_3^-$ are higher for the mixed salts $K_2Mg_2(SO_4)_3$ or $K_2Na(SO_4)_2$ than in the simple

5: *Inorganic and Organometallic Radicals*   223

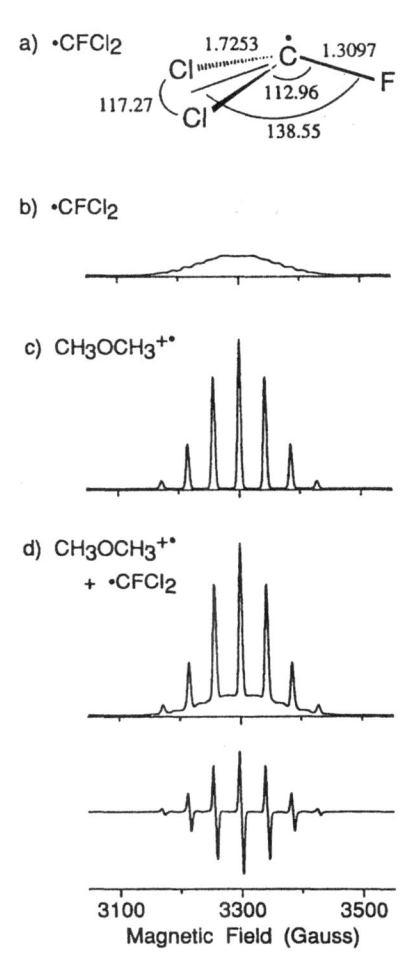

a) •CFCl₂

b) •CFCl₂

c) CH₃OCH₃⁺•

d) CH₃OCH₃⁺•
   + •CFCl₂

3100   3300   3500
Magnetic Field (Gauss)

Figure 8 (a) The optimised geometry of the ·CFCl$_2$ radicals by GAUSSIAN 86 with the basis set of Dunning DZ plus polarisation functions. Bond lengths are in Å and angles are in degrees. (b) Theoretical absorption spectrum for randomly oriented ·CFCl$_2$ radicals. Both isotopes ^{35}Cl and ^{37}Cl were taken into account. (c) Theoretical absorption spectrum for randomly oriented radical cations of dimethyl ether. (d) Theoretical absorption and its first derivative spectra for the admixture of randomly oriented CFCl$_2$ radicals and radical cations of dimethyl ether [taken from Ref. 118].

sulphates.[126] This may be because back reaction with the $SO_4\cdot^-$ hole centres is more inhibited. The effects of doping with La or Eu was studied both on the thermoluminescence and on the thermal esr response.

The $\cdot SeO_3^-$ radical was formed in irradiated ammonium selenate,[127] irradiated at room temperature. This gave $\cdot SeO_2^-$ either directly, or from electronically excited selenite ions.

6.5 Phosphorus Centred ·AB₃ Radicals

Phosphoryl radicals, $R_3P\cdot$, have been the subject of several studies. Janssen and his coworkers, who have worked extensively in this field, have been especially interested in chiral radicals.[128,129] Using enantiomerically pure $Ar_2P(O)Cl$ species exposed to X-rays at 77K, they generate pure $Ar_2\dot{P}O$ phosphoryl radicals. On warming to 120K these radicals undergo stereoinversion in the crystal matrix, as evidenced by clear changes in the esr spectra, aided by X-ray crystallography. The structure of these radicals has been further probed by *ab initio* calculations.

The observation that $Ph_2PCH_2CH_2\dot{P}Ph_2^+$ radical cations formed by radiolysis of the parent molecules in freon matrices can cyclise to form internal σ* 'dimers' (III)[130] which have a bent P–P bond has been nicely confirmed in a very careful study by Janssen and coworkers.[131] The esr spectra give clear indication of the bent bond, and this is supported by theoretical calculations.

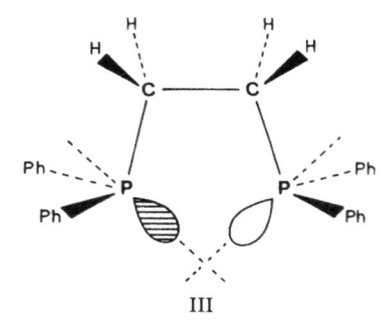

III

When phosphoryl radicals are rapidly generated in the liquid-phase they can be strongly spin-polarised.[132] Addition to carbon-carbon double bonds then results in new carbon centred radicals which also exhibit CIDEP effects, so the polarisation can be transferred. In some cases, reaction proceeded to give a third radical, also with transfer of polarisation.

Finally, some work has been done on the interesting species formed by electron addition to trivalent phosphorus compounds, R_3P:.[133] This gives a system having an extra 'lone-pair' of electrons, and there has been controversy regarding their effect on conformation. The species studied in the present work by Cattani-Lorente and Geoffroy involves electron addition to 2-chloro-1,3,2-dithiaphospholane. In this case trapping may occur in a Cl–P σ* orbital rather than giving a phosphoranyl type radical. The structure is discussed. On annealing, Cl^- is lost to give phosphinyl radicals.

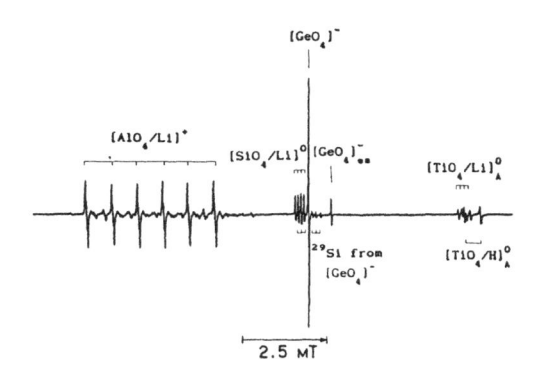

Figure 9 ESR spectrum of rose quartz crystal (175J) at 35K showing the major species present after irradiation at 77K, at ≈9.957 GHz with B‖c and B₁‖a₁ [taken from Ref. 135].

7 Penta-atomic ·AB₄ Radicals

7.1 Introduction

With increase in complexity, there is a concomitant decrease in diversity. As in previous reviews, it is only necessary to consider two major classes of radicals, those formed by electron-loss (7.2) and those formed by electron-gain (7.3). I end with a section on transition-metal centres, which, in this case, are all concerned with chromium (7.4).

7.2 Electron-loss ·AB₄ Centres

I start by considering the AlO₄· and SiO₄· type centres formed in zeolites and in silica.

A careful ENDOR study has been made of a long-lived radical that grows in during certain catalytic reactions on MNaX and MNaY zeolites (M = Li, Na, K, Rb, Cs).[134] Proton coupling to distant protons is observed, and weak coupling to ^{133}Cs. Various arguments lead to the conclusion that the tetrahedral aluminium centre, written as [AlO₄⁻] is converted into [AlO₄·] which is responsible for the ENDOR responses. However, such a species would surely show coupling to ^{27}Al nuclei in the esr spectra, and also considerable g-value variation, but this is not mentioned.

A somewhat similar centre, writtten as [SiO₄/Li]⁰ has been formed by X-irradiation at 77K in natural rose quartz.[135] Coupling to ^{29}Si in natural abundance establishes the nature of this centre and superhyperfine coupling to ^{7}Li establishes the involvement of lithium. In fact the irradiated crystal contains an abundance of centres of this type as can be seen in Fig.9.

A completely different silicon centre is Si(CH₃)₄⁺, prepared in a range of tetrachlorides by γ-radiolysis.[136] This cation had previously been studied in CFCl₃ and was found to be strongly distorted, with two strongly and two weakly coupled sets of methyl protons. This has been confirmed, and the dynamic distortion has been studied, giving an interconversion barrier. Calculations support the data and help to confirm the proposed structure.

Loss of an electron from dimethylphosphonate using the freon matrix technique gave a series of radicals, the most interesting being the parent cation (IV).[137] Electron loss evidently occurs primarily from the unique oxygen, which then forms a weak 3-electron σ-bond to one chlorine of a solvent molecule. In fact the spectrum is dominated by the chlorine splitting

IV

(A_{max} ≈ 78G), the ^{31}P splitting being only ca. 22G. This is because the chlorine is directly involved in the σ* bond, whilst the phosphorus gains negative spin-density via spin-polarisation of the P–O σ electrons.

Electron-loss, formally from fluorine, in methyl fluoride to give H3C–Ḟ$^+$ has been detected by Knight and his coworkers despite the high ionisation potential.[138] This was achieved in a neon matrix by photoionisation (16.8 eV) and also by electron bombardment. Because of the high electron affinity of fluorine this formal structure, which is perhaps reasonable for other alkyl halides, is clearly not suitable in this case, and the major loss is in fact from the methyl group. Thus for H$_3$CF·$^+$ the protons are equivalent with an isotropic coupling of 317MHz. The relatively large g-shift shows that there is delocalisation onto fluorine as does the large ^{19}F coupling (965; −130, −166 MHz). An interesting ^{2}H isotope effect was also observed which prevents methyl proton averaging and fixes the structure such that coupling to the two protons is greatly increased relative to the H$_3$CF·$^+$ derivative. This effect was also found for CH$_4$·$^+$ and CH$_2$D$_2$·$^+$, when, for the former, the four protons were equivalent on the esr time scale, but the distorted structure was fixed in CH$_2$D$_2^+$, with maximum delocalisation onto the two protons.[139]

7.3 Electron Gain Centres

There is still considerable interest in electron capture by phosphorus derivatives, and fortunately, the powder spectra are often easy to analyse. Buck, Janssen and their coworkers are very interested in reactions of stereoisomers. Using isomers of 2-chloro−3,4-dimethyl−5-phenyl−1,3,2-oxazaphospholidine 2-sulphide (V).[140] In this case electron capture occurs primarily into the P−Cl bond giving a σ* derivative rather than a phosphoranyl radical, for one of the isomers, but, curiously, no radical formation was detected for the other isomer. This was explained in terms of the different environments of the P–Cl unit in the two crystals. This theory is nicely supported by the fact that identical spectra of the ∸P∸Cl species were obtained using MTHF as a solvent.

V

These workers have also studied electron-addition into P–S σ bonds for R_3P–S compounds, where again, this occurs rather than distortion to give a normal phosphoranyl radical. They compared the three species $R_3P \dotdiv S^-$, $R_3P \dotdiv SR$ and $R_3P \overset{+}{\dotdiv} SR_2$ and found that the σ^* species was always favoured, with remarkably little change in the esr spectra. This result is not in accord with simple theory, nor with quantum mechanical calculations.

Of three papers describing the ever popular AsO_4^{4-} radical, formed in irradiated arsenates, one is concerned with the use of ESEEM (electron spin echo envelope modulation) relative to ENDOR to pick up coupling to protons and hence measure disorder,[141] another is concerned with dynamic effects in γ-irradiated cesium dihydrogen arsenate,[142] and a third with spin-lattice relaxation measurements via AsO_4^{4-} centres in antiferromagnetic ammonium dihydrogen arsenate.[143]

7.4 Transition Metal AB₄ Centres

The Cr^{III}—Cr^V system has become both popular and important during the past few years. Many years ago, our demonstration that CrO_4^{2-} ions in strongly alkaline solution decomposed to give CrO_4^{3-} ions was just an oddity. The CrO_4^{3-} ions were incorporated into vanadate crystals and studied by esr spectroscopy, establishing the d^1 configuration.[144,145]

Chromate is now recognised as an important toxin and carcinogen, and the formation of Cr^V complexes is accepted as often being of considerable importance. These can be detected readily by esr spectroscopy provided the ligand field is strong and asymmetric. It is important, however, to recognise that Cr^{III} can be confused with Cr^V. Both give "green" complexes, and both, in solution, can give singlets in the $g = 1.9 - 2.0$ range, possibly with detectable ^{53}Cr satellites. For good esr spectra, the Cr^V complexes need to be strongly distorted, or very low temperatures may be required. However the Cr^{III} centres need to be symmetrical, to avoid large zero-field splittings and consequent loss of the central line(s).

The present references are largely concerned with the detection of Cr^V intermediates formed from CrO_4^{2-} ions, rather than with their structure. It is likely that CrO_4^{3-} ions are intermediates, but these would not be detected, and are expected to protonate and undergo ligand changes, probably via chromyl-type species with a very strong $Cr-O$ bond as in vanadyl or ferryl. One issue that is discussed extensively is—does CrO_4^{2-} react with H_2O_2 by a "Fenton" mechanism? In particular, does reaction (8) occur:

$$CrO_4^{2-} + H_2O_2 \rightarrow CrO_4^{3-} + OH^- + \dot{O}H \tag{8}$$

Again, it is interesting for me to recall that the first paper I ever published asked a similar question for permanganate—is it a "source of free hydroxyl radicals"?[147] Some claim that the unique species CrO_8^{3-} [ie $Cr(OO)4^{3-}$] is an important first product (9)

$$Cr(VI) + H_2O_2 \; CrO_8^{3-} \xrightarrow{\text{cell reductant}} Cr(VI) + \cdot OH \tag{9}$$

whilst others consider that the Cr(V) centre is responsible for the generation of ·OH radicals.[148-151]

Finally, returning to chromate as such, the triplet state of CrO_4^{2-} in $CaSO_4$ has been probed using electron-spin-echo methods.[152] The centre relaxes rapidly to an orbitally non-degenerate state with well-defined D- and g-values.

8 Other Inorganic Radicals

8.1 Introduction

In this section I treat a miscellaneous collection of radicals that failed to fit in readily with the classification used above. I start by discussing the interesting ·AB$_5$ centre, ·IO$_5^{2-}$, formed by reaction of periodate ions with O·$^-$ radicals, because of the link with section 7.

8.2 The ·IO$_5^-$ Centre

Addition of O·$^-$ or ·OH to aqueous periodate species has been studied using pulse radiolysis and flash photolysis, the product having $\lambda_{max} \approx 525$ nm. One problem is that IO_4^- reversibly hydrates to give $H_4IO_6^-$, $H_3IO_6^{2-}$ etc. in water, so the structure of this intermediate is unknown. By incorporating IO_4^- into $KClO_4$ crystals, one is sure that only IO_4^- is present. Byberg has found that irradiation of such doped crystals at room temprature generates O·$^-$ ions which selectively add to IO_4^- giving ·IO$_5^{2-}$.[153] As with other ·AB$_5$ radicals of this type, a distorted trigonal bipyramid is suggested, with a single weaker I–O bond containing considerable spin-density. Thus, a limiting structure is simply $\equiv$I $\dotdiv$ O^{2-} as a σ* radical. The optical band for this centre (σ→σ*?) is at 530 nm, which agrees nicely with the pulse-radiolysis work and confirms the structure of this intermediate.

8.3 σ* Radical Dimers

After many years of study by solid-state physicists and esr spectroscopists, chemists are just beginning to realise that 3-electron σ-bonds, or σ* radicals, are major intermediates in radical chemistry.[154] In many cases, if a compound X:, having a "lone-pair" of electrons, undergoes electron-loss, the resulting radical-cation, X·$^+$ will add to X to give X $\dotdiv$ X$^+$, the unpaired elctron being in the σ* orbital and equally shared between the two X units. Examples are: hal $\dotdiv$ hal$^-$ (V$_k$ centres) Rhal $\dotdiv$ halR$^+$, RS $\dotdiv$ SR$^-$, R$_2$S $\dotdiv$ SR$_2^+$ and R$_3$P $\dotdiv$ PR$_3^+$. This mode of reaction is relatively rare for first row species (F $\dotdiv$ F$^-$ is formed, but RO $\dotdiv$ OR$^-$, R$_2$O $\dotdiv$ OR$_2^+$ and R$_3$N $\dotdiv$ NR$_3^+$ are unknown, except for H$_3$N $\dotdiv$ NH$_3^+$;[155] however, mixed 1st. and 2nd. row species are well known, see below).

Also, so far as I know, there are no clear-cut examples of R$_3$Si $\dotdiv$ SiR$_3^-$ centres, although, judging by the stability of the isoelectronic R$_3$P $\dotdiv$ PR$_3^+$ species, they should be detectable. In this regard, Tada has carried out *ab initio* MO calculations on the parent anion, H$_3$Si $\dotdiv$ SiH$_3^-$. Not surprisingly, he concludes that the excess electron would be primarily in the Si–Si σ* orbital. Also, in agreement with esr findings, the results suggest that *d* orbital participation is not important.[156] Units of this type could be important in silicon chemistry, if impurity centres were to make the stretching of one particular type of Si–Si bond more favourable than others.

These dimers can also be prepared by electron addition to X–X molecules. For example, various R$_1$R$_2$(S)P–P(S)R$_3$R$_4$ add electrons generated by radiolysis into the P–P σ* orbital.[157] However, there is an alternative choice, and that is for the added electron to become localised on one phosphorus only, either by distortion to give a phosphoranyl radical, or by stretching a P–S bond to give the σ* $\supseteq$P $\dotdiv$ S$-$ unit. In a range of chiral compounds, Janssen,

Buck and coworkers have beautifully demonstrated the effect of environmental control on the course of these reactions.[157] The *racemic* compounds gave the P–P type anions but the *meso*-derivatives gave addition to one side only. In frozen glasses, only the P−P derivatives were formed. Some other examples of σ^*-dimer formation, including cyclic dimer formation have been discussed in section 6.[129,130]

8.3 Other σ^* Radicals

As stressed above, first row dimers are very rare. However, compounds containing $\geqq$N-hal bonds readily give $\geqq$N-hal$^-$ σ^* radicals. For $\geqq$C-hal units electron addition sometimes occurs, but for simple alkyl halides, the units fall apart to give R· + hal$^-$. Examples in which σ^* electron capture is favoured are various halogenoimidazoles.[158] Irradiation of various bromo- and iodo-derivatives in CD$_3$OD or methyltetrahydrofuran at 77K gave well defined spectra with very large hyperfine coupling to one halogen nucleus with hyperfine parameters close to expectation for C $\dot{-}$ hal$^-$ centres. In these cases, electron-addition into the π^* system seems to be less favourable.

Because nitrogen compounds have a higher electron-affinity than corresponding carbon derivatives, σ^* radicals of the type R$_3$N $\dot{-}$ hal, isoelectronic with R$_3$C $\dot{-}$ hal$^-$, are far more likely to be stable, and we have established this by esr spectroscopy for a range of derivatives.[123] A simple orbital energy diagram illustrates this point, as well as showing how the extent of orbital hybridisation on carbon or nitrogen can affect the strength of σ^* bonds (Fig.10).

The esr spectra for these R$_3$N-hal centres showed good ^{14}N and halogen hyperfine coupling with values strongly suggesting colinearity. The results show that the R$_3$N· units are pyramidal, in contrast with the parent R$_3$N·$^+$ radicals which are clearly planar. The degree of bending increased on going from Cl to I, in parallel with the increase in spin-density on the halogen.[123]

One area in which organic chemists have invoked σ^* radical formation is in reactions of pyridine with halogen atoms.[159] Normally, for aromatic derivatives halogen atoms add to the π-system, but for pyridine the kinetic results seem to require initial addition to nitrogen to give the $\geqq$N $\dot{-}$ Cl σ^* centre. To show that this suggestion was plausible, we prepared pyridine radical cations ⟨N⟩·$^+$ in CFCl$_3$ by low temperature radiolysis. These are well identified by esr spectroscopy and clearly have a SOMO which is largely the in-plane *s-p* hybrid on nitrogen. On allowing the system to warm towards the softening point these cations moved to the chloride anions generated from CFCl$_3$ by electron capture and gave the required σ^* species.[160]

A rather similar reaction was carried out by Janssen and coworkers using a range of freon-like solvents in which two potential cation-forming molecules, PCl$_3$ and Me$_2$S, had been dissolved.[161] Irradiation should then generate ·PCl$_3$$^+$ and Me$_2$S·$^+$ cations in competition and, on annealing, a range of possible reactions can occur, including that with Cl$^-$ described above. Thus the adducts Cl$_3$P $\dot{-}$ PCl$_3$$^+$, Me$_2$S $\dot{-}$ SMe$_2$$^+$ and Me$_2$S $\dot{-}$ PCl$_3$$^+$ are expected to form unless there is any major preference, or unless interchange can occur.

One interesting aspect of this study is that a novel species with an extrememly large ^{31}P splitting (5115 MHz) was also observed. This is thought to be the trimer Cl$_3$P(SMe$_2$)$_2$·$^+$. A rather similar study to this was carried out some time ago using frozen glasses containing organic phosphines and sulphides, which gave R$_3$P $\dot{-}$ SR$_2$ similar to the mixed species detected in the present study.[162]

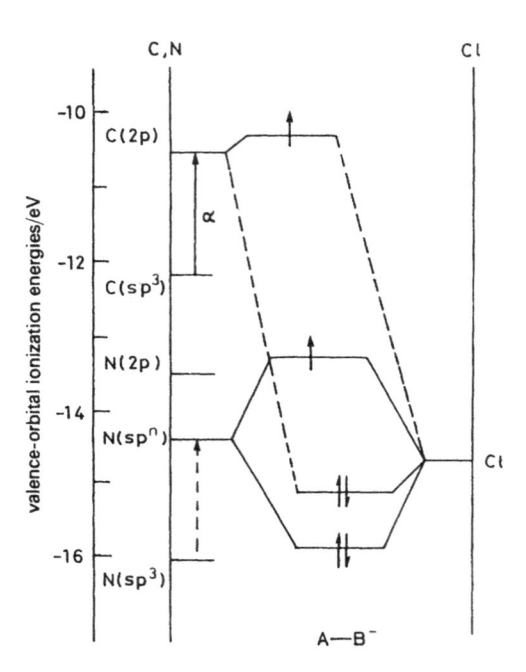

Figure 10 Atomic orbial energies for carbon, nitrogen and chlorine, with an indication of possible bonding interactions; α represents the change in energy for CH₃ on moving from pyramidal (sp³) to planar (p). Hybridisation for Cl of sp⁹ is used to indicate the small admixture of s-character detected by ESR [taken from Ref. 123].

Another example in which Cl⁻, ions formed from CFCl₃ by electron-capture, react on annealing with parent radical cations to give halide σ* centres is for various $R_3P \cdot ^+$ radical cations. In addition to $R_3P-PR_3^+$ dimers, good yields of the R_3P-Cl derivatives were detected when the samples were warmed close to the melting point.[163] This result confirms that 'free' chloride ions are available which supports the ·CFCl₂ structure for the electron-capture centre.

8.4 Carbon and Carbon–Sulphur Centres

Exposure of 1,3-butadiene in neon at 4K to vacuum UV-photolysis gives the C₄ molecule.[164] This has a triplet ground state, but there has been considerable debate as to whether or not it is linear. A careful study of the ¹³C hyperfine splitting and the zero-field splitting shows clear splittings into x and y features, thus fully establishing non-linearity.

Various radicals have been detected in the radiation products from cyanamide and dimethylcyanamide.[165] These include the electron-loss centres, HṄCN and Me₂ṄCN⁺. We were unable to detect H₂ṄCN⁺, which acts as a strong acid. The latter cations were formed by irradiating dilute solutions in $CFCl_3$. Electron capture by H_2NCN gave a centre, identified by a large ^1H splitting, as $\begin{smallmatrix}H\\HN\end{smallmatrix}{>}C=\dot{N}$. This is thought to be formed from the parent anion by proton transfer from N to C. In contrast, Me_2NCN gave only Me_2N radicals by dissociative electron capture with loss of CN^-. In alkaline glasses, the $O\cdot^-$ radical formed in abundance, is thought to react to give the species $\cdot N_2CO^{3-}$, with the unpaired electron largely shared between the two nitrogen ligands [$\cdot\begin{smallmatrix}N\\N\end{smallmatrix}{>}C{-}O$]. The oxygen ligand may well be protonated, but the esr spectrum showed well defined coupling to two equivalent ^{14}N nuclei, sharing common axes. This suggests a π-type orbital rather than the in-plane non-bonding orbital on nitrogen.

A new class of radicals, the 1,2,3-trithiolium cations (VI) has been fully characterised as its AsF_6^- salt by X-ray crystallography.[166] These are 7π-electron radical cations, which are

VI   VII   VIII

clearly remarkably stable. It seems that species previously thought to be "1,2-dithiete" radical-cations (VII), studied extensively by Bock and coworkers and Russell and coworkers, will prove to be these 1,2,3,-trithiolium cations (VI). The esr spectrum for the $R = -CO_2Me$ derivative shows ^{33}S satellite lines with two different sulphur splittings of 8.9G and 8.0G in intensity ratios of 1:2, which clearly supports (VI) over (VII). It is interesting that M.O. calculations suggest almost equal spin-densities on the three sulphur atoms. These cations can be compared with the neutral 7π-electron centres (VIII), previously reported.

8.5 Nitrogen and Sulphur Centres

The most studied group of inorganic "aromatic" ring structures are the sulphur-nitrogen derivatives starting from S_2N_2, and including $S_3N_2^+$ (also with 7π-electrons) $S_3N_3^-$, $S_4N_3^+$, and $S_5N_5^+$.[167] Structures, stabilities and esr spectra of several of these species have been examined, and M.O. calculations given in support of the conclusions.

8.6 Metal Clusters

This remains a much studied and important field which has been touched on already in some of the previous sections. Of especial interest is the point at which the "localised electron" cluster merges into the normal metal structure. This issue has been looked at in a very useful semi-review by Mile et al.[168] Based largely on esr observations, they have come to the following general conclusions:

Trimers of alkali metals, such as Na_3 are triangular, either with a static obtuse-angled ground state or with a dynamic Jahn – Teller distortion. The Cu, Ag and Au trimers are all in static distorted states at 77K, but there is a very interesting matrix induced switch in structure. Thus in hydrocarbon matrices the obtuse angled triangle is favoured, whilst in nitrogen matrices the acute angle form is favoured. In the former, the spin density is high on the two terminal nuclei whilst for the latter it is high on the 'central' atom. The trimers of aluminium and gallium are reported to have quartet ground states. (But see below.)

The septamers of all the first row metals have pentagonal bipyramidal structures, with high spin-density on the two apical atoms. The total s-atomic orbital character is considerably reduced relative to the trimers, and is further reduced as the size of the cluster increases.

These authors also report on the reactions of various clusters with dioxygen, ethene and carbon monoxide, and on the importance of their results to catalysis.

Zeolites are very useful for preparing small clusters because of the limited sizes of holes that they offer. Sodium clusters have been widely studied, and Kuranova has reported detecting an $(Na)_7$ centre in NaA zeolites and an $(Na)_{13}$ centre in NaX and NaY zeolites.[169]

In a second study of boron and aluminium trimers, B_3 and Al_3, Hamrick et al. have decided that in rare-gas matrices, these molecules have doublet ground states[170] rather than the quartet states postulated by Mile et al.[171] Calculations indicate that these alternatives are very close in energy and hence it is quite possible that there is a switch on going from rare gases ($^2A_{1'}$) to adamantane ($^4A_2$). The doublet state is found for Al_3 in the gas phase, which supports the rare-gas results. In both studies, almost isotropic hyperfine splitting is found: for $^{11}B_3$, $A_\parallel = a_\perp = 130$ MHz and for $^{27}Al_3$ $A_\parallel = 109$ and $A_\perp = 100$ MHz. For $^{27}Al_3$, the calculated 3s character is ca. 8%. Mile et al. find the Al_3 spectrum to be also almost isotropic, and indeed very similar to that in rare-gases. Weltner and coworkers are convinced that B_3 has a doublet spin-state and argue cogently for the same structure for Al_3. They stress that it is surprising that no features for the expected zero-field splitting have ever been detected in any matrix and that it is clearly in a doublet state in the gas-phase.

There is no such problem with the Cu_7 cluster. All agree that it is pentagonal bipyramidal (D_{5h}) with a $^2A_{2'}$ ground state.[168,171]

Several reports on silver centres in zeolites have appeared. For example, Blatter and Blazey obtained a strong conduction esr signal at $g = 1.96$ in zeolite Y, indicating metallic centre formation.[172] Wasowicz and Michalik, using γ-irradiated Ag-Na A zeolites detected much smaller clusters such as $Ag_3{}^{2+}$ and an $Ag_2{}^+ \cdots Ag^+$ unit using hydrated zeolites, and an $Ag_6{}^{n+}$ (n unknown) in the dehydrated form. The factors that dominate the sizes of these clusters is a matter of considerable interest.[173]

Finally some diaryl diphosphine radical cations, $(Ar_2P - PAr_2) \cdot {}^+$ are of interest.[174] When $R_2N - NR_2$ molecules lose an electron both the pyramidal nitrogen centres become planar, to give complete planarity to the radical-cation which then has a $(\pi)_1{}^2 (\pi)_2{}^1$ configuration. It is well established that although $R_3N \cdot {}^+$ units are planar, $R_3P \cdot {}^+$ units remain pyramidal—presumably for the same reason that makes R_3N: molecules have considerably larger bond-angles than corresponding R_3P: molecules.

Hence it is not surprising that $(Ar_2P-PAr_2) \cdot {}^+$ cations are non-planar, as established by the large A_{iso} (^{31}P) values. Our solid-state results show that as well as being equivalent, the two phosphorus centres share common axes, so that the trans-configuration appears to be favoured.

9 Paramagnetic Centres in Inorganic Materials

9.1 Introduction

This is now an enormous field, with a "language" of its own. This especially applies to the field of silicon and amorphous silicon, when esr spectroscopy and ODMR play a role, which needs to be understood in the light of all the other techniques used to probe these materials, as well as the methods of preparation. All I can do is to give a 'potted' overview, and apologise to the experts both for my ignorance and my omissions. Part of the field has been reviewed by Ikeya[175] and, with respect to preparation methods, by Almond[176]. In particular, there is a useful review on metal-organic chemical vapour deposition. This is widely used in industry and it is essential to understand the chemical reactions involved. Matrix isolation coupled with spectroscopy can help.

I have arbitrarily divided the section into carbon (9.2), crystalline silicon (9.3), amorphous silicon (9.4), silica centres (9.5), centres in III–V compounds (9.6), centres in II–VI materials (9.7), and centres in I–VII materials (9.8), which are entirely silver halide centres.

9.2 Carbon

Synthetic diamond is becoming an extremely important material for devices of various types, and esr spectroscopists are once more taking a great interest in impurity centres such as nitrogen. It seems that a model for this that we proposed some years ago[177] is still used and is strongly supported by theory.

Nitrogen remains the major paramagnetic impurity in synthetic as well as natural diamonds. However, two other single-line defects are detected having about the same g-values but different widths. Possible structures are discussed.[178] In another study of diamond films, two similar defects were found, but with slightly different g-values (2.0028 and 2.0019).[179] Again, possible structures are discussed.

It seems that nickel is also a common impurity in synthetic diamonds, though its incorporation presents a puzzle. Ikeya and his coworkers have used their esr microwave imager to show that there are nickel rich regions in some parts of the (III) sectors only. Also the linewidth of the N-centre correlates with the [Ni] implying that they are quite close together.[180]

Amorphous carbon and amorphous hydrogenated carbon, similar to their silicon analogues, are also of importance. Optical spectra of thin films have been measured and correlated with esr measurements. These films clearly have amorphous and graphitic regions and possibly diamond like regions, so interpretation is difficult.[181]

9.3 Crystalline Silicon

Most studies still concentrate on the well known intrinsic centres that are formed in crystalline silicon, and which are of obvious importance. These are less well defined, but far more abundant in amorphous silicon unless this incorporates hydrogen, which serves to convert such "dangling bonds" (if that's what they are) into $-SiH$ units.

Having achieved techniques in which impurity centres are extremely low, experiments are now being conducted where controlled addition of specific impurities is used to see how the properties are modified. In addition to the well known N-centre (deep binding) and P centre (shallow binding), sulphur centres have been studied.[182] ODMR studies indicate the presence of triplet states. Clearly sulphur is involved, but precise structures are as yet unknown.

Nickel is also an important additive, as with carbon. Using ^{61}Ni ($I = 3/2$) a dihedral bonding model is proposed on the basis of the derived hyperfine coupling.[183] Another impurity usually found in silica is aluminium. A new defect in thermally treated Czochralski-grown silicon (C_z) silicon, which always contains a high proportion of oxygen, has been shown by ENDOR spectroscopy to involve aluminium.[184]

One interesting way of probing silicon based materials is to bombard them with muons. This gives rise to two very interesting centres known to physicists as normal and anomalous muonium centres. In fact, even the normal centre is highly anomalous in the sense that the hyperfine coupling is far less than that for a gas-phase muonium atom. Sometime ago I proposed two models to explain these centres, which are of particular interest since, unlike most muonium centres, there are no clear hydrogen analogues. To explain the large reduction in hyperfine splitting for "normal" muonium, it was suggested there was weak bond formation within an intrinsic cavity, with fast migration between equivalent sites.[185a] To explain the 'anomalous' centre, where the hyperfine coupling is small and anisotropic, I proposed a 'bond-centred' model, with marked stretching of one unique Si–Si bond, which traps the muon. This would have a node through the muonium atom, thereby explaining the small hyperfine parameters. It seems that this model, described more fully by Cox and Symons[185b] has finally been accepted now that it is realised that Si – Si bond stretching is perfectly possible within bulk silicon. A detailed calculation of this centre has been reported.[186]

As a bridge between crystalline and amorphous silicon, a comparative study of these two classes of silicon using implanted muons has been reported.[187] The major result is that both the diamagnetic signal from μ^+ in non-radical sites, and the anomalous muonium centres, are substantially increased in the amorphous materials. This seems quite reasonable in view of the range of Si–Si bonds in the amorphous material. We were unable to get positive identification of the 'normal' muonium centres in amorphous silicon.

9.4 Amorphous and Hydrogenated Silicon

Because they are prepared from $\geq$SiH precursors, amorphous silicon films usually contain residual hydrogen.

The effect of high pressures on the ODMR for amorphous silicon has been studied. Quenching of ODMR is from the paramagnetic "dangling bonds" and this is increased at high pressures. Photolysis gave an increase in the number of dangling bonds.[188]

The origin of this optically induced esr signal has been further probed using ^{29}Si. Again, the extra signal is assigned to an increase in the number of dangling bonds.[189] The g-value assigned to dangling bonds shifts from 2.0047 to 2.0052 on thermal annealing.[190] This is interpreted in terms of changes in the environment of these dangling bonds.

It has been argued that in the photolysis of hydrogenated amorphous silica, the generated dangling bonds are charged rather than neutral. Only the neutral centre is detected directly with esr. Both centres increase on photolysis.[191] The effect of doping with boron on hydrogen diffusion in hydrogenated amorphous silicon has been probed.[192] Boron was introduced via B_2H_6 in the gas phase. The results are discussed in terms of the depth of trapping sites and the dangling bond models.

Oxygen and nitrogen are common impurities in silicon. Using purified materials, these impurities have been re-introduced in a controlled manner.[193] Both O and N increase the dark conductivity, and delay the photoresponse. Carbon, however, has no such effect. Centres O_3^+ and N_4^+ are postulated.

As a bridge between this section and that on silica centres, I mention two papers

5: *Inorganic and Organometallic Radicals* 235

concerned with silicon-silica interfaces, which are of major importance in devices of all kinds. The major radiation induced centre seems to resemble the 'dangling bond' centre in being a trivalent Si defect on the surface (P_b centre). It exhibits strong ^{29}Si hyperfine coupling.[194] These P_b defects have also been studied by Brower and Meyers.[195]

9.5 Silica Centres

There have been two major reviews by one of the masters in this field. Griscom has described the relationships between optical properties and structures of defects in silica glass,[196] and also the use of esr spectroscopy to study self-trapped holes in these materials.[197] These complementary reviews are of great help to anyone entering this important field.

The mechanism of radiation damage to high purity silica has been probed by Tsai and Griscom.[198] Simple ionisation and trapping is evidently not enough to explain the results, and it is suggested that oxygen vacancies together with interstitial pairs can explain the esr results when damage is induced with 6.4 eV highly focused laser light.

A useful series of *ab initio* MO calculations on a range of radiation induced defects has been published.[199] These include Al^{3+} substituted ions, i.e. the $[AlO_4]^0$ centre. Also oxygen vacancy centres such as the E'_1 and E'_4 defects were studied. A comparison of the effects of neutral and ionic beam bombardment in SiO_2 film has been made.[200] Both generate E' centres, but the yields are much higher for the ion beams. The results are discussed in terms of bond-breaking versus ionisation events.

The Ti^{4+} ion can, of course, readily substitute for Si^{4+} in silica. This is unusually represented as $[TiO_4]$, and this can add an extra electron giving $[TiO_4]\cdot^-$, when the electron is strongly confined to the $3d$ orbital of titanium. Because this is degenerate there is a Jahn–Teller distortion which favours occupancy of the $3d_{z^2}$ orbital. Nevertheless, there is considerable line-broadening even at low temperatures.[201]

In some ways, TeO_2 is related to SiO_2 and so I include it here. Spaeth *et al.*, using esr and ENDOR, have discovered two new intrinsic defects induced by electron-bombardment.[202] One is modelled as a double oxygen vacancy bearing a net positive charge, depicted as $(VO)_2^0$. The second new centre is said to be a "small peroxy radical" depicted as $(Te-O_2-Te)$.

9.6 III–V Compounds

There has been a "Spaeth" of papers on gallium arsenide, but before turning to these, I mention work on AlN.

Neutron irradiation of AlN gives a very broad esr sprectrum attributed to electrons trapped at nitrogen vacancies. The coupling to the surrounding ^{27}Al nuclei responsible for this broad line is given a splitting of 10.9×10^{-4} cm^{-1}, with a g-value of 2.007.[203] What are described as "anion antisite" centres have been studied in GaP using ODMR spectroscopy.[204] These sites are formed if a group V atom occupies a group III atomic site and vice-versa. For GaP doped with vanadium two types of P antisite defects were found, both with tetrahedral symmetry. The major difference between these centres is in the spin-lattice relaxation times. Hence these defects are thought to be connected with other such defects forming a complex antisite unit.

Defects in III – V semiconductors have been reviewed by Spaeth *et al.* from the view-point of ODMR.[205] This includes ODEPR, ODENDOR and magnetic circular dichroism (MCDA) of the optical absorption and comprises an extremely powerful combined structural probe. This is a very helpful review for workers in this general field. In particular, these workers have concentrated on GaAs and its intrinsic antisite defects. These include electron-irradiated systems,[206] deep level defects in high-resistivity material[207] and anti-structure defects produced by rapid-quenching of semi-insulating GaS.[208]

The situation becomes even more complex when these III–V compounds are doped. This may be with like charged ions—such as (Al^{3+}), or (P^{3-}), or it may be ions from other groups such as group IV. In one such study, Fockele et al.[208] have doped GaAs with Al, and then with silicon or tin. In the latter, splittings from ^{117}Sn and ^{119}Sn isotopes were detected, but ^{29}Si satellites were not observed. Another such system comprises GaAs – AlAs superlattices.[209] ODMR of defects show hyperfine coupling from Ga that suggests an interstitial Ga rather than an anti-site Ga.

9.7 II–VI Materials

Again, ODMR and related techniques can probe the nature of defects in these important materials with great 'insight'. This has been reviewed by Davies.[210] Zinc oxide ceramics, doped with lead, gave signals assigned to Pb^{3+} ions on photolysis.[211] It is argued that these ions occupy Zn^{2+} sites in ceramic grains. Thus they are formed from Pb^{2+} by loss of an electron into the conduction band.

Zinc oxide doped with ruthenium has also been studied.[212] The role of the ruthenium seems to be to increase the yields of point defects such as Zn^+, oxygen vacancies ($V_o{}^+$) and $O_2{}^-$ superoxide defects. Switching to cadmium telluride, Sun et al. have studied the effect of phosphorus implantation into this compound. Rapid freeze-thaw techniques force the phosphorus into Te lattice sites as judged by esr spectroscopy.[213]

9.8 I–VII Materials: Silver Halides

Some mention of silver cluster centres has been given above. Despite all sorts of modern developments, silver halides still dominate the photographic market and hence remain important. Hole trapping at certain vecancies in doped irradiated Ag halides has been probed by esr spectroscopy.[214] The results are discussed in terms of two centres which are close enough to perturb each other magnetically. Silver in TiO_2 has been studied. Many centres were formed by hydrogen reduction, which may help to unravel the results for silver halides themselves.[215]

Reference has already been made to the use of zeolites when sudying metal clusters. Silver clusters, of course, are of special interest with respect to latent image formation in the silver halides. Xu and Kevan[216] have formed an $Ag_4{}^{n+}$ cluster by hydrogen reduction of silver-charged zeolites at room temperature using esr spectroscopy. On heating to 373K *in vacuo* much larger metal particles are formed. One, very large particle is formed on the external surface and the other smaller particles are formed in the α-cages of the zeolites.

10 Spin-Trapping of Inorganic Radicals

10.1 Introduction

As before, I consider the trapping of inorganic radicals in chemical systems (10.2) separately from trapping them in biological systems (10.3). Because of its importance and novelty I have treated the radical NO separately (10.4). This is a huge field, and it is quite impossible to cover everything. Much of the work is routine, and of quantitative interest. Such work is not covered herein.

A book by Rice-Evans, Diplock and Symons entitled "Techniques in Free Radical

Research" contains an extensive review of spin-trapping, stressing its utility and its limitations.[217] An extensive and useful review on spin-trapping of inorganic radicals has been written by Rehorek.[218] This compares the use of nitrones and nitroso compounds as traps, and stresses that they are complementary. The nitrones are best for small inorganic radicals, whilst the nitroso compounds are good for trapping metal centred "radicals" (see section 11). An interesting background history is provided, as are some extremely useful tables.

A further review by Buettner and Mason on the trapping of $\cdot OH$ and $\cdot O_2^-$ both *in vivo* and *in vitro* is critical and most instructive,[219] and a short review on the uses of 5,5-dimethyl-1-pyrroline-N-oxide (DMPO) has appeared.[220]

10.2 Trapping Inorganic Radicals in Chemical Systems
One of the problems in spin-trapping is the correct identification of the trapped species. Changes in $A(^{14}N)$ and $A(^1H)$ can be quite small between different radicals, despite large steric constraints. A new trap containing the α-2,6-difluorophenyl group is recommended because the ^{19}F coupling seems to have far higher variability.[221]

Exposure of aqueous NO_2^- ions to 360 nm. light results in the induced oxidation of many organic compounds.[222] The $\cdot OH$ radical is strongly implicated, generated according to reaction (10).

$$(NO_2^-)^* + H_2O \rightarrow \cdot OH + \cdot NO + OH^- \qquad (10)$$

Using DMPO, the "characteristic" 4-line signal from DMPO–OH was detected. The OH radicals can also attack NO_2^- to give $\cdot NO_2$, which may also be involved. This is an interesting new method for generating a range of interesting radicals.

The "Fenton" reaction, or rather the first step which gives rise to the ubiquitous $\cdot OH$ radical, is still extensively discussed. The situation with copper is even less clear. However, it has now been suggested that the copper complex $Cu(en)_2^{2+}$ reacts with H_2O_2 to give $\cdot OH$ which was characterised by various traps.[223] Presumably the other product is a Cu^{III} complex since electron-donation is required.

Chlorine atoms generated by photolysis from C_2Cl_6 have been trapped by PBN traps to give stable adducts.[224] However, it seems that Cl^- ions are also generated and these compete effectively, to give $Cl_2\cdot^-$ radicals which do not donate chlorine to these traps.

Spin-traps can also be useful to study intermediates in the gas-phase. Photoactivation of the system $CH_4 + NH_3 + H_2O$ in the presence of PBN traps gave evidence for CH_3O radicals, $\cdot CH_2NH_2$ radicals and $CH=NH$ radicals.[225]

10.3 Biological Uses of Spin-Traps
Singlet oxygen, like hydrogen peroxide, is frequently formed in biological systems, and can do damage if not quenched or scavenged rapidly. The problems that arise in attempts to spin-trap $^1O_2$ during the photolysis of hypocrellin A have been discussed.[226] Since this species can be thought of as having no unpaired elecrons, the reactions with spin-traps leading to doublet state radicals are not immediately obvious.

The use of dibromonitrosobenzene sulphonate in biological systems has been probed.[227] It is thought to be useful for trapping $O_2\cdot^-$ which, being stable, is difficult to trap. The difficulty seems to be that this trap can form adducts, but is rapidly metabolised by cellular systems.

The variation in OH production by Fe^{2+} ions in the presence of a range of ligands has been studied using DMPO. The ligand makes a large difference to the rate of reaction, in the order phosphate ≈ ADP < EDTA < DETAPAC, and only the last gives ·OH radicals clearly, without interference from species like ferryl.[228]

Finally the xanthine oxidase + hydrogen peroxide system converts sulphite ions into ·SO$_3^-$ radicals very effectively.[229] This was identified using DMPO.

11 Transition-Metal "Radical" Centres

11.1 Introduction

The reason for including certain transition-metal complexes in this section is that certain complexes do have "radical" character, in contrast with the majority of complexes. Complexes that have strongly bonded ligands, and an unpaired electron in an orbital that is available for bonding tend to behave in some respects as if they were radicals. For example, they can be "spin-trapped". Rehorek, in his review[218] includes a range of complexes that react with nitroso spin-traps to give relatively stable adducts. These include complexes such as ·Co(CN)$_5^{2-}$, ·Mo(CO)$_3$Cp, ·Re(CO)$_5$, ·Mn(CO)$_5$, ·Fe(CO)$_2$Cp, ·Cr(CO)$_4$Cl etc. (Cp = cyclopentadienyl anion).

Many of these types of centre are studied by matrix-isolation methods, which have been reviewed by Almond.[4] The major techniques used are infra-red and Raman spectroscopy, but esr spectroscopy has played a major part in solving electronic structures. In addition, it is a convenient method for studying photochemical processes, especially in different types of matrix. Also, their reactions with atoms in the matrix or with free ligands are conveniently studied.

11.2 Some Chromium Complexes

A range of techniques, including esr spectroscopy, cyclic voltammetry, and low temperature IR spectroscopy have been brought to bear on the oxidation chemistry of RSCr(CO)$_5^-$ thiolate anions and of their dimers, RS[Cr(CO)$_5$]$_2^-$.[230] The former rapidly decompose to give thiyl radicals, RS· + Cr(CO)$_5$ on electron-loss, giving eventually the disulphide derivative, (RS–SR)Cr(CO)$_5$. Electron-loss from the RS[Cr(CO)$_5$]$_2^-$ dimer is more complex, although the neutral radicals are initially formed. The unpaired electron in these centres is largely on sulphur rather than on chromium.

A combination of X-ray crystallography, esr spectroscopy and MO theory has been used to probe the structures of a range of chromium complexes, η^5-C$_5$R$_5$Cr(CO)$_2$L, when R = H or Me, and L = CO, or PR$_3$.[231] Manganese derivatives have also been studied. The 18 electron Mn complexes have the expected "piano-stool" structure. However, the 17-electron chromium derivatives have a remarkably small OC – Cr – Co angle of ca. 80°. Esr studies support a ^{2}A′ ground state.

A chromium carbonyl Cr(CO)$_5^-$ has been prepared by ionising radiation or photolysis at 77K. The species is formed by electron capture at Cr(CO)$_6$, but the parent 19-electron Cr(CO)$_6$·$^-$ was not detected.[232]

11.3 Some Mo, W and Re complexes

The same workers[232] have also studied the corresponding Mo and W derivatives formed from Mo(CO)$_6$ and W(CO)$_6$ by electron capture. In all cases the g-tensors indicate square pyramidal C$_{4v}$ structures, with the major spin density in the d$_{z^2}$ orbital pointing towards the vacant site.

A 19-electron rhenium complex, Re(CO)$_4$L derived from Re$_2$(CO)$_{10}$ has been prepared, where L = 2,3-bis(diphenylphosphino) maleic anhydride. The unpaired electron in this case is largely localised on the ligand.[233]

References

1. P.W. Atkins, M.C.R. Symons, "The Structure of Inorganic Radicals" 1967, Elsevier, Amsterdam.
2. J.R. Pillbrow, abstract (Transition Ions Epr conference, Oxford 1991)
3. A. Bencini, D. Gatteschi, "EPR of Exchange Coupled Systems" Springer–Verlag, Berlin, 1990
4. M.J. Almond, R.H. Orrin, *Ann.Rep. Chem.Soc. C.* 1991 in press
5. A. Lund and M. Shiotani, (Eds.) "Radical Ionic Systems" 1991 Kluwer Academic Publishers, Netherlands.
6. L.B. Knight in "Radical Ionic Systems" 1991 Kluwer Academic Publishers, p.73
7. L.B. Knight, J.M. Bostick, R.W. Woodward, J. Steadman, *J.Chem.Phys.* 1983, **78** 6415
8. L.B. Knight, M. Winiski, P. Miller, C.A. Arrington, D. Feller, *J.Chem.Phys.* 1989, **91** 4468
9. L.B. Knight, E. Earl, A.R. Ligon, D.P. Cobranchi, *J.Chem.Phys.* 1986, **85** 1228
10. S. Suzer, L. Andrews, *J.Chem.Phys.* 1988, **88** 916
11. M. Chanon, M. Rajzmann, F. Chanon, *Tetrahedron* 1990, **46** 6193
12. J.-M. Spaeth, F.K. Koschnick, *J.Phys.Chem.Solids* 1991, **52** 1
13. J.-M. Spaeth, F. Lohse, *J.Phys.Chem.Solids* 1990, **51** 861
14. J.-M. Spaeth, *J.Appl.Magn.Res.* 1992 In press
15. M. Ikeya, M. Furusawa, M. Kasuya, *Scanning Microscopy* 1990 **4** 245
16. M. Bowman in: "Modern Pulsed and Continuous-wave Esr", eds. L. Kevan, M. Bowman (Wiley, New York, 1990) ch.1,
17. B.R. Patyal, R.H. Crepeau, D. Gamliel, J.H. Freed, *Chem.Phys.Lett.*, 1990, **175** 445
18. B.R. Patyal, R.H. Crepeau, D. Gamliel, J.H. Freed, *Chem.Phys.Lett.*, 1990, **175** 453
19. A.V. Astashkin, A. Schweiger, *Chem.Phys.Lett.*, 1990, **174** 595
20. J. Forrer, S. Pfenninger, J. Eisenegger, A. Schweiger, *Rev.Sci.Instrum.* 1990, **61** 3360
21. S. Taleb, T.W. Hijmans, W.G. Clark, *Rev.Sci.Instrum.* 1990, **61** 2354
22. P. Vikram, B.P. Roberts, C.R. Willis, *J.Chem.Soc. Perkin Trans.* II 1989 1953
23. D.J. Lurie, I. Nicholson, J.R. Mallard, *J.Magn.Reson.* 1991, **94** 197
24. J.A. Brivati, A.D. Stevens, M.C.R. Symons, *J.Magn.Reson.* 1991, **92** 480
25. M. Alecci, S. Colacicchi, P.L. Indovina, F. Momo, P. Pavone, A. Sotgiu, *Magn.Res.Imag.* 1990, **8** 59
26. H.J. Halpern, D.P. Spencer, J. van Polen, *Rev.Sci.Inst.* 1989, **60** 1040
27. J.F. Gibson, D.J.E. Ingram, M.C.R. Symons, M.G. Townsend, *Trans.Faraday Soc.*, 1957, **53**, 914
28. J.A. Brivati, N. Keene, M.C.R. Symons, *J.Chem.Soc.* 1962 237
29. C. Ferradini, J.-P. Jay-Gerin (Eds.) "Excess Electrons in Dielectric Media" CRC Press London.
30. K.H. Shmidt, P. Han, D.M. Bartels, *J.Phys.Chem.*, 1992, **96** 199
31. H. Abramczyk, J. Kroh, *J.Phys.Chem.*, 1992, **96**, 3653
32. A.S. Jeevarajan, R.W. Fessenden, *J.Phys.Chem.*, 1989, **93**, 3511
33. A.S. Jeevarajan, R.W. Fessenden, *J.Phys.Chem.*, 1992, **96**, 1520
34. H. Sothe, M. Jordan, J.-M. Spaeth, *Radiation Effects and Defects in Solids*, 1991, *119–121* 931
35. F.K. Koschnik, Th. Hangleiter, J.-M. Spaeth, R.S. Eachus, *Radiation Effects and Defects in Solids*, 1991, *119–121*, 837
36. F.K. Koschnik, Th. Hangleiter, J.-M. Spaeth, R.S. Eachus, *J.Phys.: Condens. Matter* 1992, **4** 3001
37. F.K. Koschnik, J.-M. Spaeth, R.S. Eachus, *J.Phys.: Condens. Matter* 1992, **4** 3015
38. F.K. Koschnik, J.-M. Spaeth, R.S. Eachus, W.G. McDugle, R.H.D. Nuttal, *Phys.Rev.Lett.* 1991, **67** 3571
39. H. Sothe, J.-M. Spaeth, F. Luty, *Radiation Effects and Defects in Solids*, 1991 *119–121* 269
40. N. Akiyama, S. Tahara, H. Ohkura, *Okayama Rika Daigaku Kiyo,A* 1990, **26A** 79
41. M. Zorita, J. Casa-Gonzalez, *J.Phys.Chem.Solids* 1990, **51** 1019
42. D.H. Shin, A.S. Ellaboudy, J.L. Dye, M.G. DeBacker, *J.Phys.Chem.*, 1991, **95** 7085
43. P. Gurtler, J. LeCalve, D. Raasch, *Chem.Phys.Lett.* 1991, **177** 585
44. P.-O. Astrand, G. Karlstrom, *Chem.Phys.Lett.* 1990, **175** 624
45. J.B. Raynor, I.J. Rowland, M.C.R. Symons, *J.Chem.Soc. Dalton Trans.* 1987 421

46. T. Miyazaki, H. Morikita, K. Fueki, T. Hiraku, *Chem.Phys.Lett.* 1991, **182** 35
47. H. Muto, K. Matsuura, K. Nunome, *J.Phys.Chem.* 1992, **96** 5211
48. P. Han, D.M. Bartels, *J.Phys.Chem.* 1992, **96** 4899
49. J.M. Baker, A. Cox, A.J. O'Connell, R.C.C. Ward, *J.Phys.:Condens.Matter* **3** 1991 6189
50. J.H.B. Chenier, J.A. Howard, H.A. Joly, B. Mile, *J.Chem.Soc. Faraday Trans.* 1990, **86** 2169
51. J.H.B. Chenier, H.A. Joly, J.A. Howard, B. Mile, *J.Chem.Soc. Faraday Trans.* 1990, **86** 3329
52. M. Histed, J.A. Howard, H.A. Joly, B. Mile, *Chem.Phys.Lett.* 1990, **174** 411
53. M. Histed, J.A. Howard, R. Jones, M. Tomietto, *J.Phys.Chem.* 1992, **96** 1141
54. M. Histed, J.A. Howard, R. Jones, M. Tomietto, H.A. Joly, *J.Phys.Chem.* 1992, **96** 1144
55. J.H.B. Chenier, J.A. Howard, H.A. Joly, B. Mile, P.L. Timms, *J.Chem.Soc. Chem.Comm.* 1990 581
56. V. Luca, D.R. Brown, L. Kevan, *J.Phys.Chem.* 1991, **95** 10065
57. J.A. Howard, H.A. Joly, P.P. Edwards, R.J. Singer, D.E. Logan, *J.Am.Chem.Soc.* 1992, **114** 474
58. J.A. Howard, H.A. Joly, B. Mile, *J.Chem.Soc. Faraday Trans.* 1990, **86** 219
59. J.H.B. Chenier, J.A. Howard, H.A. Joly, M. LeDuc, B. Mile, *J.Chem.Soc.* Perkin Trans.[234] in press
60. J.A. Howard, H.A. Joly, B. Mile, *J.Phys.Chem.* 1992, **96** 1233
61. J.A. Howard, H.A. Joly, B. Mile, R. Sutcliffe, *J.Phys.Chem.* 1991, **95** 6819
62. I. Heynderickx, E. Goovaerts, S.V. Nistor, D. Schoemaker, *Phys.Rev.* B 1986, **33** 1559
63. E.S. Grinberg, V.N. Efimov, *Ferroelectrics* 1990, **107** 331
    J.-M. Spaeth, R.H. Bartram, M. Rac, M. Fockele, *J.Phys.: Condens.Matter* 1991, **3** 5013
64. E. Goovaerts, S.V. Nistor, D. Schoemaker, 1992 *J.Phys.Chem.* In press
65. E. Goovaerts, S.V. Nistor, D. Schoemaker, *Phys.Rev.* B 1990, **42** 3810
66. See, for example, A. Begum, M.C.R. Symons, *J.Chem.Soc.*A 1971 2062; I.S. Ginns, M.C.R. Symons, *J.Chem.Soc. Faraday Trans.* **2** 1972 **68** 631
67. R.S. Eachus, W.G. McDugle, R.H.D. Nuttall, M.T. Olm, F.K. Koschnik, Th. Hangleiter, J.-M.Spaeth, *J.Phys.:Condens.Matter* **3** 1991 9327
68. R.S. Eachus, W.G. McDugle, R.H.D. Nuttall, M.T. Olm, F.K. Koschnik, Th. Hangleiter, J.-M.Spaeth, *J.Phys.: Condens.Matter* **3** 1991 9339
69. J. Zhu, K. Petit, A.O. Colson, S. DeBolt, M.D. Sevilla, *J.Phys.Chem.* 1991, **95** 3676
70. K.S. Kim, H.S. Kim, J.H. Jang, H.S. Kim, B.-J. Mhin, Y. Xie, H.F. Schaefer III *J.Chem.Phys.* 1991, **94** 2057
71. S.H. Ashworth, J.M. Brown, *J.Chem.Soc. Faraday Trans.* 1990, **86** 1995
72. G. Ohanessian, M.J. Brusich, W.A. Goddard III, *J.Am.Chem.Soc.* 1990, **112** 7179
73. T.D. Varberg, R.W. Field, A.J. Merer, *J.Chem.Phys.* 1990, **92** 7123
74. C.F. Kernizan, D.M. Lindsay, *J.Phys.Chem.* 1990, **94** 7445
75. I. Heynderickx, E. Goovaerts, D. Schoemaker, *Phys.Rev.* B 1987 **36** 1843
76. I. Heynderickx, H. De Raedt, D. Schoemaker, *J.Magn.Res.* 1986, **70** 134
77. L.B. Knight, J.O. Herlong, R. Babb, E. Earl, D.W. Hill, C.A. Arrington, *J.Phys.Chem.* 1991, **95** 2732
78. L.B. Knight, J.O. Herlong, T.J. Kirk, C.A. Arrington, *J.Phys.Chem.* 1992, **96** 5604
79. D. Feller, *J.Chem.Phys.* 1990, **93** 579
80. P. Mordvintcev, A. Mulsch, R. Busse, A. Valin, *Analyt.Biochem.* 1991, **199** 142
81. T. Maksumoto, K. Kuwata, *Chem.Phys.Lett.* 1990, **171** 314
82. E. Garrone, E. Giamello, M. Ferraris, G. Spoto, *J.Chem.Soc. Faraday Trans.* 1992, **88** 333
    G.R. Bradburn, H.V. Lilenfield, *J.Phys.Chem.* 1991, **95** 555
83. N. Harrison, M.C.R. Symons, *J.Chem.Soc. Faraday Trans.* 1992 In Press
84. F. Maes, F. Callens, P. Matthys, E. Boesman, *Phys.Status Solidi* B 1990, **161** K1
85. L.B. Knight, R. Babb, D.W. Hill, A.J. McKinley, *J.Chem.Phys.* 1992 In Press
86. Y.M. Hamrick, W. Weltner, *J.Chem.Phys.* 1991, **94** 3371
87. R.J. Van Zee, Y.M. Hamrick, S. Li, W. Weltner, *Chem.Phys.Lett.* 1992, **195** 214
88. M. Cheeseman, R.J. Van Zee, H.L. Flanagan, W. Weltner, *J.Chem.Phys.* 1990, **92** 1553
89. M. Cheeseman, R.J. Van Zee, W. Weltner, *J.Phys.Chem.* 1990, **94** 7808
90. Y.M. Hamrick, R.J. Van Zee, W. Weltner, *Chem.Phys.Lett.* 1991, **181** 193
91. C. Bordas, P.C. Cosby, H. Helm, *J.Chem.Phys.* 1990, **93** 6303
92. Z. Peng, A. Kupperman, *Chem.Phys.Lett.* 1990, **175** 242
93. L.B. Knight, J.R. Woodward, T.J. Kirk, C.A. Arrington, *J.Phys.Chem.* 1992 in press

94. K. Funken, B. Engels, S.D. Peyerimhoff, *Chem.Phys.Lett.* 1990, **172** 180
95. Y. Gauduel, S. Pommeret, A. Migus, A. Antonetti, *Chem.Phys.* 1990, **149** 1
96. T.A. Claxton, I.S. Ginns, M.J. Godfrey, K.V.S. Rao, M.C.R. Symons, *J.Chem.Soc. Faraday Trans.* 2 1973 **69** 217
97. L.B. Knight, K. Kerr, M. Villanueva, A.J. McKinley, *J.Chem.Phys.* 1992 In Press
98. M.C.R. Symons and coworkers, unpublished results.
99. M.J. Almond, A. Raqabah, D.A. Rice, M.C.R. Symons, C.A. Yates, *J.Chem.Soc.* Dalton Trans. 1992 1
100. L.B. Knight, B. Gregory, J. Cleveland, C.A. Arrington, *Chem.Phys.Lett.* 1992 In Press
101. J.M. Brown, H.E. Radford, T.J. Sears, *J.Mol.Spectrosc.* 1991, **148** 20
102. M.D. Sevilla, D. Becker, M. Yan, *J.Chem.Soc. Faraday Trans.* 1990, **86** 3279
103. Y.M. Hamrick, R.J. Van Zee, W. Weltner, *J.Chem.Phys.* 1992, **96** 1767
104. J.A. Howard, H.A. Joly, B. Mile, *J.Phys.Chem.* 1992 In Press
105. L.B. Knight, D.W. Hill, T.J. Kirk, C.A. Arrington, *J.Phys.Chem.* 1992, **96** 555
106. Y.M. Hamrick, R.J. Van Zee, J.T. Godbout, W. Weltner, W.J. Lauderdale, J.F. Stanton, R.J. Bartlett, *J.Phys.Chem.* 1991, **95** 2840
107. M.E. Jacox, *J.Chem.Phys.* 1990, **93** 7622
108. P. Moens, F. Callens, P. Matthys, F. Maes, R. Verbeek, D. Naessens, *J.Chem.Soc. Faraday Trans.* 1991, **87** 3137
109. R. Debuyst, M. Bidiamambu, F. Dejehet, *Bull.Soc.Chim.Belg.* 1990, **99** 535
110. F. Ramondo, N. Sanna, L. Bencivenni, F. Grandinetti, *Chem.Phys.Lett.* 1991, **180** 369
111. K.M. Johnson, B.P. Roberts, *J.Chem.Res.(S)* 1989 352
112. U. Bley, P.B. Davies, T.J. Sears, F. Temps, *Chem.Phys.* 1991, **152** 281
113. R.J. Van Zee, S. Li, Y.M. Hamrick, W. Weltner, *J.Chem.Phys.* 1992 In Press
114. R.J. Van Zee, Y.M. Hamrick, S. Li, W. Weltner, *J.Phys.Chem.* 1992 In Press
115. J. Wang, R.J. Boyd, *J.Chem.Phys.* 1992, **96** 1232
116. A.J. McKinley, J. Michl, *J.Phys.Chem.* 1991, **95** 2674
117. M. Iwasaki, C. Miyazawa, A. Kubota, E. Suzulki, K. Sato, J. Naoi, A. Katoh, K. Niwa, *Radioisotopes* 1991, **40** 421
118. T. Shida, T. Kato, T. Momose, M. Matsushita, *J.Chem.Phys.* 1991, **95** 4725
119. L. Bonazzola, J.P. Michaut, J. Roncin, *Chem.Phys.Lett.* 1988, **149** 316
120. M.C.R. Symons, J.L. Wyatt, *J.Chem.Soc. Chem.Comm.* 1989 362
121. R. Krzyminiewski, J. Kasprzak, *J.Mol.Struct.* 1990, **240** 127
122. R. Cammack, R. Williams, K.K. Rao, M.C.R. Symons, *J.Chem.Soc.* Perkin Trans. 2 1992 1417
123. J.B. Raynor, I.J. Rowland, M.C.R. Symons, *J.Chem.Soc. Faraday Trans.* 1991, **87** 571
124. I.G. Green, K.M. Johnson, B.P. Roberts, *J.Chem.Soc.* Perkin Trans. II 1989 1963
125. S.I. Bannov, V.A. Nevostruev, B.A. Khisamov, *Khim.Vys.Energ.* 1991, **25** 426
126. Y. Kawada, M. Sakaguchi, *Radiat.Eff.Defects Solids* 1990, **112** 55
127. M. Contineanu, O. Constantinescu, *Rev.Roum.Phys.* 1990, **35** 495
128. B.F.M. de Waal, O.M. Aagaard, R.A.J. Janssen, *J.Amer.Chem.Soc.* 1991, **113** 9471
129. O.M. Aagaard, R.A.J. Janssen, B.F.M. de Waal, H.M. Buck, *Heteroatom Chemistry* 1991, **2** 39
130. C.J. Rhodes, M.C.R. Symons, *J.Chem.Soc. Chem.Comm.* 1989 1393
131. R.A.J. Janssen, O.M. Aagaard, M.J.T.F. Cabbolet, B.F.M. de Waal, *J.Phys.Chem.* 1991, **95** 9256
132. K.A. McLauchlan, N.J.K. Simpson, *J.Chem.Soc. Perkin Trans.* 2 1990 1371
133. M. Cattani-Lorente, M. Geoffroy, *Chem.Phys.Lett.* 1990, **167** 460
134. A. Ponti, O. Cesare, L. Forni, *J.Chem.Soc. Faraday Trans.* 1991, **87** 3151
135. P. Bailey, J.A. Weil, *J.Chem.Soc. Faraday Trans.* 1991, **87** 3143
136. L. Bonnazzola, J.P. Michaut, J. Roncin, *J.Phys.Chem.* 1991, **95** 3132
137. R. Janes, M.C.R. Symons, *J.Chem.Soc. Faraday Trans.* 1990, **86** 2173
138. L.B. Knight, B.W. Gregory, D.W. Hill, C.A. Arrington, T. Momose, T. Shida, *J.Chem.Phys.* 1991, **94** 67
139. L.B. Knight, J. Steadman, D. Feller, E.R. Davidson, *J.Am.Chem.Soc.* 1984 **106** 3700
140. O.M. Aargaard, R.A.J. Janssen, B.F.M. de Waal, H.M. Buck, *J.Amer.Chem.Soc.* 1990, **112** 938
141. R.A.J. Janssen, O.M. Aagard, M.J. van der Woerd, H.M. Buck, *Chem.Phys.Lett.* 1990, **171** 127
142. G.J. Gerfen, D.J. Singel, *J.Chem.Phys.* 1990, **93** 4571
143. P.K. Kahol, D. Scoular, *Solid State Commun.* 1990, **75** 205

144. S. Waplak, Z. Trybula, J.E. Drumheller, V.H. Schmidt, *Phys.Rev. B*: *Condens.Matter* 1990, **42** 7777
145. N. Bailey, M.C.R. Symons, *J.Chem.Soc.* 1957 202
146. N. Bailey, A. Carrington, K.A.K. Lott, M.C.R. Symons, *J.Chem.Soc.* 1960 290
147. M.C.R. Symons, *Research*, 1953 **6** 55
148. X. Shi, N.S. Dalal, *Arch.Biochem.Biophys.* 1990, **281** 90
149. X. Shi, N.S. Dalal, V. Vallyathan, *Arch.Biochem.Biophys.* 1991, **290** 381
150. R.H. Liebross, K.E. Weyyerhahn, *Chem.Res.Toxicol.* 1990, **3** 401
151. C. Coudray, P. Faure, S. Rachidi, A. Jeunet, M.J. Richard, A.M. Roussel, A. Favier, *Biol.Trace Elem.Res.* 1992, **32** 161
152. C.J.M. Coremans, E.J.J. Groenen, J.H. van der Waals, *J.Chem.Phys.* 1990, **93** 3101
153. J.R. Byberg, *J.Phys.Chem.* 1992, **96** 4220
154. See, for example, S. Stinson, *Chem.and Eng.News*, 1987 **12** 32
155. N. Ganghi, J.L. Wyatt, M.C.R. Symons, *J.Chem.Soc. Chem.Comm.* 1986 1424
156. T. Tada, *Chem.Phys.Lett.* 1990, **173** 15
157. O.M. Aagaard, R.A.J. Janssen, B.F.M. de Waal, J.A. Kanters, A. Schouten, H.M. Buck, *J.Amer.Chem. Soc.* 1990, **112** 5432
158. M.C.R. Symons, W.R. Bowman, *J.Chem.Soc. Perkin Trans.* 2 1990 975
159. R. Breslow, M. Brandl, J. Hunger, N. Turro, K. Cassidy, K. Krogh-Jesperson, J.D. Westbrook, *J.Amer.Chem.Soc.* 1987, **109** 7204
160. A. Abu-Raqabah, M.C.R. Symons, *J.Amer.Chem.Soc.* 1990, **112** 8614
161. O.M. Aagaard, B.F.M. de Waal, M.J.T.F. Cabbolet, R.A.J. Janssen, *J.Phys.Chem.* 1992, **96** 614
162. M.C.R. Symons, R.L. Peterson, *J.Chem.Soc. Faraday Trans.* 1 1979 216
163. A. Abu-Raqabah, M.C.R. Symons, *J.Chem.Soc. Faraday Trans.* 1990, **86** 3293
164. Q. Jiang, W.R.M. Graham, *J.Chem.Phys.* 1991, **95** 3129
165. A.B. O'Connell, H. Chandra, S.P. Mishra, A. Hasegawa, M.C.R. Symons, *J.Chem.Soc. Faraday Trans.* 1991, **87** 3129
166. T.S. Cameron, R.C. Haddon, S.M. Mattar, S. Parsons, J. Passmore, A.P. Ramirez, *J.Chem.Soc. Chem.Comm.* 1991 358
167. B.M. Gimarc, D.S. Warren, *Inorg.Chem.* 1991, **30** 3276
168. B. Mile, J.A. Howard, M. Histed, H. Morris, C.A. Hampson, *Faraday Discuss.* 1991, **92** 129
169. G.A. Kuranova, *Khim.Vys.Energ.* 1991, **25** 110
170. Y.M. Hamrick, R.J. Van Zee, W. Weltner, *J.Chem.Phys.* 1992, **96** 1767
171. R.J. Van Zee, W. Weltner, *J.Chem.Phys.* 1990, **92** 6976
172. F. Blatter, K.W. Blazey, *Z.Phys.D: At., Mol.Clusters* 1991, **18** 427
173. T. Wasowicz, J. Michalik, *Radiat.Phys.Chem.* 1991, **37** 427
174. M.C.R. Symons, P. Tordo, J.L. Wyatt, *J.Organomet.Chem.* 1992 in press
175. M. Ikeya, *Annu.Rev.Mater.Sci.* 1991, **21** 45
176. M. Almond, *Ann.Rep.Chem.Soc.* C 1991 in press
177. H. Bower, M.C.R. Symons, *Nature*, 1966 210 1037
178. T.A. Karpukhina, L.L. Builov, V.F. Chuvaev, *Surf.Coat.Technol.* 1991, **47** 538; N.D. Samsonenko, V.V. Tokii, S.V. Gorban, V.I. Timchenko, *Surf.Coat.Technol.* 1991, **47** 618
179. M. Fanciulli, T.D. Moustaka, *Diamond Relat.Mater.* 1992, **1** 773
180. M. Furusawa, M. Ikeya, *J.Phys.Soc.Japn.* 1990, **59** 2340
181. D. Dagupta, F. Demichelis, C.F. Pirri, A. Tagliaferro, *Phys.Rev.B*: *Condens.Matter* 1991, **43** 2131
182. W.M. Chen, A. Henry, E. Janzen, B. Monemar, M.L.W. Thewalt, *Mater.Res.Soc.Symp.Proc.* 1990, **163** 303
183. N.T. Son, A.B. Van Oosten, C.A.J. Ammerlaan, *Solid State Commun.* 1991, **80** 439
184. N. Meilwes, J.R. Niklas, J.-M. Spaeth, *Mater.Sci.Forum* 1990 **65–66** 247
185. a M.C.R. Symons, *Hyp.Int.* 1984 **17–19** 771
   b S.F.J. Cox, M.C.R. Symons, *Chem.Phys.Lett.* 1986, **126** 516
186. T.A. Claxton, Dj.M. Maric, P.F. Meier, *Chem.Phys.Lett.* 1992, **192** 29
187. S.F. J.Cox, E.A. Davis, W. Hayes, M.C.R. Symons, A. Wright, A. Singh, F.L. Pratt, T.A. Claxton, F. Jansen, *Hyperfine Interactions* 1990, **64** 551
188. M. Kondo, W. Utsumi, T. Yagi, K. Morigaki, *J.Non-Cryst. Solids*, 1991, **137** 275

189. S. Yamasaki, H. Okushi, A. Matsuda, K. Tanaka, J. Isoya, *Phys.Rev.Lett.* 1990, **65** 756
190. Y. Wu, A. Stesmans, *Solid State Commun.* 1991, **79** 741
191. P.J. McElheny, H. Okushi, S. Yamasaki, A. Matsudea, *J.Non-Cryst.Solids* 1991 **137–138** 243
192. S. Mitra, *Report* 1991, from *Energy Res.Abstr.* 1991, **16** Abstr.No. 15816
193. A. Morimoto, M. Matsumoto, M. Yoshita, M. Keneda, T. Shimizu, *Appl.Phys.Lett.* 1991, **59** 2130
194. M.A. Jupina, P.M. Lenahan, *IEEE Trans.Nucl.Sci.* 1990, **37** 1650
195. K.L. Brower, S.M. Myers, *Appl.Phys.Lett.* 1990, **57** 162
196. D.L. Griscom, *J.Ceram.Soc.Japn.* 1991, **10** 923
197. D.L. Griscom, Proc.Int.Symp.on Point Defects in Glasses, *J.Non-Cryst.Solids* 1992, in press
198. T.E. Tsai, D.L. Griscom, *Phys.Rev.Lett.* 1991, **67** 2517
199. F. Sim, C.R.A. Catlow, M. Dupois, J.D. Watts, *J.Chem.Phys.* 1991, **95** 4215
200. K. Yokogawa, Y. Yajima, T. Mizutani, S. Nishimatsu, K. Ninomiya, *Jpn.J.Appl.Phys.* Part 1 1991, **30** 3199
201. P. Bailey, T. Pawlik, H. Sothe, J.-M. Spaeth, J.A. Weil, *J.Phys.:Condens.Matter* 1992, **4** 4063
202. G. Corradi, A. Watterich, I. Foldvari, R. Voszka, J.R. Niklas, J.-M. Spaeth, O.R. Gilliam, L.A. Kappers, *J.Phys.:Condens.Matter* 1990, **2** 4325
203. M. Honda, K. Atobe, N. Fukuoka, M. Okada, M. Nakagawa, *Jpn.J.Appl.Phys.* Part 2 1990, **29** L652
204. J.M. Spaeth, J.J. Lappe, *Appl.Magn.Res.* 1991, **2** 311
205. J.-M. Spaeth, M. Fockele, K. Krambrock in "Non-Stoichiometry in Semiconductors" (eds. K.J. Bachmann, H.-L. Hwang, C. Schwab) 1992 Elsevier p.193
206. K. Krambrock, J.-M. Spaeth, C. Delerue, G. Allan, M. Lannoo, *Phys.Rev.B* 1992, **45** 1481
207. M. Jordan, M. Linde, T. Hangleiter, J.-M. Spaeth, *Semicond.Sci.Technol.* 1992, **7** 731
208. M. Fockele, J.-M. Spaeth, H. Overhof, P. Gilbart, *Semicond.Sci.Technol.* 1991, **6** B88 .
209. J.M. Trombetta, T.A. Kennedy, W. Tseng, D. Gammon, *Phys.Rev.B: Condens. Matter* 1991, **43** 2458
210. J.J. Davies, *NATO ASI Ser.*, *Ser.B* 1989 200
211. A. Poeppl, G. Voelkel, *Phys.Status Solidi A* 1990, **121** K69
212. F. Morazzoni, R. Scotti, S. Volonte, *J.Chem.Soc. Faraday Trans.* 1990, **86** 1587
213. C.Y. Sun, Y.J. Hsyu, H.L. Hwang, *J.Cryst.Growth* 1990, **101** 479
214. C.T. Kao, L.G. Rowan, L.M. Slifkin, *Phys.Rev.B: Condens.Matter* 1990, **42** 3142
215. Y. Wang, C. Yeh, *J.Chem.Soc. Faraday Trans.* 1991, **87** 345; B. Xu, L. Kevan, *J.Phys.Chem.* 1991, **95** 1147
216 B. Xu, L. Kevan, *J.Phys.Chem.* 1991, **95** 1147
217. C.A. Rice-Evans, A.T. Diplock, M.C.R. Symons, "Techniques in Free Radical Research" Elsevier, 1991
218. D. Rehorek, *J.Chem.Soc. Perkin Trans.* 2 1992 in press
219. G.R. Buettner, R.P. Mason, *Methods Enzymol.* 1990, **186** 127
220. K. Makino, T. Hagiwara, A. Murakami, *Radiat.Phys.Chem.* 1991, **37** 657
221. E.G. Janzen, Y.-L.E. Wang, *Free Rad.Res.Comms.* 1990, **10** 63
222. P. Bilski, C.F. Chignell, J. Szychlinski, A. Borkowski, E. Oleksy, K. Reszka, *J.Am.Chem.Soc.* 1992, **114** 549
223. T. Ozawa, A. Hanaki, *J.Chem.Soc. Chem.Comm.* 1991 330
224. D. Rehorek, E.G. Janzen, Y. Kotake, *Can.J.Chem.* 1991, **69** 1131
225. K. Ogura, C.T. Migita, M. Nakayama, *J.Chem.Soc. Faraday Trans.* 1990, **86** 2565
226. L.Y. Zang, Z. Zhang, H.P. Misra, *Photochem.Photobiol.* 1990, **52** 677
227. A. Samuni, A. Samuni, H.M. Swartz, *Free Radical Biol. Med.* 1989, **7** 37
228. I. Yamazaki, L.H. Piette, *J.Biol.Chem.* 1990, **265** 13589
229. X. Sun, X. Shi, N.S. Dalal, *FEBS Lett.* 1992, **303** 213
230. J. Springs, C.P. Janzen, M.Y. Darensbourg, J.C. Calabrese, P.J. Krusic, J.-N. Verpeaux, C. Amatore, *J.Am.Chem.Soc.* 1990, **112** 5789
231. S. Fortier, M.C. Baird, K.F. Preston, J.R. Morton, T. Ziegler, T.J. Jaeger, W.C. Watkins, J.H. MacNeil, K.A. Watson, K. Hensel, Y. LePage, J.-P. Charland, A.J. Williams, *J.Am.Chem.Soc.* 1991, **113** 542
232. A. Buttafava, A. Faucitano, F. Martinotti, *Radiat.Phys.Chem.* 1991, **37** 567
233. J. Ko, Y.C. Park, *Bull.Korean Chem.Soc.* 1989, **10** 606

Author Index

In this index the number given in parenthesis is the Chapter number of the citation and this is followed by the reference number or numbers of the relevant citations within that Chapter.

Aagaard, O.M. (5) 128, 129, 131, 140, 141, 157, 161
Aasa, R. (1) 57; (2) 279, 340
Abasov, Ya.M. (1) 272
Abou-Kais, A. (1) 519-524
Abragam, A. (3) 2, 21
Abraham, M.M. (3) 100, 147, 148
Abram, U. (1) 424-427, 429-433
Abramczyk, H. (5) 31
Abu-Raqabah, A. (1) 398, 417; (5) 99, 160, 163
Ackrell, B.A.C. (2) 183
Adams, J.M. (1) 393
Adams, M.W.W. (2) 169-171, 193, 194, 196, 240, 241
Addison, A.W. (2) 117, 120, 122
Adrian, F.J. (1) 285
Adriani, A. (2) 53
Aebli, G. (1) 29
Affleck, I. (1) 220
Agalidis, I. (2) 311
Aguilar, M. (1) 614
Agullo-Lopez, F. (1) 586
Ainscough, E.W. (1) 551; (2) 95
Aisen, P. (2) 93
Aissi, C.F. (1) 520-524
Ajayakumar, K.S. (1) 324
Ajiro, Y. (1) 217; (3) 94
Akagi, K. (1) 364, 365
Akiyama, N. (5) 40
Alagna, L.A. (2) 100
Al-Basseet, J. (2) 150
Albracht, S.P.J. (2) 195, 205, 206, 211, 215, 242
Alcalá, R. (1) 90, 580

Alecci, M. (5) 25
Alekseevski, N.E. (3) 98
Alimov, A.I. (1) 292
Allakhverdiev, S.I. (2) 325
Allan, G. (5) 206
Allmann, R. (1) 104, 559
Allsopp, R.A. (3) 48, 49
Almond, M.J. (5) 4, 99, 176
Alonso, P.J. (1) 90, 396, 562, 580
Alvarez, S. (1) 168
Alves Marinho, L. (1) 461
Amatore, C. (5) 230
Amiell, J. (1) 253-255
Ammerlaan, C.A.J. (1) 625; (5) 183
Amoretti, G. (1) 313, 349
Ananyev, G. (2) 329
Anderson, C.H. (3) 39
Anderson, J.R. (1) 386
Andersson, A. (1) 392
Andersson, B. (2) 333
Andersson, K.K. (2) 59, 60, 82, 140, 173
Andre, J.-J. (1) 19, 265
Andréasson, L.-E. (2) 355
Andrews, K.M. (2) 287
Andrews, L. (5) 10
Andrews, S.C. (2) 94
Andriesson, J. (3) 45
Anemüller, S. (2) 296
Angelova, O. (1) 550
Angelucci, F. (1) 535
Anger, G. (2) 219
Anikeenok, O.A. (3) 50
Antholine, W.E. (1) 66, 455; (2) 26, 30, 40, 48, 76, 138
Antipin, A.A. (3) 125, 126

Antonetti, A. (5) 95
Antonyuk-Barynina, S.V. (2) 256
Aoki, H. (1) 630
Aono, S. (2) 170, 196
Aquilar, M. (1) 275
Arakawa, M. (1) 630, 645, 649, 650; (3) 100
Aramburu, J.A. (1) 48
Araullo-McAdams, C. (1) 472, 473, 502
Arena, G. (1) 542
Arizmendi, L. (1) 614
Armstrong, F.A. (2) 157, 163, 164, 172
Arnaud, S. (2) 295
Arp, D.J. (2) 235
Arrington, C.A. (5) 8, 77, 78, 93, 100, 105, 138
Ascenzi, P. (2) 115, 116
Ashworth, S.H. (5) 71
Assis, M.D. (1) 136
Asso, M. (2) 204, 227
Astashkin, A.V. (1) 33; (5) 19
Astrand, P.-O. (5) 44
Atakishiev, S.M. (1) 272
Atamian, M. (2) 315
Atanasov, M. (1) 559
Atimirova, T.V. (1) 644
Atkins, P.W. (5) 1
Atobe, K. (5) 203
Atta-Asafo-Adjei, E. (2) 285
Aubke, F. (1) 570
Auburn, P.R. (1) 125, 129
Augier, V. (2) 227
Auteri, F.P. (1) 25
Averill, B.A. (2) 90
Avigliano, L. (2) 36, 37

Axe, J.D. (3) 41
Azzoni, C.B. (1) 303, 371

Baar, D.J. (1) 293, 294, 311
Babb, R. (5) 77, 85
Babcock, G.T. (1) 132; (2) 16, 131, 314-316
Baberschke, K. (3) 46, 89
Backes, G. (2) 72
Bacquet, G. (3) 88
Bag, N. (1) 127, 128, 459, 460
Bagchi, R.N. (1) 317, 438
Bagguley, D.M.S. (3) 32
Bagyinka, C. (2) 201, 202
Bai, G. (1) 53
Baidya, N. (1) 493, 504; (2) 202
Baiker, A. (1) 387
Bailey, N. (5) 145, 146
Bailey, P. (1) 367; (5) 135, 201
Bain, A. (1) 17
Baird, M.C. (5) 231
Bakak, A. (1) 469
Baker, E.N. (2) 95
Baker, H.M. (2) 95
Baker, J.E. (2) 276
Baker, J.M. (1) 647; (3) 20, 22, 34, 38, 47-49, 51, 57, 58, 61, 63, 64, 68, 69, 73, 76, 77, 83, 99, 115-117, 128, 129, 139, 144, 147, 148; (5) 49
Bakou, A. (2) 356
Balagoplakrishna, C. (1) 229
Balakrishnan, I. (1) 659
Baldas, J. (1) 428
Balestrino, G. (1) 336
Ball, J.M. (1) 150
Ballinger, M.D. (2) 190
Ban, T. (1) 166
Banci, L. (2) 19
Bandoh, T. (1) 89
Bandow, S. (1) 346
Bandyopadhyay, D. (2) 128
Bank, J.F. (2) 303
Bannov, S.I. (5) 125
Bansal, R.S. (1) 376, 620, 632
Barak, J. (3) 91
Barassi, C.A. (2) 305
Barbaro, P. (1) 160
Barber, M.J. (2) 224-226, 230
Barbieri, R. (2) 99
Barkigia, K.M. (1) 503
Barnham, R.J. (1) 312
Barone, V. (1) 52
Barra, A.L. (1) 114, 213
Barra, D. (1) 61
Barrette, W.C., jun. (2) 280
Barry, B.A. (2) 314, 326, 331, 342
Bartels, D.M. (5) 30, 48
Bartlett, R.J. (5) 106
Bartley, S.L. (1) 141
Bartolino, R. (1) 394
Bartram, R.H. (1) 584; (5) 63
Barynin, V.V. (2) 255, 256
Bashkin, J.S. (1) 203
Basosi, R. (1) 541, 552
Bass, J.S. (1) 483
Bastian, N.R. (2) 230
Basu, P. (1) 128, 459, 600, 601
Baszynski, J. (1) 306
Bates, C.A. (1) 107; (3) 30
Batra, G. (1) 547
Battaglia, L.P. (1) 170, 555
Baugher, J.F. (4) 2
Baumgarten, M. (2) 358
Bauminger, R. (2) 94
Bear, J. (1) 147, 318, 319, 322, 328
Beauchamp, A.L. (1) 476
Bechara, R. (1) 519-523
Beck, W.F. (2) 319, 336
Becker, D. (5) 102
Bedioui, F. (1) 131
Bednarek, J. (1) 530-532
Begum, A. (5) 66
Beinert, H. (2) 1, 161, 162, 166, 167, 186
Bejjit, L. (3) 92
Belford, R.L. (4) 20
Belougne, P. (1) 419
Belt, L.F. (1) 366
Beltrán, D. (1) 413
Beltrán-Lopez, V. (1) 72
Beltran-Porter, D. (1) 104
Bencini, A. (1) 1, 55, 142, 160, 212; (5) 3
Bencivenni, L. (5) 110
Bender, C. (1) 484; (2) 44
Benelli, C. (1) 113, 138
Benetis, N.P. (1) 73
Beneto, M. (1) 238
Beneto Borja, M. (1) 169
Bennani, A. (1) 524
Benner, H. (1) 9
Bennett, B. (2) 222
Bennett, R.L.H. (4) 24
Bensimon, Y. (1) 419
Ben Taarit, Y. (1) 363
Bentsen, J.G. (1) 196; (2) 80
Beranger, M. (1) 19
Berardi, E. (2) 249
Berardinelli, G. (1) 246
Berger, R. (1) 63, 420
Bergonia, H.A. (2) 105
Berlinger, W. (1) 98, 275; (3) 107
Bernardo, M. (1) 13; (2) 194, 234
Bernhardt, P.V. (1) 180, 441, 442, 598
Bersuker, G.I. (1) 102, 103
Bertin, P.Y. (1) 298
Bertini, I. (1) 160; (2) 19, 165
Bertrand, P. (2) 143, 204, 227, 295
Bertran-Lopez, V. (1) 70, 71
Beth, A.H. (1) 25
Beurskens, P.T. (1) 163, 206
Beveridge, K. (1) 181
Bhagat, S.M. (3) 91
Bhanja Choudhury, S. (1) 488, 491, 492
Bhat, S.V. (1) 350
Bhattacharya, S. (1) 458
Bhattacharyya, L. (2) 257
Bhirud, R.G. (1) 540
Bhula, R. (1) 154, 159
Bianchini, C. (1) 123, 478
Bidiamambu, M. (5) 109
Bied-Charreton, C. (1) 131
Biensan, P. (1) 253-255
Bienvenue, A. (2) 134
Biggins, J. (2) 362, 364, 372
Bill, E. (1) 3, 130, 134, 194, 209, 250, 452; (2) 99, 126
Bill, H. (1) 109, 289, 506
Bilski, P. (5) 222
Bilton, R.F. (1) 174
Birgeneau, R.J. (3) 139
Blackmore, R.S. (2) 146
Blake, W.B.J. (3) 38, 63, 67
Blankenship, R.E. (2) 14
Blasco, F. (2) 227
Blatter, F. (5) 172
Blazey, K.W. (1) 275; (5) 172
Bleaney, B. (1) 5; (3) 3, 2, 16, 33-36, 51, 56, 86, 128, 146-148; (4) 21
Bley, U. (5) 112
Blonk, H.L. (1) 163
Blubaugh, D.J. (2) 315
Bluck, L.J.C. (3) 48, 72, 99
Boas, J.F. (1) 428
Boatner, L.A. (3) 100
Böcher, R. (2) 215
Boeker, A. (1) 658
Boerner, R.J. (2) 331, 342
Boesman, E. (5) 84
Böttcher, R. (1) 430-432, 589
Bogdanov, V.L. (1) 525
Bogomolova, L.D. (1) 525
Bogumil, R. (2) 244
Bohandy, J. (1) 285

Bohr, A. (3) 66
Bois, C. (1) 173
Bojanowski, B. (1) 99, 256
Bollinger, J.M., jun. (2) 71
Bolognesi, M. (2) 115
Bolscher, B.G.J.M. (2) 131
Bominaar, E.L. (1) 3, 134, 452; (2) 126
Bonaccorsi di Patti, M.C. (2) 55
Bonamartini-Corradi, A. (1) 555
Bonazolla, L. (5) 119
Bond, A.M. (1) 438
Bond, M.R. (1) 235
Bonicel, J. (2) 143
Bonnazzola, L. (5) 136
Bonnelle, J.P. (1) 519-521, 523
Bonner, J.C. (1) 215
Bonomaar, E.L. (1) 130
Bonomi, F. (2) 87, 283
Bonomo, R.P. (1) 542, 546, 595, 596; (2) 249
Bonori, M. (1) 26
Bontemps, N. (1) 296, 298
Boone, S.R. (1) 117, 124
Boorman, P.M. (1) 150
Booth, K.S. (2) 132
Borchardt, D. (2) 15, 57
Bordas, C. (5) 91
Boricha, A.B. (1) 464
Borkowski, A. (5) 222
Borovik, A.S. (1) 195; (2) 79
Borrás-Almenar, J.J. (1) 183, 205, 242, 243
Bossa, F. (2) 61
Bostick, J.M. (5) 7
Bote, P.R. (1) 626
Bottin, H. (2) 363
Bottomley, L.A. (1) 401
Boucher, J.P. (1) 9
Boulard, B. (1) 582
Bourbonnais, C. (1) 266, 267
Boussac, A. (2) 332, 350, 351, 353, 354
Bouwens, E.C.M. (2) 206
Bouwman, E. (1) 206
Bowden, S.J. (2) 317, 323
Bower, H. (5) 177
Bowers, C.R. (1) 21
Bowlby, N.R. (2) 316
Bowman, M. (5) 16
Bowman, W.R. (5) 158
Boyd, P.D.W. (1) 203; (4) 15
Boyd, R.J. (5) 115
Bracete, A.M. (2) 155
Bradburn, G.R. (5) 82
Bradbury, M.I. (3) 10
Brader, M.L. (2) 15, 57
Bramley, R. (1) 405-407; (2) 341

Branca, M. (2) 247
Brandl, M. (5) 159
Brauer, S.L. (1) 408
Bravo, D. (1) 614, 615
Bray, P.J. (4) 17
Bray, R.C. (2) 219, 222
Bredenkamp, G. (2) 371
Breitbarth, F.W. (1) 370
Breivogel, C.S. (1) 23
Brennan, B.A. (1) 135; (2) 58, 60
Brenner, M.C. (2) 217
Breñosa, A.G. (1) 48
Breslow, R. (5) 159
Bressan, M. (1) 546
Breton, J. (2) 157, 163, 164
Brettel, K. (2) 365, 368
Brewer, F. (2) 257
Brewer, J.C. (1) 434
Briant, R.G. (1) 20
Briganti, F. (1) 160; (2) 165
Britt, R.D. (2) 288, 337, 338, 339
Brivati, J.A. (5) 24, 28
Brnicevic, N. (1) 357
Broderick, J.B. (2) 62
Brodie, A.M. (1) 551; (2) 95
Brodie, S. (1) 544
Brook, D.J.R. (2) 9
Brouet, G. (1) 638, 640
Brower, E.O. (3) 145
Brower, K.L. (3) 145; (5) 195
Brown, D.R. (1) 573, 574; (5) 56
Brown, J.M. (5) 71, 101
Brudvig, G.W. (1) 157, 201, 202, 434; (2) 310, 319, 336, 356
Brune, D.C. (2) 14
Brunel, L. (1) 114, 213
Brunori, M. (2) 149
Bruschi, M. (2) 143, 218
Brusich, M.J. (5) 72
Bryant, D.A. (2) 366, 367
Buchanan, R.M. (1) 176, 549
Buck, H.M. (5) 129, 140, 141, 157
Buckmaster, H.A. (3) 81
Buettner, G.R. (5) 219
Builov, L.L. (5) 178
Bukowska-Strzyzewska, M. (1) 172
Buluggiu, E. (1) 313, 349
Bunker, G. (2) 51
Buratto, S.K. (1) 21
Burbaev, D.Sh. (2) 142, 277
Burgess, B.K. (2) 172

Burn, J.L.E. (1) 185
Burns, G. (3) 41
Burnside, S.D. (1) 309
Burriel, R. (1) 243
Busca, G. (1) 390
Buser, C. (2) 356
Busse, R. (2) 106; (5) 80
Butt, J.N. (2) 157, 163, 164, 172
Buttafava, A. (5) 232
Butzlaff, C. (1) 194; (2) 99
Buzare, J.Y. (3) 28, 48, 71, 106-109
Byberg, J.R. (5) 153
Byeon, S.H. (1) 495, 496, 604
Bylina, E.J. (2) 313
Byrt, D. (1) 317

Cabbolet, M.J.T.F. (5) 131, 161
Cabelli, D. (2) 19
Cabello, C.I. (1) 259
Cabral-Prieto, A. (1) 67-69
Cai, J. (1) 164
Calabrese, J.C. (5) 230
Calabrese, L. (2) 53, 55
Calafat, A.M. (1) 597
Calamiotou, M. (1) 341; (3) 96, 97
Calco, R. (1) 230
Calhoun, M.W. (2) 308
Callens, F. (5) 84, 108
Calvo, R. (1) 225-227, 231-234; (3) 23
Camenish, T.G. (1) 22
Cameron, J. (1) 474
Cameron, T.S. (5) 166
Cammack, R. (2) 156, 177, 181, 182, 200, 223, 228; (4) 3; (5) 122
Campagna, C.F. (1) 493
Campillos, E. (1) 562
Caneschi, A. (1) 137, 138, 213, 259, 260
Cannons, A.C. (2) 226
Canters, G.W. (2) 7, 11
Capelletti, R. (1) 399
Carbonaro, M. (2) 53
Carlin, R.L. (1) 259; (5) 85
Caron, L.G. (1) 267
Carrano, C.J. (2) 246
Carrington, A. (5) 146
Carroll, P.J. (1) 544
Carroll, R.T. (2) 64, 67
Carson, M.R. (1) 412
Carson, P.J. (1) 21
Carter, S.F. (3) 114
Casas-Gonzales, J. (1) 576, 580;

Author Index

(5) 41
Casassas, E. (1) 560
Casella, L. (1) 385, 542; (2) 35
Casellato, U. (1) 596
Cassidy, K. (5) 159
Cassoux, P. (1) 139, 265, 487
Castan, P. (1) 246
Castellano, E.E. (1) 225, 226, 233
Castro, I. (1) 173
Castro-Tello, J. (1) 70, 71
Catlow, C.R.A. (5) 199
Cattani-Lorente, M. (5) 133
Caughey, W.S. (2) 132
Cecchini, G. (2) 183-185
Centi, G. (1) 391
Cerny, V. (1) 79
Cervilla, A. (1) 413
Cesare, O. (5) 134
Chakravorty, A. (1) 127, 128, 458-460, 488, 491, 492, 600, 601, 603
Cham, P. (1) 376
Chambers, D.N. (3) 76
Champion, P.M. (2) 110
Chan, S.I. (2) 297, 305
Chand, P. (1) 620
Chandra, H. (5) 165
Chandra, S.K. (1) 600, 603
Chandrasekhar, S. (1) 479
Chang, C.K. (1) 132; (2) 155
Chang, C.S.J. (1) 415, 416
Chanon, F. (5) 11
Chanon, M. (5) 11
Chapellier, M. (3) 21
Chapman, A. (2) 223
Charland, J.-P. (5) 231
Chary, K.V.R. (1) 389
Chasteen, N.D. (2) 86, 91-93, 96
Chaston, S.H.H. (4) 18
Chattopadhyay, S. (1) 459
Chaudhuri, P. (1) 209
Chaudhury, M. (1) 126, 410
Che, M. (1) 6, 423
Cheeseman, M. (2) 50, 140; (5) 88, 89
Chehab, S. (1) 253, 254, 255
Chen, C. (1) 164
Chen, S.J. (1) 186
Chen, V.J. (2) 68-70
Chen, W.M. (5) 182
Chen, X. (1) 513-517, 638, 640
Chen, Y. (2) 91; (3) 100
Chen, Z. (1) 345
Cheng, C.Y. (1) 252
Cheniae, G.M. (2) 315
Chenier, J.H.B. (1) 571; (5) 50, 51, 55, 59
Cheong, S.W. (1) 307, 323
Chibotaru, L.F. (1) 102, 103
Chignell, C.F. (5) 222
Chikira, M. (2) 30
Chiorino, A. (1) 402
Chitambar, C.R. (2) 76
Chmielewski, P.J. (1) 470
Cho, H.M. (1) 21, 31
Choh, S.H. (1) 616, 627
Choi, D. (1) 653
Choi, N. (1) 548
Chottard, G. (2) 114, 143, 218
Choy, J.H. (1) 495, 496, 604
Christidis, T. (3) 73, 77
Christou, G. (1) 158, 203
Chung, N.S. (3) 75
Chuvaev, V.F. (5) 178
Cimino, A. (1) 403
Ciriolo, M.R. (2) 22
Ciurli, S. (1) 161
Civitareale, P. (2) 259
Clark, K. (2) 17
Clark, P.A. (2) 41
Clark, W.G. (1) 40; (5) 21
Claxton, T.A. (5) 96, 186, 187
Cleary, D.A. (1) 606
Clemens, J.M. (3) 133, 137
Cleveland, J. (5) 100
Cobranchi, D.P. (5) 9
Cocco, T. (2) 283
Colacicchi, S. (5) 25
Colaneri, M.J. (2) 25
Coldea, M. (1) 361
Cole, J.L. (2) 36, 41
Coletta, M. (2) 115, 116
Collier, S. (1) 154
Collins, T.J. (1) 404, 480, 494
Collison, D. (1) 64, 150, 377, 415; (4) 19
Colmanet, S.F. (1) 428
Colpas, G.J. (2) 201, 202
Colson, A.O. (5) 69
Colton, R. (1) 438
Comba, P. (1) 180, 441, 442, 598
Condò, S.G. (2) 115, 116
Connelly, N.G. (4) 22
Conover, R.C. (2) 170, 171
Constantinescu, O. (5) 127
Conti, A.J. (1) 158
Contineau, M. (5) 127
Cook, M.I. (3) 116, 117
Cook, R.J. (3) 62
Cooper, C.E. (2) 133, 298, 299
Copland, G.M. (3) 38, 67
Corbett, J.A. (2) 198
Cordischi, D. (1) 402, 403
Coremans, C.J.M. (1) 38; (5) 152
Coremans, J.M.C.C. (2) 205, 206, 211
Cornelissen, J.P. (1) 144
Coronado, E. (1) 183, 205, 242, 243
Corradi, G. (1) 585, 617; (5) 202
Corradi-Bonamartini, A. (1) 170
Corrie, A.R. (2) 322
Cosby, P.C. (5) 91
Costa, C.C. (2) 147
Costa, L. (1) 535
Costa, R. (1) 168
Costes, J.P. (1) 171, 454
Cote, C.E. (2) 33
Cotton, F.A. (1) 187-190
Coudray, C. (5) 151
Cowan, J.A. (2) 148
Cowling, R.A. (2) 65
Cox, A. (5) 49
Cox, S.F.J. (5) 185, 187
Crabtree, R.H. (1) 157, 201, 202, 501
Creece, I. (1) 438
Crepeau, R.H. (5) 17, 18
Crofts, A.R. (2) 287
Crowder, M.W. (2) 90
Crowther, J.A. (1) 548
Cucurou, C. (2) 66
Cugunov, L. (3) 122
Cui, Y. (1) 533
Culvahouse, J.W. (3) 53
Currell, G. (3) 22
Curti, B. (2) 188
Curtis, J.F. (2) 154
Curtis, N.F. (1) 598
Cusanovich, M.A. (2) 14
Cuvier, S. (1) 315, 318, 319
Czechowski, M. (2) 191
Czyzak, B. (1) 286, 287, 306

Dagupta, D. (5) 181
Dahan, F. (1) 169, 171, 238, 454
Dahl, C. (2) 178
Dalal, N.S. (1) 282, 288, 297; (2) 251, 261-264; (5) 148, 149, 229
Daldal, F. (2) 285, 286, 291, 292
Dalton, H. (2) 80
Dalton, L.R. (1) 17
Dance, J.M. (1) 495, 496, 581
Dandulakis, G. (2) 356
Dannefaer, S. (1) 281

Darensbourg, M.Y. (5) 230
Dartiguenave, M. (1) 476
Dartiguenave, Y. (1) 476
Das, M. (1) 93, 94
Das, T.P. (3) 45
Date, M. (1) 217-219, 247, 301
David, S.S. (2) 88
Davidov, D. (1) 296, 298
Davidson, E. (2) 285
Davidson, E.R. (5) 139
Davies, E.R. (3) 58, 76
Davies, G. (1) 120, 207
Davies, J.J. (5) 210
Davies, M.J. (2) 125
Davies, P.B. (5) 112
Davis, E.A. (5) 187
Davoust, C.E. (2) 8; (3) 34
Davydov, R. (2) 135
Dawson, J.H. (1) 453; (2) 155
Day, C.R. (3) 114
Day, E.P. (2) 49, 86, 208, 253
Day, P. (1) 269
Deák, Z. (2) 318
Dean, D.R. (2) 234, 237
DeBacker, M.G. (5) 42
De Boer, E. (1) 81, 430, 431, 439, 440
DeBolt, S. (5) 69
Debrunner, P.G. (2) 82
Debus, R.J. (2) 342
Debuyst, R. (5) 109
de Castro, B. (1) 467, 486
de CastroMartins, S. (1) 461
Dege, J.E. (2) 84
De Graaf, R.A.G. (1) 144
DeGray, J.A. (2) 154; (4) 22, 25
Deguenon, D. (1) 246
Dei, A. (1) 113-116
Dejehet, F. (5) 109
De Jonge, W.J.M. (1) 271
Dekker, J.P. (2) 316
Delerue, C. (5) 206
De Ley, M. (2) 56
Della Vedova, B.P.C. (1) 209
DeLooze, J.P. (1) 288
DeMaio, R. (2) 123
Demazeau, G. (1) 495, 496, 604
Demeter, S. (2) 325
Demichelis, F. (5) 181
De Montauzon, D. (1) 487
Denariaz, G. (2) 139
Deng, Y. (1) 118
Dengel, A.C. (1) 435
Dengler, J. (1) 578
Denis, M. (2) 295
De Raedt, H. (5) 76
de Rezende, N.M.S. (1) 461
Deroide, B. (1) 419

DeRose, V.J. (2) 339
De Rossi, S. (1) 402, 403
DerVartanian, D.V. (2) 191, 207
Des Courieres, T. (1) 397
Desideri, A. (2) 22, 115, 116
Dessí, A. (2) 247
Deville, A. (3) 92
Devynck, J. (1) 131
de Waal, B.F.M. (5) 128, 129, 131, 140, 157, 161
DeWitt, J.G. (2) 80
de Wolf, I. (3) 35
Dexheimer, S.L. (2) 334
D'Huysser, A. (1) 519, 521
Diab, J. (1) 363
Diaz, A. (1) 552
Diaz, D. (2) 218
Di Bilio, A.J. (1) 595, 596; (2) 249
Didiuk, M. (1) 157
Dietrich, M. (2) 89
Dietzsch, W. (1) 427, 456, 457, 564, 565
DiFeo, T.F. (2) 117
Ding, H. (2) 292
Ding, X.-Q. (1) 130, 134; (2) 126
Dinse, K.P. (1) 295
Diplock, A.T. (5) 217
Dismukes, G.C. (2) 335, 348, 349, 358
Dmitriev, A.V. (1) 101
Dobbert, O. (1) 295
Dodsworth, E.S. (1) 129
Doetschman, D.C. (1) 312; (2) 321
Domenech, M.V. (1) 168
Donlevy, T.M. (1) 179
Donnerberg, H. (1) 658
Dooley, D.M. (2) 33, 49, 50
Doppelt, P. (1) 452
Dordick, J.S. (2) 130
Douce, R. (2) 168
Drabent, K. (1) 470
Drago, R.S. (1) 471
Drapier, J.-C. (2) 197
Driessen, W.L. (1) 206
Drumheller, J.E. (1) 235; (5) 144
Du, M.L. (1) 588
Ducruet, J.-M.R.C. (2) 332
Duffy, N.V. (1) 456, 457
Duine, J.A. (2) 47
Dujaux, M. (1) 363
Dulcic, A. (1) 290, 291, 357
Dumont, M.F. (1) 59
Dunbar, J.B. (2) 174
Dunbar, K.R. (1) 141, 186, 484

Dunham, W.R. (1) 663; (2) 64, 67, 174, 175, 221, 340
Dunn, B. (1) 518
Dunn, J.L. (1) 107
Dunn, M.F. (2) 15, 57
Dupois, M. (5) 199
Duret, D. (1) 19
Durocq, C. (2) 199
Durosay, P. (2) 61
Duroy, H. (1) 582
Dusausoy, Y. (1) 646
Dutel, J.F. (1) 363
Dutta, S. (1) 601
Dutton, P.L. (2) 291, 292
Dye, J.L. (5) 42
Dyer, A. (1) 393
Dyer, R.B. (2) 138
Dyrek, K. (1) 41, 421, 423

Eachus, R.S. (5) 35-38, 67, 68
Eady, R.R. (2) 232
Eagle, A.A. (1) 145
Earl, E. (5) 9, 77
Eaton, G.R. (1) 7, 24
Eaton, S.S. (1) 7, 24
Ebihara, M. (1) 211
Ebisu, H. (1) 649, 650; (3) 100
Edmonds, D.T. (3) 60
Edmondson, D.E. (2) 71, 188
Edwards, G.J. (1) 366, 399
Edwards, P.P. (1) 309, 320, 325; (5) 57
Effelman, E.S. (1) 404
Efimov, V.N. (5) 63
Egawa, Y. (5) 342
Egeberg, K.D. (2) 110
Ehara, S. (1) 356
Ehrenberg, A. (2) 219
Einarsdóttir, O. (2) 138
Eisenegger, J. (5) 20
Eisenstein, J.C. (3) 138
Eivazov, E.A. (1) 272
Ekbote, S.N. (1) 300
El Ajouz, N. (2) 114
El-Deeb, M.K. (2) 16, 314
Elgren, T.E. (2) 73
Elias, H. (1) 558
Eling, T.E. (2) 151, 154
Ellaboudy, A.S. (5) 42
Ellerman, J. (2) 216
Elliott, R.J. (3) 33
El-Maradne, A.A. (1) 119, 120
El-Sayed, M.A. (1) 119, 120, 207
Elschner, B. (3) 90
El-Sherbini, S. (2) 32
El-Toukhy, A. (1) 119, 120

Author Index

Emanuelsson, P. (1) 280, 610, 621, 622
Emery, J. (3) 112, 113
Enemark, J.H. (1) 415, 416
Engels, B. (5) 94
Eremin, M.V. (3) 50, 54, 132
Erkelens, W.A.C. (1) 216
Ernst, R.R. (1) 29, 32
Escriva Monto, E. (1) 169, 238
Escuer, A. (1) 487
Espenson, J.H. (1) 469
Evans, M.C.W. (1) 10; (2) 317, 322, 352, 371
Evans, R.W. (2) 97
Even, R. (1) 296, 298

Fabiane, M.S. (1) 225, 233
Fabre, F. (3) 88
Fait, J.F. (1) 150
Fajer, J. (1) 503
Fajer, P.G. (4) 24
Falin, M.L. (1) 99, 256; (3) 50, 54, 74
Faller, J.W. (1) 157
Fan, C. (2) 138, 161, 162, 169, 203, 212
Fanciulli, M. (5) 179
Fargin, E. (1) 581
Farkas, E. (1) 543
Farnum, M.F. (2) 127
Farrar, J.A. (2) 50
Farrell, R.D. (1) 407
Faucitano, A. (5) 232
Faulmann, C. (1) 487
Fauque, G. (2) 178, 191, 207
Faure, D. (1) 397
Faure, P. (5) 151
Faus, J. (1) 173
Favier, A. (5) 151
Fayet, J.C. (1) 80; (3) 71, 106, 107, 112, 113
Fayet-Bonnel, M. (3) 71, 106
Fecke, W. (2) 275
Fedorov, Ya.K. (3) 126
Fedorushkova, E.B. (1) 525
Fee, J.A. (2) 138, 174, 253
Feezel, L.L. (2) 362
Feiler, U. (2) 312, 370, 373
Feller, D. (5) 8, 79, 139
Feltz, A. (1) 370
Feng, H. (1) 655
Feng, X. (1) 656-658
Ferguson-Miller, S. (2) 301
Fernando, Q. (1) 239
Ferrari, R.P. (1) 385; (2) 250
Ferraris, G. (1) 402
Ferraris, M. (5) 82

Ferraro, F. (1) 137
Fessenden, R.W. (5) 32, 33
Feuerstein, J. (2) 269
Fey, F. (2) 265
Field, R.W. (5) 73
Fields, R.A. (3) 80
Figgis, B.N. (1) 51
Finazzi-Agrò, A. (2) 32
Findsen, E. (1) 397
Finel, M. (2) 307
Finkel'stein, A.M. (1) 354
Finnen, D.C. (1) 663
Fiol, J.J. (1) 597
Fischer, J. (1) 130
Fish, K.M. (2) 221
Fishel, L.A. (2) 127
Fisher, D.S. (1) 284
Fisher, J. (1) 452
Fisher, K. (2) 236
Fisher, M.P.A. (1) 284
Fisk, Z. (1) 323
Flanagan, H.L. (5) 88
Flandrois, S. (1) 253-255
Flassbeck, C. (1) 194
Flatmark, T. (2) 59, 60
Fleischhauer, P. (1) 209
Flores-Llamas, H. (1) 67, 68
Fockele, M. (5) 63, 205, 208
Foldvari, I. (5) 202
Folgado, J.V. (1) 104, 559
Folting, K. (1) 203
Fontecave, M. (2) 74
Forbes, M.D. (1) 23
Forman, A. (1) 503
Formicka, G. (2) 265
Formoso, V. (1) 394
Forni, L. (5) 134
Forrer, J. (1) 31; (5) 20
Fortier, S. (5) 231
Foster, D.L. (3) 53
Fournes, L. (1) 496
Fourquet, J.L. (1) 582
Fox, B.G. (1) 197; (2) 70, 82-84
Francangeli, O. (1) 394
France, P.W. (3) 114
Francis, A.H. (1) 618
Franco, R. (2) 208
Franco, W.O. (1) 481
Franconi, C. (1) 26
Franzen, M.M. (1) 473, 502
Frediani, P. (1) 123
Freed, J.H. (1) 16; (5) 17, 18
Freeman, A.J. (3) 5, 44
Freire, C. (1) 486
Frey, P.A. (2) 189, 190
Fricke, R. (1) 660
Fridén, H. (2) 140
Friedrich, T. (2) 275

Frolik, C.A. (2) 69
Fromme, P. (2) 368
Froncisz, W. (1) 22, 27
Fu, G. (1) 568
Fu, W. (2) 170
Fu, Y. (1) 568
Fueki, K. (5) 46
Fuji, H. (2) 108
Fujii, S. (1) 662; (2) 141
Fujii, T. (2) 98
Fujimoto, Y. (2) 137
Fujita, T. (1) 331
Fujiyasu, H. (1) 279, 628
Fukudome, A. (2) 29
Fukui, K. (1) 607; (2) 180
Fukui, M. (1) 331
Fukumori, Y. (2) 136
Fukuoka, N. (5) 203
Fukuzawa, K. (2) 98
Funk, M.O., jun. (1) 663; (2) 64, 67
Funken, K. (5) 94
Furdyna, J.K. (1) 624
Furenlid, L.R. (1) 503
Furniss, D. (3) 123
Furrer, R. (3) 135
Furusawa, M. (1) 34, 35, 422; (5) 15, 180

Gabe, E.J. (1) 177
Gabriel, N.E. (2) 305
Gacheru, S.N. (2) 34
Gadsby, P.M. (2) 101, 146
Gahan, L.R. (1) 179, 180
Gaillard, B. (3) 92
Gaillard, J. (2) 168
Gaite, J.M. (1) 646
Galiazzo, F. (2) 259
Galindo, S. (1) 67, 69
Galtieri, A. (2) 53
Galuppi, P. (1) 26
Gamari-Searle, H. (1) 341, 344; (3) 96, 97
Gamliel, D. (5) 17, 18
Gammon, D. (5) 209
Ganapathi, L. (1) 350
Ganghi, N. (5) 155
Ganguli, A.K. (1) 350
Gannett, P.M. (2) 264
Garcia, A. (1) 168
Garcia Herbosa, G. (4) 22
Garcia-Lozano, J. (1) 169, 238
Garcia-Mora, J. (2) 218
Garifullin, I.A. (3) 98
Garif'yanov, N.N. (3) 98
Garrett, R.C. (2) 97
Garrone, E. (5) 82

Gatteschi, D. (1) 55, 113-116, 137, 138, 183, 213, 242, 243, 259, 260, 394, 566; (5) 3
Gau, T.S. (1) 505
Gaudemer, A. (1) 131
Gauduel, Y. (5) 96
Gayda, J.-P. (2) 295
Gazzoli, D. (1) 403
Gebhard, M.S. (2) 63
Geck, M. (1) 578
Gehlhoff, W. (1) 111, 112, 280, 610, 621, 622
Gehring, S. (1) 209
Geick, R. (1) 245
Geifman, I.N. (3) 19
Geiger, W.E. (4) 5
Gelerinter, E. (1) 456, 457
Gelin, P. (1) 363
Gelsomini Giusti, J. (1) 555
Gemperle, C. (1) 14, 29
Gennis, R.B. (2) 287, 305, 307, 308
Geoffroy, M. (5) 133
George, S.J. (2) 157, 163
Georgiou, C.D. (2) 305
Gerardi, G.J. (1) 332
Gerber, A. (1) 109
Gerez, C. (2) 74
Gerfen, G.J. (2) 92; (5) 142
Geschwind, S. (1) 223
Ghanotakis, D.F. (2) 316, 356
Gharbage, S. (1) 420
Ghedini, M. (1) 394, 563
Ghiotti, G. (1) 402
Ghosh, S. (1) 569
Ghoussoub, D. (1) 523
Giamarchi, T. (1) 224
Giamello, E. (1) 6, 390, 391; (5) 82
Giardina, B. (2) 115, 116
Gibson, J.F. (5) 27
Gilbart, P. (5) 208
Gilliam, O.R. (1) 399, 584; (5) 202
Gimarc, B.M. (5) 167
Gingras, G. (2) 324
Ginley, D.S. (1) 332
Ginns, I.S. (5) 66, 96
Giordano, G. (2) 227
Giori, D.C. (1) 349
Gippius, A.A. (1) 110
Giraudon, J.M. (1) 418
Gismelseed, A. (1) 130, 452
Gladney, H.M. (4) 1
Glarum, S.H. (1) 223
Glasbeek, M. (1) 108
Glerup, J. (1) 151, 153
Glinchuk, M.D. (1) 644

Gochev, G. (1) 550
Godbout, J.T. (5) 106
Goddard, W.A., II (5) 72
Godfrey, M.J. (5) 96
Godlewski, M. (1) 625
Goerger, A. (1) 278
Goff, H.M. (1) 446
Golbeck, J.H. (2) 315, 366, 367, 369
Gold, A. (1) 134; (2) 126
Goldfarb, D. (1) 639
Goldman, M. (3) 21
Golic, L. (1) 427, 430, 564
Gómez-Garcia, C.J. (1) 205, 242, 243
Gómez-Romero, P. (1) 413
González-Tovany, L. (1) 67, 69, 72
Good, M. (2) 266
Goodgame, D.M.L. (1) 409, 597
Goodgame, M. (1) 637
Goodson, P.A. (1) 152, 153
Goodwin, J.A. (1) 444
Goody, R.S. (2) 269
Goovaerts, E. (5) 62, 64, 65, 75
Gorban, S.V. (5) 178
Gorst, C.M. (2) 212, 213
Goslar, J. (1) 240
Goswami, S. (1) 458
Goswitz, V.C. (2) 305
Goto, T. (1) 217
Gounaris, K. (2) 333
Gourier, D. (1) 368, 369, 518
Grace, D. (2) 173
Grachev, V.G. (3) 104
Gradshteyn, I.S. (4) 11
Graeslund, A. (2) 75
Graham, L.A. (2) 284
Graham, S. (1) 474
Graham, W.R.M. (5) 164
Gralla, E.B. (2) 20, 21
Grand, A. (1) 52
Grandinetti, F. (5) 110
Grantz, M. (5) 112
Grasshoff, P. (1) 28
Gray, H.B. (1) 434
Grayevski, A. (1) 289
Graziani, M.T. (2) 37
Graziani, R. (1) 596
Grebenko, A.I. (2) 255
Green, I.G. (5) 124
Green, J. (2) 80
Greenway, F.T. (2) 34
Greenwood, C. (2) 146
Greenwood, R.J. (1) 414; (2) 220
Gregorkiewicz, T. (1) 625
Gregory, B.W. (5) 100, 138

Greiner, S.P. (1) 553
Greneche, J.M. (1) 162
Gribnau, M.C.M. (1) 59, 60, 431
Grieb, T. (1) 245
Gries, B.L. (1) 249
Griffith, W.P. (1) 435
Griffiths, J.H.E. (3) 32
Grimblot, J. (1) 519
Grimmeiss, H.G. (1) 280, 610
Grinberg, E.S. (5) 63
Griscom, D.L. (3) 119; (4) 7; (5) 196-198
Groenen, E.J.J. (1) 38; (5) 152
Gromovoj, Yu.S. (1) 623
Gros, F. (2) 60
Grover, T.A. (2) 248
Guarini, G.G.T. (1) 555
Guelton, M. (1) 519-524
Guener, S. (2) 282
Guenzburger, D. (1) 49
Guerchais, J.E. (1) 418
Guerrisi, M. (1) 26
Guiglarelli, B. (2) 204, 218, 227, 295
Guilard, R. (1) 397, 661
Guillot, M. (1) 213
Guissani, A. (2) 61, 199
Gullotti, M. (1) 542; (2) 35
Gultneh, Y. (1) 200; (2) 359
Gundu Rao, T.K. (1) 302
Gunsalus, R.P. (2) 183-185
Gupta, G.P. (1) 449, 450
Gupta, L.C. (3) 91
Gupta, N.S. (1) 455
Gurbiel, R.J. (2) 138, 286
Gurtler, P. (5) 43
Guscos, N. (3) 24, 96, 97
Guskos, N. (1) 339, 341, 344
Guzzi, R. (2) 12

Haasnoot, J.G. (1) 144
Haavik, J. (2) 59, 60
Haddad, M. (1) 63, 420
Haddon, R.C. (5) 166
Haddy, A. (2) 340
Haedicke, U. (1) 660
Haefner, S.C. (1) 484
Hägerhäll, C. (2) 279
Haga, M. (1) 124, 129
Haga, S. (1) 192
Hage, J.P. (1) 471
Hagen, W.R. (2) 174, 175, 192, 209, 214, 239
Hagino, T. (2) 29
Hagiwara, M. (1) 220-222
Hagiwara, T. (5) 220

Author Index

Haladjian, J. (2) 143
Haldane, F.D.M. (1) 214
Hallahan, B.J. (2) 317, 323, 352
Hallewell, R.A. (2) 19
Halliburton, L.E. (1) 366, 654
Halperin, B.I. (1) 220
Halpern, H.J. (5) 26
Haltia, T. (2) 47, 294, 302
Haltiwanger, R.C. (2) 9
Halut-Desporte, S. (1) 261
Hamada, A. (1) 384
Hambley, T.W. (1) 179, 441, 598
Hampson, C.A. (5) 168
Hamrick, Y.M. (5) 86, 87, 90, 103, 106, 113, 114, 170
Han, B.C. (1) 473, 502
Han, C.S. (1) 654
Han, H. (2) 272
Han, J. (2) 14
Han, P. (5) 30, 48
Han, S. (1) 655
Hanaki, A. (5) 223
Handa, M. (1) 257
Hang, H.G. (1) 548
Hangleiter, T. (5) 35, 36, 67, 68, 207
Hanley, J.A. (2) 371
Hanna, P.M. (2) 91-93
Hansen, S. (1) 651
Hanson, G.R. (1) 179, 548; (2) 309
Hantehzadeh, M.R. (1) 654
Harada, N. (2) 104
Harding, C. (1) 146
Harpel, M.R. (2) 70
Harriman, J.E. (4) 10
Harris, E.A. (3) 123
Harrison, N. (5) 83
Hasegawa, A. (5) 165
Hasegawa, E. (1) 475
Hashimoto, D. (1) 364, 365
Hashizume, N. (1) 356
Hasnain, S.S. (2) 97
Hatano, K. (1) 451
Hatchikian, E.C. (2) 157, 163, 164
Hatfield, W.E. (1) 143, 554; (2) 126
Hauska, G. (2) 361
Hayashi, Y. (1) 331
Hayes, W. (5) 187
He, L.P. (1) 147
Heathcote, P. (2) 371
Hedderich, R. (2) 195
Hederstedt, L. (2) 140, 279
Hedman, B. (2) 80
Heidler, R. (1) 589

Heinz, B. (1) 150
Helm, H. (5) 91
Hendorfer, G. (1) 37, 276
Hendrich, M.P. (1) 193, 195, 197-199; (2) 73, 78, 79, 82, 83, 86
Hendrickson, D.N. (1) 158, 176, 203
Henke, W. (1) 104
Henriques, R.T. (1) 266
Henry, A. (5) 182
Henry, Y. (2) 197, 199
Hensel, K. (5) 231
Herbst-Irmer, R. (1) 417
Herlach, F. (1) 18
Herlong, J.O. (5) 77, 78
Hermes, C. (2) 99
Herring, F.G. (1) 570
Herve, G. (2) 114
Hessler, J.P. (3) 133
Heynderickx, I. (5) 62, 75, 76
Hiczer, W. (1) 241
Hijmans, T.W. (1) 40; (5) 21
Hikita, H. (1) 583
Hilczer, W. (1) 240, 306
Hill, D.W. (5) 77, 85, 105, 138
Hill, J.J. (2) 305
Hill, N.J. (3) 85
Hille, R. (2) 187
Hiller, W. (2) 23
Hipkin, C.R. (2) 226
Hirabayashi, I. (1) 305
Hirai, S. (2) 29
Hiraku, T. (5) 46
Hirose, T. (1) 645
Hirota, N. (1) 607, 662
Hirsh, D.J. (2) 319
Histed, M. (1) 572; (5) 52-54, 168
Hitchman, M.A. (1) 442
Hock, K.-H. (3) 90
Hodgson, D.J. (1) 151-153
Hodgson, K.O. (2) 80
Höhne, M. (1) 111, 112
Hoffman, B.M. (2) 8, 138, 161, 162, 166, 167, 169, 203, 212, 286
Hoffman, K.R. (1) 576
Hoffman, R.A. (2) 105
Hoffmann, S.K. (1) 75, 240, 241
Hoitink, C.W.G. (2) 7, 11
Hollenstein, R. (2) 266
Hollocher, T.C. (2) 51, 52
Holm, R.H. (1) 161, 489
Holman, T.R. (1) 193, 195, 198, 199; (2) 78, 79
Holmes, S. (2) 246
Honda, M. (1) 301; (5) 203

Hoogland, H. (2) 131
Horai, K. (1) 630; (3) 100
Hori, H. (1) 217; (2) 107, 109, 137, 260
Hornack, J.P. (1) 20
Horvath, L.I. (1) 76
Hoshino, M. (1) 182
Hoshino, N. (1) 395
Hosono, H. (1) 362
Hou, B.H. (1) 351
Houlier, B. (3) 111
Houseman, A.L.P. (2) 162, 166
How, H. (1) 321, 336
Howard, J.A. (1) 571, 572; (5) 50-55, 57-61, 104, 168
Howell, M.L. (2) 77
Howes, B.D. (2) 219
Hoyer, E. (1) 565
Hrabanski, R. (1) 78, 88, 619
Hsyu, Y.J. (5) 213
Huang, M.X. (3) 91
Huang, Y.H. (1) 277
Huang, Y.S. (1) 273
Hubert, A.S. (3) 112
Hubert, S. (3) 113
Hudd, R.J. (1) 405
Hüttermann, J. (1) 61; (2) 135, 244, 265
Huff, A.M. (2) 155
Huffman, J.C. (1) 158, 203
Hughes, D.L. (1) 149, 577
Hughes, R.K. (2) 222
Hui-Su, W. (1) 46
Hulsberger, F.B. (1) 477
Humblet, C. (2) 174
Hundley, M.F. (1) 323
Hunger, J. (5) 159
Hurrell, J.P. (3) 68
Hurst, J.K. (2) 280
Huse, D.A. (1) 284
Hussain, A. (1) 372
Hussain, I. (1) 597, 637
Hutchison, C.A., jun. (3) 34, 57, 78-80, 115-117, 133-137, 140
Huttner, G. (1) 412
Hutton, S.L. (1) 235
Huynh, B.H. (1) 445; (2) 71, 147, 159, 176, 179, 203, 207
Hwang, G. (1) 570
Hwang, H.L. (5) 213
Hwang, J.Y. (1) 21
Hyde, J.S. (1) 22, 27, 66; (3) 70
Hyman, M.R. (2) 235

Iamamoto, Y. (1) 136
Ianelli, S. (1) 170

Ibric, S. (1) 594
Ichikawa, Y. (2) 137
Iimura, Y. (1) 182
Iismaa, S.E. (2) 172
Iizuka, T. (2) 141
Ikeda-Saito, M. (2) 107, 109, 111, 129, 132
Ikenoue, S. (1) 257
Ikeya, M. (1) 34, 35, 422; (5) 15, 175, 180
Imhof, W. (1) 412
Impellizzeri, G. (1) 542
Inaba, S. (1) 447
Inami, T. (1) 217
Indovina, P.L. (5) 25
Indovina, V. (1) 402, 403
Ingemarson, R. (2) 75
Ingle, R.T. (2) 301
Ingledew, W.J. (2) 306
Ingram, D.J.E. (5) 27
Innes, J.B. (2) 319
Inoue, H. (1) 475
Inoue, M. (1) 239
Inoue, M.B. (1) 239
Inoue, Y. (2) 343-347
Ioannidis, N. (2) 299
Iosphe, D.M. (1) 274
Iqbal, Z. (1) 327, 330
Irkhin, V.Yu. (1) 358
Irmscher, K. (1) 622
Isaacson, R.A. (3) 23
Ishchenko, S.S. (3) 104
Ishida, T. (1) 310
Ishii, T. (1) 237
Ishizu, K. (1) 448; (2) 113
Ismael, K.Z. (1) 119, 120
Isobe, K. (1) 124, 211
Isoya, J. (1) 364, 365, 500; (5) 189
Iton, L.E. (3) 118
Ivanenko, I. (3) 26
Iwahashi, K. (1) 331
Iwaizumi, M. (1) 133, 175, 534, 536; (2) 10
Iwasaki, H. (2) 144
Iwasaki, M. (5) 117
Iyer, R.M. (1) 324
Izotov, V.V. (1) 99, 256
Izquierdo-Ridorsa, A. (1) 560

Jacobsen, J.H. (1) 599
Jacobsen, S.M. (1) 576
Jacobson, R.J. (1) 557
Jacox, M.E. (5) 107
Jaeger, T.J. (5) 231
Jain, V.K. (1) 96, 499, 611
Jakob, W. (2) 123

Jakobs, C. (1) 28
Janda, P. (1) 165
Janes, R. (1) 309, 320, 325, 326; (5) 137
Jang, J.H. (5) 70
Jang, M.S. (1) 616
Janossy, A. (3) 93
Jansen, F. (5) 187
Jansen, S. (1) 392
Janssen, P. (3) 35, 84
Janssen, R.A.J. (5) 128, 129, 131, 140, 141, 157, 161
Jantsch, W. (1) 37, 276
Janzen, C.P. (5) 230
Janzen, E.G. (5) 182, 221, 224
Jaquinot, J.F. (3) 21
Jayaraj, K. (1) 134, 568; (2) 126
Jeannin, S. (1) 261
Jeannin, Y. (1) 261
Jeevarajan, A.S. (5) 32, 33
Jeffries, C.D. (3) 15
Jegerschoeld, C. (2) 318
Jenkins, A.A. (1) 647; (3) 143, 147
Jensen, G.M. (2) 172
Jérome, D. (1) 266
Jesmanowicz, A. (1) 66
Jetten, M.S.M. (2) 209, 214
Jeunet, A. (5) 151
Jezierska, J. (1) 561
Jezowska-Trzebiatowska, B. (1) 561
Jha, N.K. (1) 455
Ji, J.Y. (1) 405-407
Jiang, F. (2) 24, 68
Jiang, J. (2) 303
Jiang, Q.J. (5) 164
Jiang, X. (1) 656
Jiménez-Domínguez, H. (1) 67-69
Jin, H. (2) 58
Jin, S. (1) 345
Jin, S.Z. (1) 351
Jing-Fen, L. (2) 121
Jinno, J. (1) 448; (2) 113
Johnson, K.M. (5) 111, 124
Johnson, M.K. (2) 170, 171, 183, 184, 185, 217
Jollie, D.R. (2) 231
Joly, H.A. (1) 571, 572; (5) 50-52, 54, 55, 57-61, 104
Jones, R. (1) 572; (5) 53, 54
Jordan, M. (5) 34, 207
Jordanov, J. (1) 162, 163
Journaux, Y. (1) 204, 208
Juarez-Garcia, C. (1) 198, 199; (2) 78
Judd, B.R. (3) 65

Julve, M. (1) 173, 208
Jun-Kai, X. (1) 45, 47
Junod, A. (1) 289
Jupina, M.A. (5) 194
Juppe, S. (1) 609
Juszczak, A. (2) 196

Kabayashi, T. (1) 556
Kadam, R.M. (1) 324
Kadir, F.H.A. (2) 150
Kadish, K.M. (1) 118, 397, 472, 473, 502, 661
Kadkhodayan, S. (2) 155
Kagen, H.M. (2) 34
Kahn, O. (1) 139, 204, 261
Kahol, P.K. (1) 282, 288; (5) 143
Kahziri, M. (1) 91, 340
Kaise, M. (1) 352
Kajiwara, A. (1) 411
Kalbitzer, H.R. (2) 268, 269
Kalinichenko, N.B. (1) 425, 426, 429
Kalyanaraman, B. (2) 26, 276
Kamenskiy, Yu.A. (1) 142
Kaminskii, A.A. (3) 4
Kanazawa, D. (1) 269
Kanda, H. (1) 500
Kanegasaki, S. (2) 141
Kang, B. (1) 164
Kang, L. (2) 191
Kang-Wei, Z. (1) 45, 47
Kanoda, K. (1) 310
Kanters, J.A. (5) 157
Kao, C.T. (5) 214
Kaplan, M.L. (1) 223
Kapoor, V. (1) 96
Kappers, L.A. (1) 399, 584; (5) 202
Kappl, R. (2) 135, 244
Karhunen, E. (2) 47
Karim, R. (1) 321, 336
Karlin, K.D. (1) 557
Karlstrom, G. (5) 44
Karpukhina, T.A. (5) 178
Karthein, R. (1) 594; (2) 152
Kasamatsu, K. (1) 257
Kasprzak, J. (5) 121
Kasuga, K. (1) 257
Kasuya, M. (5) 15
Kataev, V.E. (1) 354
Kataoka, S. (1) 356
Kato, T. (5) 118
Katoh, A. (5) 117
Katsnelson, M.I. (1) 358
Katsuda, H. (1) 89
Katsumata, K. (1) 217, 220-222,

Author Index

248
Katti, K.V. (1) 398, 417
Kau, D. (2) 226
Kauffman, G.B. (1) 537-539
Kawada, Y. (5) 126
Kawamori, A. (2) 320
Kawano, K. (1) 636; (2) 29
Kawata, S. (1) 175
Kay, C.J. (2) 224-226, 230
Kear, J. (2) 371
Keeble, D.J. (1) 332
Keene, N. (5) 28
Keijzers, C.P. (1) 54, 59, 431
Keizer, P.N. (1) 42
Keller, R.J. (2) 248
Kellner, E. (2) 360
Kempinski, W. (1) 348
Keneda, M. (5) 193
Kennedy, M.C. (2) 161, 162, 166, 167, 186
Kennedy, T.A. (5) 209
Kera, Y. (1) 422
Kernisant, K. (4) 12
Kernizan, C.F. (5) 74
Kerr, K. (5) 97
Kevan, L. (1) 210, 299, 315, 316, 318, 319, 322, 328, 465, 466, 483, 497, 498, 511-517, 573, 574, 638, 640; (4) 8; (5) 56, 215, 216
Khalilev, V.D. (1) 525
Khaliullin, G.G. (3) 98
Khan, M.M.T. (1) 462-464
Khan, N.H. (1) 462, 463, 464
Khangulov, S.V. (2) 255, 256
Khasanov, R.I. (1) 335
Khisamov, B.A. (5) 125
Khylboy, E.P. (3) 98
Kieslich, A. (1) 245
Kikuchi, H. (1) 217; (3) 94
Kim, B.F. (1) 285
Kim, D.H. (2) 337, 338
Kim, H.C. (1) 360
Kim, H.S. (5) 70
Kim, K.S. (5) 70
Kim, Y. (2) 131
Kim, Y.I. (1) 143
Kim, Y.O. (1) 446
Kime-Hunt, E. (2) 246
Kimura, H. (1) 315
Kimura, M. (2) 29
Kimura, S. (2) 129
Kimura, Y. (1) 583
Kindo, K. (1) 218, 219, 301
Kinoshida, M. (1) 268
Kintner, E.T. (1) 453
Kirilovsky, D.L. (2) 332
Kirk, M.L. (1) 554; (2) 126

Kirk, T.J. (5) 78, 93, 105
Kirk, W.P. (1) 249
Kirmse, R. (1) 424-427, 429-433, 457, 564, 565
Kitagawa, H. (1) 166
Kitai, T. (1) 89
Kito, M. (2) 29
Kjøsen, H. (2) 100
Kleeman, W. (1) 644
Klein, M.P. (2) 288, 334, 337-339
Kliava, J. (3) 122; (4) 7
Klimm, C. (1) 589
Klimov, A.A. (3) 104
Klimov, V.V. (2) 325, 329
Klipp, W. (2) 233
Knaff, D.B. (2) 282, 288
Knight, L.B. (5) 6-9, 77, 78, 85, 93, 97, 100, 105, 138, 139
Knoch, F. (1) 578
Knoch, R. (1) 558
Knüttel, K. (2) 239
Ko, J. (5) 233
Kobayashi, N. (1) 165
Kobayashi, T. (1) 89
Koch, J. (2) 195
Koch, T.H. (2) 9
Kochelaev, B.I. (3) 98, 132
Kodama, A. (1) 395
Kodera, Y. (2) 320
Koehler, J. (1) 74
Köhler, K. (1) 424-427, 429-432, 564, 565
Köngeter, D. (1) 266
Koga, K. (1) 310
Kohashi, T. (1) 301
Kohzuma, T. (2) 144
Kojima, K. (1) 605
Kojima, N. (1) 166, 270, 607
Koka, S. (1) 334
Koksharov, Yu.A. (1) 110
Kolaczkowski, S.V. (2) 313
Koley, A.P. (1) 569
Kolodziej, A.F. (2) 51, 52
Kolopus, L.A. (3) 100
Kolpakova, N.N. (1) 95
Komatsu, T. (1) 475
Kondo, M. (5) 188
Kondoh, A. (1) 294
Kondoh, H. (1) 352
Konishi, S. (1) 182
Konkin, A.L. (3) 50
Konstantinov, A.A. (2) 142
Konstantinov, V.N. (4) 4
Kool, T.W. (1) 108
Koppenol, W.H. (2) 254
Korczak, S.Z. (1) 631
Korecz, L. (3) 93

Korovina, L.A. (1) 623
Korp, J.D. (1) 147
Korytowski, W. (2) 28
Kosaka, H. (2) 104
Koschnik, F.K. (5) 12, 35-38, 67, 68
Koshta, A.A. (1) 292, 358
Kosuge, K. (3) 94
Kotake, Y. (5) 224
Kotlinski, J. (1) 91, 340
Kotlyar, A.B. (2) 277
Koufoudakis, A. (1) 341, 344; (3) 96, 97
Kovalyuk, Z.D. (3) 104
Kowal, A.T. (2) 170
Kozlova, I.V. (3) 19
Kraatz, H.B. (1) 150
Kraft, K.A. (1) 553
Krambrock, K. (5) 205, 206
Krasil'nikova, N.A. (1) 525
Kraus, D.W. (2) 121
Kraut, J. (2) 127, 128
Kreilick, R.W. (1) 553; (4) 13
Kreissl, J. (1) 610, 621, 622
Kreiter, A. (1) 61
Kriauciunas, A. (2) 69, 70, 288
Krishna, R.M. (1) 507, 509
Kriz, H.M. (4) 2
Krogh-Jesperson, K. (5) 159
Kroh, J. (5) 31
Kroneck, P.M.H. (2) 48, 66, 123, 186
Krueger, H.-J. (1) 489
Krupski, M. (1) 240; (3) 27, 28
Krusic, P.J. (5) 230
Krygin, I.M. (3) 130, 131
Krzyminiewski, R. (5) 121
Kubota, A. (5) 117
Kukkadapu, R.K. (1) 497, 498
Kukovitski, E.F. (1) 335, 354
Kulawiec, R.J. (1) 202
Kulmacz, R.J. (2) 151, 153
Kumar, S.B. (1) 126, 410
Kumeda, M. (1) 359
Kupperman, A. (5) 92
Kuppusamy, P. (2) 118
Kuranova, G.A. (5) 169
Kureshy, R.I. (1) 462-464
Kuriata, J. (1) 99, 256; (3) 24, 25, 28
Kurmoo, M. (1) 269
Kuroda, N. (1) 262
Kurtz, D.M., jun. (2) 87
Kurzak, B. (1) 543
Kurzawa, M. (1) 99
Kusunoki, M. (1) 411; (2) 346
Kutoglu, A. (1) 559
Kutsumitzu, S. (1) 270

Kuwata, K. (1) 422; (5) 81
Kwei, G.H. (1) 304
Kwik, W.L. (1) 548

Labanowska, M. (1) 421
Lacour, C. (1) 298
Lacroix, P. (1) 139
Lagutin, A.S. (1) 101
Lah, M.S. (1) 602
Lahiri, G.K. (1) 127, 128, 458-460
Lakshman, S.V.J. (1) 373, 374, 507-509, 526, 527, 633; (3) 120
Lakshmana Rao, J. (3) 120
Lakshmi, P.S. (3) 121
Lakshmi Reddy, S. (1) 648
Lam, H. (1) 165
Lampreia, J. (2) 178, 179
Lancaster, J.R., jun. (2) 105, 198
Landin, J.A. (2) 49
Lang, G. (2) 107
Lange, R. (2) 134
Langrehr, J.M. (2) 105
Lania, A. (2) 53
Lannoo, M. (5) 206
Lappe, J.J. (5) 204
Larroque, C. (2) 134
Larsen, R. (2) 297
Lasage, S. (2) 184
Laschi, F. (1) 123, 206, 478
Lassmann, G. (1) 28; (2) 154
Latour, J.M. (1) 176
Latwesen, D.G. (2) 258
Lauderdale, W.J. (5) 106
Lauraeus, M. (2) 294
Laurent, J.P. (1) 171, 454
Laurenti, E. (1) 385
Lavergne, J. (2) 351
Lawangar-Pawar, R.D. (1) 626
Lawrance, G.A. (1) 180, 441, 598
Lay, P.A. (1) 405-407
Lazuta, A.V. (1) 308
Lea, K.R. (3) 9
Leask, M.J.M. (3) 9, 34, 57, 116
Lebedev, Y.S. (1) 36
LeCalve, J. (5) 43
Lech, J. (1) 78, 619
Lecomte, C. (1) 661
LeDuc, M. (5) 59
Lee, C.W. (1) 148, 515, 516, 638, 640
Lee, E.Y.S. (3) 75
Lee, F.L. (1) 177

Lee, H.C. (2) 111, 132
Lee, H.K. (1) 360
Lee, J.C. (1) 147
Lee, K.C. (1) 570
Lee, K.M. (1) 223
Lee, Y.H. (1) 627
Lefebvre, R. (4) 23
LeGall, J. (2) 139, 147, 158, 159, 176, 179, 191, 203, 207
Legros, J.-P. (1) 169, 238, 265
Lehmann, G. (1) 611, 651
Lei, G.D. (1) 465, 466
Lei, X. (1) 164
Leif, H. (2) 275
Leigh, J.S. (2) 258
Leising, G. (1) 357
Leising, R.A. (1) 135
Lejus, A.M. (1) 369
Lelenfield, H.V. (5) 82
Lelj, F. (1) 121, 566, 567
Leluk, M. (1) 561
Lenahan, P.M. (5) 194
Lenfant, M. (2) 199
LePage, Y. (5) 231
Leroy, G. (2) 143
Lestienne, B. (1) 581
Levanyuk, A.P. (3) 19
Levasseur, A. (1) 420
Lever, A.B.P. (1) 125, 129, 165, 568
Levi, Z. (1) 639
Levstein, P.R. (1) 225, 226, 232
Levy, A. (2) 118
Levy, L.P. (1) 224
Lewis, W.B. (1) 592
Leznoff, C.C. (1) 165
Li, C.H. (1) 252
Li, F.Z. (1) 613
Li, G. (2) 371
Li, J.-B. (2) 40
Li, L. (1) 319
Li, N. (2) 366, 367
Li, Q. (1) 203
Li, S. (5) 87, 113, 114
Li, X. (1) 105, 381; (3) 29, 110
Li, Z. (2) 297
Li, Z.M. (1) 613
Liang, R. (1) 342, 351
Libby, E. (1) 203
Licci, F. (1) 349
Liebl, U. (2) 312, 360
Liebross, R.H. (5) 150
Lifshitz, E. (1) 618
Ligon, A.R. (5) 9
Lii, K.H. (1) 252
Likodimos, L. (3) 96
Lin, S.S. (1) 273, 277
Lindahl, P.A. (2) 210

Lindberg, K. (2) 355
Linde, M. (5) 207
Linder, D. (2) 195, 216
Lindley, P.F. (2) 97
Lindner, J. (1) 245
Lindsay, D.M. (4) 12; (5) 74
Lino, A.R. (2) 191
Lippard, S.J. (1) 196; (2) 80, 81
Lipscomb, J.D. (1) 197; (2) 70, 82-84, 231
Liu, E.D. (3) 133, 134
Liu, K.E. (2) 81
Liu, M.-Y. (2) 139, 147, 159, 176
Liu, Q. (1) 164
Liu, R.-S. (1) 320, 325, 326
Liu, S.C. (1) 196
Liu, W. (1) 129
Liu, X. (2) 372
Liu, Y. (2) 84, 127, 290
Liu, Y.H. (1) 397, 661
Ljones, T. (2) 173
Llopis, E. (1) 413
Lloret, F. (1) 173, 204, 208
Lodros, N. (1) 344
Loehr, T. (2) 72, 77
Loewen, P.C. (2) 102, 155
Logan, D.E. (5) 57
Lohse, F. (1) 11; (5) 13
Lommen, A. (2) 7
London, C.A. (3) 96
Long, G.T. (1) 606
Long, R.C. (2) 87
Lontie, R. (2) 56
López, F.J. (1) 575, 586, 614, 615
Lord, K.A. (2) 245
Loreti, P. (2) 37
Lorusso, M. (2) 283
Lott, K.A.K. (5) 146
Lou, S.H. (1) 92
Lowe, D.J. (1) 577; (2) 219, 236
Lowy, D. (1) 289
Lu, J. (1) 164; (2) 44
Lu, Y. (2) 20
Luca, V. (1) 497, 498, 517, 573, 574; (5) 56
Luchinat, C. (2) 19, 165
Ludowise, P. (1) 24
Lue, J.T. (1) 314
Lueck, R. (1) 50
Luft, H. (3) 89
Lund, A. (1) 73
Lunk, H.J. (1) 50
Luo, X.L. (1) 501
Lurie, D.J. (5) 23
Lurz, C.J. (1) 250

Luty, F. (5) 39
Lutz, R.S. (2) 109
Luz, Z. (4) 9
L'vov, S.G. (1) 335

Ma, L. (2) 217
Mabbs, F.E. (1) 150, 377, 415
McAuley, A. (1) 181, 479, 485, 490
McCausland, M.A.H. (3) 37
McCormick, J.M. (2) 85
McCracken, J. (2) 24, 33, 44, 223, 238, 257
McCurley, J.P. (2) 281
McCusker, J.K. (1) 176
McDaniel, M.L. (2) 198
McDermott, A.E. (2) 339
McDonald, P.F. (1) 249
McDowell, C.A. (1) 533
McDugle, W.G. (5) 38, 67, 68
McElheny, P.J. (5) 191
McGarvey, B.R. (3) 42, 43
McGregor, K. (1) 438
McGuirl, M.A. (2) 33
Macicek, J. (1) 550
McKay, D.B. (3) 78
McKee, V. (1) 146
McKelvey, E.J. (2) 109
MacKenzie, I.S. (3) 37
Mackey, M.F. (1) 145
McKinley, A.J. (5) 85, 97, 116
Maclachlan, D.J. (2) 321
McLauchlan, K.A. (5) 132
McManus, J.D. (2) 14
McMillin, D.R. (2) 40, 42, 44, 45
MacNeil, J.H. (5) 231
McPartlin, M. (1) 548
McPherson, G.L. (1) 579
Madej, A. (1) 41
Maeder, M. (1) 598
Maes, F. (5) 84, 108
Maes, G. (2) 56
Magliozzo, R.S. (2) 111, 112
Magruder, R.H., III (1) 362
Mahal, A.T. (1) 288
Mahl, T.A. (1) 297
Mahy, M. (3) 84
Maichle, C. (2) 23
Maiya, B.G. (1) 118
Makino, K. (5) 220
Maksumoto, T. (5) 81
Malatesta, F. (2) 295
Malatesta, V. (1) 535
Maletka, K. (1) 612
Malkin, B.Z. (3) 13, 26
Malkin, R. (2) 288

Mallard, J.R. (5) 23
Malmstrom, G. (2) 6
Mandal, S.K. (1) 177
Mandon, D. (1) 134, 452; (2) 126
Manfredini, T. (1) 555
Mangrich, A.S. (1) 383
Maniukiewicz, W. (1) 172
Mann, G.J. (2) 75
Manodori, A. (2) 183, 185
Manoharan, P.T. (1) 167, 569
Mar, T. (2) 324
Marchesini, A. (2) 35
Marchettini, N. (1) 552
Marco, S.D. (2) 55
Marco de Lucas, C. (1) 629
Marcos, M. (1) 396, 562
Maric, D.M. (5) 186
Markiewicz, A. (1) 179
Markley, J.L. (2) 162
Marnett, L.J. (2) 151, 154
Marnier, G. (1) 646
Maroney, M.J. (2) 201, 202
Marov, I.N. (1) 425, 426, 429
Marracone, G. (1) 542
Marrelli, S. (1) 319, 322
Marsh, D. (1) 76
Marshall, S.A, (1) 624
Martin, A. (1) 586
Martineau, P.M. (3) 34, 57, 64, 115, 117, 140, 147
Martínez, A. (2) 59
Martini, G. (1) 388
Martinis, S.A. (2) 110
Martino, D.M. (1) 231
Martinotti, F. (5) 232
Maruani, J. (1) 73; (4) 23
Maruyama, H. (1) 337
Mascharak, P.K. (1) 493, 504; (2) 202
Masiakowski, J.T. (1) 299, 315, 316, 328
Maskos, Z. (2) 254
Mason, J.R. (2) 177
Mason, R.P. (2) 154; (5) 219
Massey, V. (2) 221
Masuda, H. (1) 305
Masui, H. (1) 125
Masumura, M.S. (1) 62
Matauda, J. (1) 605
Mathews, C.K. (2) 77
Mathur, P. (1) 547
Matsubayashi, G. (1) 263, 264
Matsuda, A. (5) 189, 191
Matsuda, S. (2) 66
Matsumoto, M. (5) 193
Matsunaga, Y. (1) 395
Matsushida, N. (1) 166

Matsushita, M. (5) 118
Matsushita, T. (2) 346
Matsuura, K. (5) 47
Mattar, S.M. (5) 166
Mattera, R. (2) 109
Matthys, P. (5) 84, 108
Matzanke, B.F. (2) 99
Maurin, L. (2) 134
Mauro, J.M. (2) 127, 128
Maurya, R.C. (1) 436
May, H.D. (2) 237
Mazur, E. (1) 41
Mazúr, M. (1) 65
Meadows, K.A. (2) 42
Mednis, A. (3) 122
Meeker, A.K. (2) 273
Mehari, T. (2) 369
Mehran, F. (3) 55
Mei, R. (2) 357
Meier, P.F. (5) 186
Meijers, S.M. (1) 439
Meiklyar, V.P. (5) 50
Meilwes, N. (5) 184
Meinhardt, S.W. (2) 274, 293
Melandri, M.C. (1) 138
Meli, A. (1) 123
Melkonyan, V.Z. (2) 13
Melník, M. (1) 65, 174
Meloni, M.G.L. (2) 249
Merer, A.J. (5) 73
Mesman, J. (1) 163
Meyer, T.E. (2) 14
Mhin, B.-J. (5) 70
Micera, G. (2) 247
Michalik, R. (5) 173
Michalski, T.J. (2) 362
Michaut, J.P. (5) 119, 136
Michel, B.R. (2) 280
Michel, H. (2) 370
Michel, J. (1) 223
Michelsen, K. (1) 153
Michelson, A.M. (2) 61
Michiels, J.J.M. (1) 81
Midollini, S. (1) 160, 212
Mieher, R.L. (3) 75
Mielczarek, E.V. (2) 94
Migita, C.T. (5) 225
Migus, A. (5) 95
Miki, T. (2) 281, 289
Mikuriya, M. (1) 257
Mildvan, A.S. (2) 271-273
Mile, B. (1) 571; (5) 50-52, 55, 58-61, 104, 168
Miller, A.-F. (2) 310
Miller, M. (2) 372
Miller, M.A. (2) 127, 128
Miller, P. (5) 8
Mims, W.B. (2) 238; (3) 18

Minard, R.A. (3) 14
Minelli, M. (1) 412
Ming, L.-J. (2) 68, 69
Minge, J. (1) 652
Minghetti, K.C. (2) 305
Minichino, C. (1) 52
Minner, E. (1) 506
Minuto, M. (2) 283
Miraglio, R. (2) 295
Mishra, D.D. (1) 436
Mishra, S.P. (5) 165
Misiak, L.E. (3) 88, 110
Misra, H.P. (5) 226
Misra, S.K. (1) 91, 105, 106, 340, 375, 378-382, 631; (3) 29, 88, 110
Mistry, F. (1) 570
Mitani, T. (1) 251
Mitin, A.V. (3) 98
Mitra, S. (5) 192
Mitrofanov, Yu.E. (3) 74
Mitros, C. (1) 341, 344; (3) 96, 97
Miyajima, S. (1) 395
Miyamoto, R. (1) 534; (2) 10
Miyazaki, T. (5) 46
Miyazawa, C. (5) 117
Mizuno, F. (1) 305
Mizuno, M. (1) 352
Mizutani, T. (5) 200
Moens, P. (5) 108
Mohammed, M.Y. (1) 149, 577
Mohan, M. (1) 455; (2) 246
Moisy, P. (1) 131
Mombourquette, M.J. (1) 652
Momo, F. (5) 25
Momose, T. (5) 118, 138
Mondovi, B. (2) 31
Monemar, B. (5) 182
Monfort, M. (1) 168
Monnereau, O. (3) 92
Monnier, A. (1) 109
Monod, P. (1) 296, 298
Moody, A.J. (2) 299
Moore, M.W. (3) 114
Moorjani, K. (1) 285
Morazzoni, F. (1) 535; (5) 212
Mordvintcev, P. (2) 106; (5) 80
More, C. (2) 143, 227
Morelli, G. (1) 121, 566, 567
Moreno, M. (1) 4, 48, 629
Morgan, T.V. (1) 544; (2) 235, 238
Morie-Bebel, M.M. (2) 42
Morigaki, K. (5) 188
Morikis, D. (2) 110
Morikita, H. (5) 46
Morimoto, A. (1) 359; (5) 193

Morishima, I. (2) 108
Moroz, I.A. (2) 277
Morpurgo, L. (2) 36, 37
Morris, H. (1) 174; (5) 168
Morris, M. (1) 393
Morrone, S. (1) 394, 563
Mortenson, L.E. (2) 235, 238
Morton, J.R. (1) 42; (5) 231
Morvillo, A. (1) 546
Moshchalkov, V.V. (1) 110, 298
Moshkov, K.A. (2) 54
Mosina, L.V. (1) 184
Mosny, K. (1) 544
Moto, A. (1) 359
Motschi, H. (1) 594
Mouli, V.C. (1) 510
Moura, I. (2) 139, 147, 158, 159, 176, 178, 179, 191, 203, 207, 208
Moura, J.J.G. (2) 147, 158, 159, 176, 178, 179, 191, 203, 207, 208
Moussavi, M. (1) 19
Moustaka, T.D. (5) 179
Moya, E. (1) 575
Mrozinski, J. (1) 172
Müller, A. (2) 233, 239, 361
Müller, K.A. (1) 80, 98, 275; (3) 107
Mueller, L.J. (1) 21
Muelsch, A. (2) 106
Münck, E. (1) 161, 195, 197-199; (2) 70, 78, 79, 83
Münstermann, D. (2) 89
Münze, R. (1) 427
Mukai, K. (2) 98
Mukund, S. (2) 240, 241
Mulsch, A. (5) 80
Murakami, A. (5) 220
Murakami, M. (1) 294
Muralheedhran, K. (1) 302
Muramatsu, H. (1) 279, 628
Murase, N. (2) 105
Murieto, H.S. (3) 101
Murray, K.S. (1) 179
Musci, G. (2) 53, 55, 149
Mus-Veteau, I. (2) 218
Muto, H. (5) 47
Myers, S.M. (5) 195

Naessens, D. (5) 108
Nagarajan, R. (1) 350
Nagata, K. (1) 342
Nagata, T. (1) 475
Naidu, Y.N. (1) 508, 633
Naito, A. (1) 533
Nakagawa, M. (5) 203

Nakajima, K. (2) 29
Nakajima, T. (1) 636
Nakamura, A. (1) 411
Nakamura, F. (3) 95
Nakamura, T. (1) 279, 342, 628
Nakanishi, A. (1) 247
Nakanishi, Y. (1) 279, 628
Nakata, R. (1) 636
Nakatani, K. (1) 261
Nakayama, M. (5) 225
Nalbandyan, R.M. (2) 13
Nandagawe, J.K. (1) 626
Naoi, J. (5) 117
Narahara, Y. (3) 95
Narasimhan, J. (2) 76
Narendra, G.L. (1) 373, 527; (3) 120
Naris, M. (1) 147
Nascimento, O.R. (1) 136, 230; (2) 119
Nasri, H. (1) 130, 444, 445, 452
Neely, F.L. (1) 401
Negri, A. (2) 87
Nehls, U. (2) 275
Nelson, J. (1) 146
Nelson, M.J. (2) 58, 65
Nerissian, A.M. (2) 13
Neto, L.M. (2) 119
Netto Grillo, M.L. (1) 499
Neuberger, M. (2) 168
Nevin, W.A. (1) 129, 165
Nevostruev, V.A. (5) 125
Newlands, M.J. (1) 177
Newman, D.J. (3) 10-12, 17, 102, 124
Newman, R.C. (3) 76
Newton, M.E. (1) 22
Newton, W.E. (2) 234, 237
Ng, B. (3) 12
Nguyen, A.P. (2) 342
Niarchos, D. (1) 341, 344; (3) 96, 97
Nichols, T.R. (1) 494
Nicholson, I. (5) 23
Nieter Burgmayer, S.J. (1) 544
Niklas, J.R. (5) 184, 202
Niku-Paavola, M.-L. (2) 47
Nilsson, F. (3) 333
Ninomiya, K. (5) 200
Nishida, Y. (1) 192
Nishide, H. (1) 475
Nishihara, C. (1) 352
Nishimatsu, S. (5) 200
Nishina, Y. (1) 262
Nistor, S.V. (5) 62, 64, 65
Nitschke, W. (2) 145, 311, 312, 360, 361, 370, 373
Niwa, K. (5) 117

Nizhankovskii, V.I. (3) 98
Nocera, D.G. (1) 258
Noel, H. (3) 92
Noglik, H. (1) 443
Nogueira, S.R. (1) 49
Noltemeyer, M. (1) 398, 417
Nolting, H.-F. (2) 99
Noren, G.H. (2) 326, 331
Norris, J.R. (2) 313, 362, 364
Novak, M.A. (1) 230
Nozoye, H. (1) 352
Nuber, B. (1) 194, 209
Nugent, J.H.A. (2) 317, 321-323, 352
Numata, M. (2) 136
Nunome, K. (5) 47
Nuttall, R.H.D. (5) 38, 67, 68

Oberhausen, K.J. (1) 176, 549
O'Brian, R.J. (1) 549
Ochiai, E.-I. (2) 75
Ochiai, Y. (3) 95
Ochiuzzi, M. (1) 403
O'Connell, A.B. (5) 165
O'Connell, A.J. (5) 49
O'Connor, C.J. (3) 85
Odell, E. (2) 133
Odenwaller, R. (2) 154
Oehlmann, G. (1) 660
Oertling, W.A. (1) 132; (2) 131, 138
O'Gara, C.Y. (2) 34
Ogoshi, H. (1) 447
Ogura, K. (5) 225
Oh, B.-H. (2) 162
Oh, M.H. (1) 627
O'Halloran, T.V. (2) 62
Ohanessian, G. (5) 72
Ohba, Y. (1) 133, 534; (2) 10
Ohi, K. (1) 97
Ohkura, H. (5) 40
Ohnishi, T. (2) 274, 275, 285, 286, 293
Ohya-Nishiguchi, H. (1) 448, 607, 662; (2) 113, 141, 180
Okada, M. (5) 203
Okamoto, H. (1) 251
Oki, A.R. (1) 151
Okibayashi, K. (1) 356
Okulov, S.M. (3) 104
Okushi, H. (5) 189, 191
Oleksy, E. (5) 222
Olk, B. (1) 564
Olk, R.M. (1) 564, 565
Olm, M.T. (5) 67, 68
Olmstead, M. (1) 504
Olnaji, O. (1) 476

Omling, P. (1) 280, 610, 621, 622
Ong, E.W. (1) 304
Ono, T.-A. (2) 343-347
Orhun, U. (1) 380
Orlinskii, S.B. (3) 125, 126
Orlowski, T.E. (3) 79
Orme-Johnson, W.H. (2) 51, 52, 238
Orosz, R.D. (1) 449
Orrin, R.H. (5) 4
Ortiz de Montellano, P.R. (2) 124
Orton, J.W. (3) 1
Orville, A.M. (2) 70
Oseroff, S.B. (1) 225, 230, 307, 323
Osgood, M. (2) 290
Osguera, V.U. (3) 101
Otaki, N. (2) 29
Ottaviani, M.F. (1) 388, 478, 567
Ovchinnikov, I.V. (4) 4
Overhof, H. (5) 208
Owen, J. (3) 40
Owens, F.J. (1) 327, 329, 330; (3) 105
Oxawa, S. (2) 108
Oyanagi, H. (2) 346
Ozawa, T. (5) 223
Ozbay, B. (3) 76

Pabst, T. (1) 245
Pace, M.D. (1) 62
Pace, R.J. (2) 341
Paci, M. (2) 21, 22
Pal, A.K. (1) 93, 94
Pal, S. (1) 600
Palanca, P. (1) 413
Paleari, A. (1) 303, 371
Pallanza, G. (2) 35
Pallavicini, G.B. (1) 303
Palmer, G. (2) 153, 300
Pan, L.P. (2) 297
Pandey, R.K. (1) 249
Papa, S. (2) 283
Papaefthymiou, G.C. (1) 202; (2) 80
Papaefthymiou, V. (1) 195; (2) 79
Parasiris, A. (1) 249
Paraskevas, S.M. (1) 344
Pardi, L. (1) 113-116
Park, J.-B. (2) 169-171
Park, Y.C. (5) 233
Parker, L.W. (1) 107
Parlitz, B. (1) 660

Paroli, P. (1) 336
Parrett, K.G. (2) 369
Parsons, S. (5) 166
Parthasarathy, S. (2) 26
Pasenkiewicz-Gierula, M. (1) 66
Passeggi, M.C.G. (1) 227, 230, 234
Passmore, J. (5) 166
Pastawski, H.M. (1) 232
Pastusiak, W. (3) 25
Patapis, S.K. (1) 339
Patil, P.K. (1) 626
Patyal, B.R. (5) 17, 18
Pavone, P. (5) 25
Pawlik, T. (5) 201
Payne, W.J. (2) 139
Peake, B.M. (4) 5
Pearce, L.L. (2) 86
Pechenyi, A.P. (1) 644
Peck, H.D., jun. (2) 147, 179, 203, 207
Pecoraro, V.L. (1) 602
Pedersen, E. (1) 599
Pedersen, J.Z. (2) 32, 259
Peiffer, W.E. (2) 301
Peisach, J. (2) 24, 25, 33, 44, 68, 111, 112, 121, 223, 238, 257
Pekker, S. (3) 93
Pelikán, P. (1) 65, 174
Pellacani, G.C. (1) 555
Pellat, C. (2) 197
Peng, G. (1) 489
Peng, Q. (2) 102
Peng, Z. (5) 92
Penner-Hahn, J.E. (2) 17
Penrose, R.P. (3) 16, 31
Perkinson, J. (1) 544
Perrot, I. (1) 661
Peruzzini, M. (1) 478
Petering, D.H. (2) 30, 76
Petersen, J. (2) 316
Peterson, J. (1) 23; (2) 102, 253
Peterson, J.A. (2) 135
Peterson, R.L. (5) 162
Petit, K. (5) 69
Petit, P. (1) 265
Petrouleas, V. (2) 335
Petrovich, R.M. (2) 189
Petruzzelli, R. (2) 115
Peyerimhoff, S.D. (5) 94
Pezennec, S. (2) 360
Pezeshk, A. (2) 27
Pezeshk, V. (2) 27
Pfeffer, J.Z. (3) 147, 148
Pfenninger, S. (1) 31; (5) 20
Phatak, G.M. (1) 324
Phillips, P.S. (1) 570

Philo, J.S. (2) 349, 358
Phor, J.S. (1) 634, 635
Picher, T. (1) 413
Pickering, S.R. (1) 179
Pickett, C.J. (1) 149, 577
Pierik, A.J. (2) 175, 192, 209, 214
Pierpont, C.G. (1) 117, 124
Pietraszko, A. (1) 95
Piette, L.H. (5) 228
Pike, R.D. (4) 6
Pilbrow, J.R. (1) 2, 39, 56, 414; (2) 220; (4) 15, 18; (5) 2
Pilkington, S. (2) 80
Pilo, M.I. (1) 541
Pinelli, D. (1) 391
Ping-Feng, W. (1) 45-47
Pinhal, N.M. (1) 577; (2) 219
Pinkerton, A.A. (1) 663
Pintar, A. (2) 35
Piro, O.E. (1) 225, 226, 231, 233
Pirri, C.F. (5) 181
Pizzarro, P.J. (1) 21
Planinic, P. (1) 357
Plant, W.J. (3) 75
Plyatsko, V.S. (1) 623
Poe, M. (2) 258
Poeppl, A. (5) 211
Pogni, R. (1) 541, 552
Pointdexter, E.H. (1) 332
Polgár, K. (1) 585, 617
Poli, S. (1) 385
Polinger, V.Z. (1) 102, 103
Politis, C. (1) 339
Polizio, F. (2) 115, 116
Pollock, W.B.R. (2) 143
Polnaszek, C.F. (4) 24
Poluektov, O.G. (1) 36
Pommeret, S. (5) 95
Pommier, J. (2) 227
Pon, G. (1) 555
Ponti, A. (5) 134
Popovic, S. (1) 357
Potel, M. (3) 92
Potenza, J.A. (2) 25
Powell, A.K. (1) 64
Powell, R.D. (1) 480
Power, P.P. (1) 122
Poyner, R.R. (2) 270
Pozek, B. (1) 290
Pozek, M. (1) 291, 357
Poz'skii, Yu.E. (3) 74
Pramanik, A. (1) 127
Prasad, L.S. (1) 569
Pratt, F.L. (5) 187
Preston, K.F. (1) 42; (5) 231
Prokhorov, A.D. (3) 130, 131

Prosperi, T. (2) 100
Puget, K. (2) 61
Puppo, A. (2) 125
Puri, M. (1) 299, 315, 316, 318, 319, 322, 328
Purohit, S. (1) 569
Puustinen, A. (2) 307

Qiu, Z.-H. (2) 282
Que, L., jun. (1) 135, 193, 195, 198, 199; (2) 58, 60, 68, 69, 73, 78, 79, 86, 88
Quiros, M. (1) 409

Raasch, D. (5) 43
Rac, M. (5) 63
Rachidi, S. (5) 151
Radford, H.E. (5) 101
Rager, H. (1) 646
Raghunathan, P. (1) 528, 529
Ragsdale, S.W. (2) 212, 213
Rai, R. (1) 347
Raitsimring, A.M. (1) 639
Rajagopalan, K.V. (2) 230
Rajasekharan, M.V. (1) 155, 191, 229
Rajavajan, A.K. (3) 91
Raju, A.R. (1) 350
Rajzmann, M. (5) 11
Raksanyi, K. (1) 584
Rakshit, D. (2) 267
Rakvin, B. (1) 357
Ramachandran, S. (2) 67
Ramakrishna, B.L. (1) 304
Ramana, M.V. (3) 121
Ramirez, A.P. (5) 166
Ramirez, J.A. (1) 413
Ramondo, F. (5) 110
Ranford, J.D. (1) 551
Rangel, M. (1) 467
Ranon, U. (3) 70
Rao, A.S. (1) 374
Rao, B.R. (1) 389
Rao, C.N.R. (1) 350
Rao, D. (1) 307, 323
Rao, J.L. (1) 373, 374, 507-509, 526, 527, 633
Rao, K.K. (5) 122
Rao, K.V.S. (5) 96
Rao, P.S. (1) 648
Rao, T.B. (1) 372
Rauer, S. (1) 564
Rausch, M.D. (4) 5
Rauschmann, H. (1) 245
Ravi, N. (2) 159, 176
Ravkin, B. (1) 290, 291, 297

Ray, D. (1) 488, 492, 600
Ray, R.K. (1) 537-539
Ray, S.N. (3) 45
Raynor, J.B. (1) 467; (2) 219; (5) 45, 123
Razavi, F.S. (1) 340
Razeghi, M. (3) 111
Real, J.A. (1) 173
Reddy, B.J. (1) 648
Reddy, T.R. (3) 58, 76
Reed, C.A. (1) 449
Reed, G.H. (2) 189, 190, 245, 258, 270
Reedijk, J. (1) 144, 206, 477
Reefman, D. (1) 144
Reem, R.C. (2) 85
Regnault, L.P. (1) 216
Rehder, D. (2) 243
Rehorek, D. (1) 437; (5) 218, 224
Reich, D.H. (1) 224
Reijerse, E.J. (1) 59, 60
Reineker, P. (1) 74
Reinen, D. (1) 104, 558, 559
Ren, Y. (2) 153
Renard, J.P. (1) 216, 217, 220, 221
Renger, G. (2) 329, 330
Renner, M.W. (1) 503
Reszka, K. (5) 222
Rettori, C. (1) 307
Rewaj, T. (3) 24, 27, 28
Rey, P. (1) 52, 137, 138, 259, 260
Reynolds, P.A. (1) 51
Reynolds, R.W. (3) 100
Rhodes, C.J. (5) 130
Ribas, J. (1) 168, 216, 265, 487
Ricciardi, G. (1) 121, 566, 567
Riccio, P. (2) 31
Rice, D.A. (5) 99
Rice-Evans, C.A. (5) 217
Rich, P.R. (2) 299
Richard, M.J. (5) 151
Richard, P. (1) 661
Richardson, J.F. (1) 150, 176, 549
Rickert, K.W. (2) 336
Ricter, M. (1) 660
Riedel, A. (2) 360, 361
Rieger, A.L. (4) 6
Rieger, P.H. (4) 5, 6, 16, 22, 25
Riester, J. (2) 48
Rietzel, M. (1) 398, 417
Rifkind, J.M. (2) 118, 272
Riggi, F. (1) 595
Riley, J.D. (3) 139, 144
Riley, M.J. (1) 442

Rillema, D.P. (1) 185
Ristori, S. (1) 388
Ritschl, F. (1) 50
Ritter, G. (1) 578
Rivero, B.E. (1) 226
Rizzarelli, E. (1) 542, 546
Roberge, P.C. (1) 423
Roberts, B.P. (5) 22, 111, 124
Robertson, D.E. (2) 282, 286, 291, 292
Robinson, B.H. (1) 25; (4) 5
Robinson, M.G. (3) 34, 57, 116, 141
Robinson, R.A. (1) 304
Robinson, W.T. (1) 154
Rockenbauer, A. (1) 117; (3) 93; (4) 14
Rodriguez, F. (1) 629
Roe, J.A. (2) 20
Roesky, H.W. (1) 398, 417
Rohlfs, R.J. (2) 187
Rojo, T. (1) 104
Rokosz, A. (1) 41
Romanelli, M. (1) 315
Romanyukha, A.A. (1) 292, 358
Romero, P. (1) 396
Roncin, J. (5) 119, 136
Rosa, A. (1) 121, 566, 567
Roseman, N.A. (2) 77
Rosen, H. (2) 280
Rosenzweig, A.C. (2) 49, 80
Rospert, S. (2) 215, 216
Rossat-Mignod, J. (1) 216
Roth, E.K.H. (1) 162
Rothery, R.A. (2) 229
Rothier Amaral, M., jun. (1) 482
Rotilio, G. (2) 21, 22, 32, 259
Rottman, G.A. (2) 93
Rousseau, J.J. (1) 582; (3) 109
Roussel, A.M. (5) 151
Rowan, L.G. (5) 214
Rowland, I.J. (5) 45, 123
Rozentuller, B.V. (1) 353
Rubins, R.S. (1) 235; (3) 83
Rubio, J. (3) 101, 103
Rudolf, M.F. (1) 470
Rudowicz, C. (1) 44
Ruebsam, M. (1) 295
Ruf, H.H. (2) 152
Ruffle, S.V. (2) 317, 323
Ruiz-Mejia, C. (3) 101
Rush, J.D. (2) 248, 254
Russo, U. (1) 115; (2) 99
Rustandi, R. (2) 362, 364
Rutherford, A.W. (2) 144, 311, 312, 328, 332, 350, 351, 353, 354, 361, 373

Ruzicka, F.J. (2) 189
Rylkov, V.V. (2) 54
Ryu, K. (2) 130
Ryzhik, I.M. (4) 11

Saadeh, S.M. (1) 602
Saalfrank, R.W. (1) 250
Sackmann, U. (2) 274
Sadlowski, L. (1) 99, 256
Sadowski, S. (1) 289
Safarov, A.F. (1) 272
Safo, M.K. (1) 449-451
Sahlin, M. (2) 72
Saito, K. (1) 211, 342
Saito, T. (2) 136
Sakaguchi, M. (5) 126
Sakai, M. (1) 262
Sakakibara, T. (1) 217
Sako, S. (1) 331
Sakurai, H. (2) 29, 113
Sakurai, T. (2) 38, 39, 43, 46
Sala-Pala, J. (1) 418
Salerno, J.C. (2) 278, 290, 298, 306, 308
Salmon, L. (1) 131, 518
Samsonenko, N.D. (5) 178
Samuni, A. (5) 227
Sanati, M. (1) 392
Sanders-Loehr, J. (2) 14, 72, 77
Sands, R.H. (1) 663; (2) 64, 174, 175, 221, 340
Sandusky, P.O. (1) 132; (2) 16, 314
Sang-Bo, Z. (1) 45, 46, 47
Sanjuán, M.L. (1) 396
Sanna, D. (2) 247
Sanna, G. (1) 170
Sanna, N. (5) 110
Sanz, V. (1) 413
Saraiva, L.M. (2) 139
Saraste, M. (2) 294
Sardanelli, A.M. (2) 283
Sarna, T. (2) 28, 304
Sarneski, J.E. (1) 157, 200-202; (2) 359
Sarra, R. (2) 97
Sasaki, Y. (1) 211
Sastry, D.L. (1) 533
Sastry, G.S. (1) 510; (3) 121
Sastry, M.D. (1) 324
Sathyanarayan, S.G. (3) 121
Sato, K. (5) 117
Sato, M. (3) 94
Sato, T. (2) 30
Satoh, K. (2) 328
Satoh, M. (1) 133, 536
Sauer, K. (2) 288, 337-339

Savini, I. (2) 37
Sazou, D. (1) 397, 473, 502
Scarrow, R.C. (2) 58
Schäfer, G. (2) 296
Schaefer, H.F., II (5) 70
Schake, A.R. (1) 158
Schechinger, T. (2) 23
Scheidt, W.R. (1) 444, 445, 449-451
Scherer, H.-J. (2) 152
Schiesser, A. (2) 259
Schinina, M.E. (2) 61
Schinke, D.P. (3) 53
Schirmer, O.F. (1) 609, 658
Schlick, S. (1) 530-532; (4) 8
Schmidt, H.G. (1) 417
Schmidt, K.H. (5) 30
Schmidt, V.H. (5) 144
Schmiede, A. (2) 275
Schmitt, E.A. (1) 158, 203
Schmitz, R.A. (2) 242
Schneider, K. (2) 233, 239
Schneider, R. (2) 275
Schobert, K. (1) 250
Schoemaker, D. (5) 62, 65, 64, 75, 76
Scholes, C.P. (1) 200; (2) 127, 290, 303, 359
Scholz, G. (1) 50
Schoonover, J.R. (2) 300
Schouten, A. (5) 157
Schramm, U. (2) 233
Schröder, I. (2) 183-185
Schugar, H.J. (2) 25
Schulte, G.K. (1) 157, 201, 501
Schulte, U. (2) 275
Schultz, S. (1) 307, 323
Schwartz, R.N. (1) 315
Schwartz, R.W. (3) 85
Schwarz, D. (1) 28
Schweiger, A. (1) 12, 14, 29-33, 594; (5) 19, 20
Scott, R.A. (2) 217
Scotti, R. (1) 535; (5) 212
Scoular, D. (5) 143
Scovil, H.E.D. (3) 33
Scripsick, M.P. (1) 366
Sczaniecki, P.B. (1) 27
Sears, J. (2) 336
Sears, T.J. (5) 101, 112
Seeber, R. (5) 541
Seed, R. (1) 336
Seefeldt, L.C. (2) 235
Seitz, S.P. (2) 65
Sellmann, D. (1) 578
Senglet, N. (1) 397
Serra, O.A. (1) 136
Serrano, J.L. (1) 396, 562

Servant, Y. (1) 420
Servent, D. (2) 199
Sessoli, R. (1) 137, 213, 259, 260
Seth, V.P. (1) 376, 620, 632
Setif, P. (2) 312, 332, 354, 363, 365
Sette, M. (2) 21
Severns, J.C. (2) 44, 45
Sevilla, M.D. (5) 69, 102
Sha, K. (1) 533
Shabarchina, M.M. (1) 353
Shah, M.A. (2) 34
Shaltiel, D. (1) 289
Sharma, V.S. (2) 103
Sheldrick, G.M. (1) 417
Shelley, D.A. (2) 109
Shen, G. (1) 53
Sheng, Z.Z. (1) 297
Shergill, J.K. (2) 181, 182
Shi, S. (1) 469
Shi, X. (2) 251, 261-264; (5) 148, 149, 229
Shibayama, H. (1) 331
Shibuya, T. (1) 395
Shida, T. (5) 118, 138
Shidara, S. (2) 144
Shiga, T. (2) 104
Shigematsu, M. (1) 448; (2) 113
Shim, H.K. (1) 503
Shimizu, H. (3) 95
Shimizu, T. (1) 359; (5) 193
Shimokoshi, M. (1) 97
Shin, D.H. (5) 42
Shin, W. (2) 210
Shin, Y.K. (1) 258
Shinde, R.F. (1) 659
Shindo, H. (1) 352
Shino, R. (1) 468
Shiohara, Y. (1) 293, 294, 311
Shiomi, D. (1) 268
Shiozaki, I. (1) 337
Shirakawa, H. (1) 364, 365
Shlenkii, V.I. (3) 125, 126
Shoner, S.C. (1) 122
Shongwe, M.S. (2) 95
Shore, R.G. (3) 144
Shrivastava, K.N. (1) 283, 333, 334
Shvachko, Yu.N. (1) 292, 358
Shykind, D.N. (1) 21
Sices, H. (2) 184
Siderer, Y. (4) 9
Sieckman, I. (2) 363, 368
Sierra, G.A. (1) 32
Sigalas, M.P. (1) 172
Silvestrini, M.C. (2) 149
Sim, F. (5) 199

Simard, M. (1) 476
Simmons, R.L. (2) 105
Simon, P. (3) 108, 109; (4) 14
Simpson, D.J. (1) 503
Simpson, J. (4) 5
Simpson, N.J.K. (5) 132
Singel, D.J. (2) 92; (5) 142
Singer, R.J. (5) 57
Singh, A. (5) 187
Singh, K.K. (1) 309
Singh, T. (1) 338
Sinn, E. (1) 456
Siriwardena, A. (1) 598
Siu, G.G. (1) 587
Sivakumar, K. (3) 121
Sivaraja, M. (2) 348, 349
Sivasubramanian, S. (1) 528, 529
Sizov, F.F. (1) 623
Sjöberg, B.-M. (2) 72
Sjöqvist, L. (1) 73
Skjeldal, L. (2) 173
Skrzypek, D. (1) 244
Slabaugh, M.B. (2) 77
Slagle, K.M. (1) 434
Slebodnick, C. (1) 404, 480
Sled, V.D. (2) 277
Sledzinska, I. (1) 612
Sletten, J. (1) 261
Slezak, A. (1) 619
Slifkin, L.M. (5) 214
Sligar, S.G. (2) 110
Smirnov, A.I. (1) 36
Smith, B.E. (2) 232
Smith, C.A. (2) 95
Smith, K.M. (1) 503
Smith, P. (2) 341
Smith, T.D. (4) 15, 18
Smith, W.L. (2) 151
Smith-Palmer, T. (1) 91
Smits, J.M.M. (1) 163, 206
Smykalla, C. (1) 206
Snetsinger, P.A. (2) 96
Snyder, S.W. (2) 362, 364, 372
So, H. (1) 148, 360
Sobral, R.R. (1) 383
Söthe, H. (1) 584, 585, 617; (5) 34, 39, 201
Sojka, Z. (1) 423
Solans, X. (1) 168
Solomon, E.I. (2) 36, 41, 63, 85
Solomonson, L.P. (2) 224-226
Solonenko, A.O. (1) 102, 103
Son, N.T. (5) 183
Son'ko, T.V. (3) 19
Sono, H. (1) 257
Sono, M. (1) 155
SooHoo, C.K. (2) 51

Sotgiu, A. (5) 25
Soto, L. (1) 238
Soto Tuero, L. (1) 169
Spacciapoli, P. (2) 34
Spacher, M. (1) 20
Spaeth, J.-M. (1) 11, 90, 278, 584, 585, 617; (5) 12-14, 34-39, 63, 67, 68, 184, 201, 202, 204-208
Spartalian, K. (2) 246
Speich, N. (2) 178
Speier, G. (1) 117
Spence, J.T. (1) 414; (2) 220
Spencer, D.P. (5) 26
Sportelli, L. (2) 12
Spoto, G. (5) 82
Springer, B.A. (2) 110
Springs, J. (5) 230
Sreedhar, B. (1) 526, 527; (3) 120
Sreehari, N. (1) 167
Srinivas, D. (1) 462, 463
Srivastava, C.M. (1) 8
Srivastava, K.K.P. (1) 161
Srivastava, T.S. (1) 540
Sroubek, Z. (3) 23
Stabler, R. (2) 244
Stafford, P.R. (2) 210
Stam, P. (1) 59
Stams, A.J.M. (2) 209
Stankowski, J. (1) 286, 287, 306, 348
Stanton, J.F. (5) 106
Stapleton, H.J. (3) 127, 145
Staren, C.A. (1) 231
Stasicka, Z. (1) 437
Steadman, J. (5) 7, 139
Stedman, G.E. (3) 14
Steggerda, J.J. (1) 163
Stehlik, D. (2) 341, 363, 368
Steinbrück, M. (1) 370
Steinitz, M.O. (1) 91, 340
Stenger, J.F. (1) 646
Stepanov, S.V. (1) 353
Stephan, D.W. (1) 493
Stephanos, J.J. (2) 117, 120, 122
Stephen, A.J. (3) 86
Stephens, P.J. (2) 172
Steren, C.A. (1) 233
Sternheimer, R. (3) 59
Stesmans, A. (5) 190
Stevens, A.D. (1) 320, 326; (5) 24
Stevens, K.W.H. (1) 43; (3) 8, 30, 55
Stevens, S.R. (3) 127
Stinson, S. (5) 154
Stirling, W.G. (1) 216

Author Index

Stoesser, R. (1) 50
Stopa, G. (1) 437
Story, T. (1) 271
Strähle, J. (2) 23
Strangeway, R.A. (1) 22
Stratemeier, H. (1) 104, 442
Streib, W.E. (1) 158, 203
Strieder, S. (2) 152
Strobel, K. (1) 245
Struck, O. (1) 250
Strzalka, K. (2) 304
Strzelecka, H. (1) 265
Stubbe, J. (2) 71
Stumm, W. (1) 594
Sturn, H. (3) 90
Stynes, D.V. (1) 443
Styring, S. (2) 318, 327, 333
Suarato, A. (1) 535
Subbanna, G.N. (1) 350
Subra, R. (1) 52
Subrahmanyam, V.S. (1) 389
Subramanian, S. (1) 181, 485, 490
Sucheta, A. (2) 164
Sudfeldt, C. (2) 244
Suerbaum, H. (2) 89
Suezawa, M. (1) 262
Sugano, T. (1) 268
Sugawara, K. (1) 293, 294, 311, 355, 356
Suhara, M. (1) 89, 556
Sulzberger, B. (1) 594
Sumino, K. (1) 262
Sumita, M. (1) 636
Sun, C.Y. (5) 213
Sun, J. (1) 106, 378-382
Sun, X. (2) 264; (5) 229
Sunder, G.S. (1) 510
Surerus, K.K. (2) 138
Sussums, A.A.F. (3) 87
Sutcliffe, R. (5) 61
Suzer, S. (5) 10
Suzuki, E. (5) 117
Suzuki, S. (2) 29, 43, 144
Svagten, H.J.M. (1) 271
Swalen, J.D. (4) 1
Swann, I.L. (1) 179
Swarnabala, G. (1) 155, 191
Swartz, H.M. (2) 304; (5) 227
Sweetland, M.A. (2) 198
Swuste, C.H.W. (1) 271
Symko, O.G. (1) 230
Symons, M.C.R. (1) 320, 326, 398, 417; (5) 1, 24, 27, 28, 45, 66, 83, 96, 98, 99, 120, 122, 123, 130, 137, 145-147, 155, 158, 160, 162, 165, 174, 177, 185, 187, 217

Szczepanska, L. (1) 95, 306
Szychlinski, J. (5) 222
Szymczak, H. (3) 30

Ta, D.T. (2) 160
Tabak, M. (1) 461; (2) 119
Tabard, A. (1) 661
Tada, T. (5) 156
Tadu, M. (1) 468
Taginov, L.R. (3) 98
Tagliaferro, A. (5) 181
Taguchi, H. (1) 605
Tahara, S. (5) 40
Tahon, J.-P. (2) 56
Tajima, K. (1) 448; (2) 113
Takahashi, T. (1) 310
Takano, M. (3) 94
Takeda, K. (1) 583
Takeda, Y. (3) 94
Takenouchi, H. (1) 268
Takeshima, S. (2) 29
Takeuchi, A. (1) 178
Takeuchi, H. (1) 630, 645, 649, 650
Takeuchi, T. (1) 217
Takeuchi, Y. (1) 279, 628
Takura, K. (2) 320
Talanov, Yu.I. (1) 335
Taleb, S. (1) 40; (5) 21
Tallon, J.L. (1) 325
Tan, J. (2) 148
Tan, X.-L. (2) 127
Tan, X.L. (1) 200; (2) 359
Tanaka, H. (1) 100
Tanaka, K. (5) 189
Tanaka, S. (1) 293, 294, 305, 311
Tang, L. (1) 656, 657
Tarasiev, M.Y. (2) 54
Tatebe, T. (1) 384
Tauler, R. (1) 560
Tavares, P. (2) 159, 176
Tawa, R. (2) 29
Taylor, D.R. (3) 87
Taylor, H. (2) 290
Taylor, P.C. (4) 2, 17
Taylor, R.H. (3) 82
Taylor, R.J. (1) 471
Tedeschi, G. (2) 87
Teitel'baum, G.B. (1) 335, 354
Teixeira, M. (2) 158, 179, 203, 207
Temperley, J. (1) 377
Temps, F. (5) 112
Terrile, M.C. (1) 230
Terron, A. (1) 597
Teslenko, V.V. (3) 104

Thanyasiri, T. (1) 456
Thauer, R.K. (2) 195, 215, 216, 242
Thelander, L. (2) 75
Thewalt, M.L.W. (5) 182
Thibault, J. (2) 60
Thomann, H. (1) 13; (2) 194, 234
Thomas, D.D. (4) 24
Thompson, D.W. (1) 443
Thompson, G.A. (4) 12
Thompson, J.D. (1) 323
Thompson, J.F. (2) 64
Thompson, L.K. (1) 139, 177
Thomson, A.J. (2) 50, 101, 140, 146, 150, 157, 163, 164
Thornely, R.F.N. (2) 236
Thornley, J.H.M. (3) 40
Thorp, H.H. (1) 156, 157, 201, 202, 434
Thova, M. (1) 307
Thurnauer, M.C. (2) 362, 364, 372
Tiberis, G.P. (3) 97
Timchenko, V.I. (5) 178
Timkovich, R. (2) 102
Timms, P.L. (1) 571; (5) 55
Tisza, S. (1) 117
Toi, H. (1) 447
Tokii, V.V. (5) 178
Tokoi, T. (1) 192
Tokushige, M. (2) 180
Tollin, G. (2) 14
Tolman, W.B. (1) 196, 545
Tomietto, M. (1) 572; (5) 53, 54
Tomlinson, A.G. (2) 100
Toney, G.E. (1) 134
Tonon, C. (1) 163
Tordi, M.G. (2) 149
Tordo, P. (5) 174
Toriumi, K. (1) 251
Torralba, R.C. (1) 187-190
Torres-Valderrama, M. (1) 67-69
Tosatti, E. (1) 98
Tosik, A. (1) 172
Toupet, L. (1) 418
Tovar, M. (1) 323
Townsend, M.G. (5) 27
Toy, A.D. (4) 15, 18
Trackman, P.C. (2) 34
Trautwein, A.X. (1) 3, 130, 134, 194, 209, 250, 452; (2) 99, 126
Traylor, T.G. (2) 103, 128
Treutmann, W. (1) 245
Triberis, G.P. (1) 341
Trifiro, F. (1) 391
Trikalinos, C. (1) 341, 344; (3)

97
Trofimenko, S. (1) 477
Trojan, K.L. (1) 554
Trombetta, J.M. (5) 209
Tronconi, A.L. (3) 116, 117, 142
Trüper, H.G. (2) 178
Trumpower, B.L. (2) 284
Trybula, Z. (1) 348; (5) 144
Tsai, A.-L. (2) 151, 153
Tsai, H.L. (1) 203
Tsai, M.N. (1) 92
Tsai, T.E. (5) 198
Tsapin, A.I. (1) 353
Tseng, W. (5) 209
Tsipis, C.A. (1) 172
Tso, J. (2) 335, 348, 349
Tsubaki, M. (2) 137, 260
Tsuchida, E. (1) 475
Tsujikawa, I. (1) 166, 270
Tu, Y. (1) 392
Tuchagues, J.P. (1) 169, 246
Tuchendler, J. (1) 248
Tucker, D.A. (1) 554
Tuel, A. (1) 363
Tulyathan, B. (4) 5
Turek, P. (1) 19
Turkevich, J. (3) 118
Turner, I.M., jun. (2) 58
Turner, N.A. (2) 219
Turro, N. (5) 159
Tusek-Bozik, L. (1) 357
Tyeklár, Z. (1) 557
Tykarski, L. (1) 612

Uchida, Y. (1) 500
Ueda, N. (2) 66
Ueda, Y. (2) 180
Ueno, I. (2) 141
Ueyama, N. (1) 411
Uffelman, E.S. (1) 480, 494
Ullrich, V. (2) 66
Urban, W. (3) 17
Urbina, J.A. (2) 271
Usachev, A.E. (1) 184
Ustinov, V.V. (1) 292, 358
Utsumi, H. (1) 384
Utsumi, W. (5) 188
Utterback, G.G. (3) 140

Vacca, A. (1) 478
Valach, F. (1) 65
Valade, L. (1) 139
Valentine, J.S. (2) 20
Valin, A. (5) 80
Valko, M. (1) 65, 174

Vallyathan, V. (2) 262; (5) 149
Van Bockstal, L. (1) 18
van de Kamp, M. (2) 7
van der Elst, A. (2) 363, 368
Van der Linden, J.G.M. (1) 163
van der Meer, R.A. (2) 47
van der Waals, J.H. (1) 38; (5) 152
van der Woerd, M.J. (5) 141
van der Zwaan, J.W. (2) 205, 206
van Garderen, C.J. (2) 211
van Gisbergen, S.J.C.H.M. (1) 625
Vanin, A.F. (2) 106
Van Kempen, R.J.T. (1) 271
van Lier, J.E. (2) 134
van Mieghem, F.J.E. (2) 328, 332
Vanngard, T. (1) 57
Vanoni, M.A. (2) 188
Van Oosten, A.B. (5) 183
van Ormondt, D. (3) 20, 45
Vanoy, T.C. (1) 579
van Polen, J. (5) 26
van Tits, J.L.C. (1) 60
van Veen, G. (1) 58
Van Vleck, J.H. (3) 6
van Vliet, P. (2) 145
van Willigen, H. (2) 96
Van Zee, R.J. (5) 87-90, 103, 106, 113, 114, 170, 171
Varberg, T.D. (5) 73
Varghese, B. (1) 167
Vašák, M. (2) 266, 267
Vass, I. (2) 318, 327
Vasson, A. (1) 107
Vasson, A.M. (1) 107
Vassort, C. (2) 60
Vàzquez, A.E. (2) 172
Venkateswarlu, M. (1) 372
Vera, A. (1) 313, 349
Verbeek, R. (5) 108
Verdaguer, M. (1) 216
Vermaas, W.F.J. (2) 332
Verniquet, F. (2) 168
Verpeaux, J.-N. (5) 230
Viana, L.F.S. (1) 383
Viana dos Santos, J.A. (1) 461
Vicente, R. (1) 265, 487
Vickery, L.E. (2) 160
Vier, D.C. (1) 307, 323
Viezzoli, M.S. (2) 19
Viikari, L. (2) 47
Vijayaraghavan, R. (1) 350; (3) 91
Vikram, P. (5) 22
Villacampa, B. (1) 580

Villaveuva, M. (5) 97
Vincent, J.B. (1) 203; (2) 90
Vinckier, C. (2) 56
Vinogradov, A.D. (2) 277
Vittier, C. (1) 216
Vittoria, C. (1) 321, 336
Vivien, D. (1) 368, 369, 518
Vizza, F. (1) 123
Voelkel, G. (5) 211
Voevodskaya, N.V. (2) 255
Voiron, J. (1) 266
Volonte, S. (5) 212
Von Dreele, R.B. (1) 304
von Wachenfeldt, C. (2) 279
Voordouw, G. (2) 143
Voronkova, V.I. (1) 110
Voronkova, V.K. (1) 184
Voszka, R. (5) 202
Vugman, N.V. (1) 49, 383, 481, 482, 499

Wabia, M. (1) 256
Wacker, T. (1) 30
Wahlen, M. (1) 295
Walczak, J. (1) 99
Walczak, T. (2) 304
Walker, E. (1) 289
Walker, F.A. (1) 445, 450, 451
Walker, P.J. (3) 77, 86
Wallenborn, L.R. (1) 392
Walther, K.L. (1) 387
Wang, C. (1) 105, 375; (3) 110
Wang, C.-P. (2) 208
Wang, D.-C. (2) 274
Wang, D.M. (1) 439, 440
Wang, H. (1) 655
Wang, J. (5) 115
Wang, P. (1) 346
Wang, S. (1) 203, 345
Wang, S.L. (1) 252
Wang, S.S. (1) 140
Wang, W.W. (1) 140
Wang, Y. (1) 345, 445; (2) 147; (4) 12; (5) 215
Wang, Y.-L.E. (5) 221
Wang, Y.X. (1) 351
Wanklyn, B.M. (3) 77
Wan-Lu, Y. (1) 44
Wannowius, K.J. (1) 558
Waplak, S. (1) 95; (5) 144
Ward, R.C.C. (1) 647; (5) 49
Warden, J.T. (2) 352, 366
Warren, D.S. (5) 167
Warren, P.V. (2) 366, 367, 369
Wasowicz, T. (5) 173
Watanabe, M. (2) 104
Waters, J.M. (1) 551

Author Index

Watkins, W.C. (5) 231
Watson, K.A. (5) 231
Watson, R.E. (3) 5, 44
Watterich, A. (1) 399, 584; (5) 202
Watts, J.D. (5) 199
Weatherburn, D.C. (1) 154, 159
Weber, D.J. (2) 273
Weber, R.T. (3) 34, 117
Webster, D.A. (2) 123
Wedd, A.G. (1) 414; (2) 220
Weeks, R.A. (1) 362
Weidner, U. (2) 275
Weihe, H. (1) 153
Weil, J.A. (1) 652, 653; (5) 135, 201
Weill, J.A. (1) 367
Weiner, J.H. (2) 181, 182, 228, 229
Weiss, H. (2) 274, 275
Weiss, J. (1) 194, 209
Weiss, R. (1) 130, 134, 452; (2) 126
Weitekamp, D.P. (1) 21
Wells, M.R. (3) 34, 57, 86, 116, 148
Weltner, W. (5) 86-90, 103, 106, 113, 114, 170, 171
Wen, J. (1) 655
Wenckebach, W.T. (1) 15
Werst, M.M. (2) 8, 162, 166, 167
Werth, M.T. (2) 183-185
Weser, U. (2) 23
Wessel, R. (1) 28
West, D.X. (1) 555
Westbrook, J.D. (5) 159
Wetterhahn, K.E. (1) 408; (5) 150
Wever, R. (2) 131
Whichard, G. (1) 362
Whiffen, D.H. (3) 62
Whisenhurst, D.W., jun. (1) 412
Whitcombe, T.W. (1) 181
White, L.K. (4) 20
White, P.S. (1) 554
Whitehead, J.P. (2) 201
Whittaker, J.W. (2) 16-18, 252
Whittaker, M.M. (2) 16-18, 252
Whittington, B.I. (1) 386
Whyte, T. (1) 438
Wicholas, M. (1) 608
Widom, A. (1) 321, 336
Wieghart, K. (1) 194
Wikström, M. (2) 294, 307
Wilamowski, Z. (1) 276
Wilk, A. (1) 558
Wilks, A. (2) 124

Willett, R.D. (1) 235, 555
Willett, R.G. (1) 228
Williams, A.J. (5) 231
Williams, F.I.B. (3) 52, 61
Williams, G.A. (1) 428
Williams, R. (5) 122
Willis, C.R. (5) 22
Willner, H. (1) 570
Wilson, G.L. (1) 414; (2) 220
Wilson, M.T. (2) 150
Windsch, W. (1) 589
Winfield, M.E. (4) 18
Winiski, M. (5) 8
Winkler, H. (1) 3, 130, 134, 452; (2) 99, 126
Winter, M. (1) 209
Winzer, K. (3) 89
Wittenberg, J.B. (2) 121
Witters, R. (2) 56
Wittinghofer, A. (2) 268, 269
Witzel, H. (2) 89, 244
Wokaun, A. (1) 387
Wolf, E. (1) 330
Wolf, W.P. (3) 9, 84
Wolny, J.A. (1) 470
Woodruff, W.H. (2) 138
Woods, C. (1) 185
Woodward, J.R. (5) 93
Woodward, R.W. (5) 7
Wrigglesworth, J.M. (2) 299
Wright, A. (5) 187
Wright, F. (1) 323
Wrobel, G. (1) 520, 524
Wu, C.C. (1) 314
Wu, C.J. (1) 92, 400
Wu, D. (1) 164
Wu, Y. (5) 190
Wyatt, J.L. (5) 120, 155, 174
Wydrzynski, T. (2) 329, 330
Wyon, C. (1) 368
Wzietek, P. (1) 266

Xie, Y. (5) 70
Xu, B. (1) 210; (5) 215, 216
Xu, C. (1) 53
Xu, L. (1) 656
Xu, Y. (1) 422, 655

Yablokov, Yu.V. (1) 184
Yachandra, V.K. (2) 339
Yachkin, V.A. (1) 525
Yagi, T. (2) 204; (5) 188
Yajima, Y. (5) 200
Yakoub, N.A. (3) 113
Yakovleva, Zh.S. (1) 256
Yamada, L. (1) 237

Yamada, S. (1) 178
Yamagishi, A. (1) 217
Yamaguchi, K. (1) 294
Yamamoto, S. (2) 66
Yamanaka, T. (2) 136
Yamasaki, S. (5) 189, 191
Yamashita, M. (1) 251
Yamauchi, S. (1) 133
Yamazaki, I. (5) 228
Yamazaki, T. (2) 136
Yan, M. (5) 102
Yang, B. (1) 394
Yao, C.L. (1) 147
Yarish, S.S. (1) 457
Yates, C.A. (5) 99
Yeh, C. (5) 215
Yen, W.M. (1) 576
Yeom, T.H. (1) 616
Yeung, Y.Y. (3) 102, 124
Ying, J. (1) 657
Yocum, C.F. (2) 316, 357
Yoder-Short, D.R. (1) 624
Yokogawa, K. (5) 200
Yokoi, H. (1) 175, 178, 536
Yokozawa, A. (1) 263, 264
Yonetani, T. (2) 107, 111, 260
Yoon, K. (1) 544
Yordanov, N.D. (1) 550
Yoshida, T. (1) 218, 630
Yoshihara, H. (2) 104
Yoshimoto, T. (2) 66
Yoshimura, T. (1) 447; (2) 144
Yoshinaga, N. (1) 411
Yoshita, M. (5) 193
Yosida, T. (3) 100
You-Ming, N. (1) 45
Young, C.G. (1) 145
Young, C.L. (1) 17
Youvan, D.C. (2) 313
Ysern, X. (2) 271
Yu, C. (1) 624
Yu, C.-A. (2) 281, 282, 288, 289
Yu, J. (2) 332
Yu, J.S. (1) 511, 512
Yu, J.T. (1) 92, 273, 277, 400, 505
Yu, L. (2) 281, 282, 288, 289
Yu, N.-T. (2) 137, 260
Yu, S. (1) 161
Yuan, S. (1) 345
Yumoto, N. (2) 180

Zacharias, E. (1) 338
Zaldo, C. (1) 575
Zamadics, M. (1) 514
Zanchetta, J.V. (1) 419

Zanchini, C. (1) 1, 142, 168, 183, 212, 228, 242, 243, 394, 563, 608
Zanello, P. (1) 123, 206, 478
Zanetti, G. (2) 188
Zang, L.Y. (5) 226
Zannoni, D. (2) 145
Zapart, M.B. (1) 82, 84
Zapart, W. (1) 82-87
Zaspel, C.E. (1) 235, 236
Zavidonov, A.Yu. (3) 132
Zeegers-Huyskens, T. (2) 56
Zehnder, A.J.B. (2) 209
Zeitler, C.M. (2) 155
Zelei, B. (1) 399
Zensen, R. (2) 275
Zeppezauer, M. (2) 265

Zhan, J.W. (1) 346
Zhang, C.-S. (2) 52
Zhang, J.-H. (2) 87
Zhang, Y. (2) 63
Zhang, Y.N. (1) 624
Zhang, Z. (5) 226
Zhao, J. (2) 367
Zhao, K. (1) 164
Zhao, M.G. (1) 587, 588
Zhao, W. (2) 303
Zheng, W. (1) 77
Zheng, W.C. (1) 590, 591, 593, 641
Zhong, Y.C. (1) 39, 56
Zhou, P. (1) 235
Zhou, X. (1) 345

Zhou, Y.Y. (1) 642, 643
Zhu, D.B. (1) 346
Zhu, J. (5) 69
Ziegler, T. (5) 231
Zimmer, M. (1) 501
Zimmermann, J.-L. (2) 350, 351
Zimpel, Z. (1) 75
Zöphel, A. (2) 186
Zorita, M. (5) 41
Zoroddu, M.A. (1) 170, 541; (2) 249
Zubieta, J. (1) 557
Zuhr, R.A. (1) 362
Zuk, T. (1) 286
Zumft, W.G. (2) 48-50
Zuotao, Z. (1) 343
Zwanenburg, G. (1) 81, 565

Milton Keynes UK
Ingram Content Group UK Ltd.
UKHW010042230624
444478UK00002B/9